INTEGRATED APPROACH TO COORDINATION CHEMISTRY

BICENTENNIAL
1807
WILEY
2007
BICENTENNIAL

THE WILEY BICENTENNIAL—KNOWLEDGE FOR GENERATIONS

*E*ach generation has its unique needs and aspirations. When Charles Wiley first opened his small printing shop in lower Manhattan in 1807, it was a generation of boundless potential searching for an identity. And we were there, helping to define a new American literary tradition. Over half a century later, in the midst of the Second Industrial Revolution, it was a generation focused on building the future. Once again, we were there, supplying the critical scientific, technical, and engineering knowledge that helped frame the world. Throughout the 20th Century, and into the new millennium, nations began to reach out beyond their own borders and a new international community was born. Wiley was there, expanding its operations around the world to enable a global exchange of ideas, opinions, and know-how.

For 200 years, Wiley has been an integral part of each generation's journey, enabling the flow of information and understanding necessary to meet their needs and fulfill their aspirations. Today, bold new technologies are changing the way we live and learn. Wiley will be there, providing you the must-have knowledge you need to imagine new worlds, new possibilities, and new opportunities.

Generations come and go, but you can always count on Wiley to provide you the knowledge you need, when and where you need it!

WILLIAM J. PESCE
PRESIDENT AND CHIEF EXECUTIVE OFFICER

PETER BOOTH WILEY
CHAIRMAN OF THE BOARD

CONTENTS

Acknowledgments

Special thanks go to Dr Emily Jarvis and undergraduate Emily Rains who contributed all Spartan-based computational projects in this book. The experiments in Chapter 2 were originally implemented at Kenyon College by Professors Russell Batt and Gordon Johnson and Dudley Thomas for our first year chemistry laboratory course. We have been using these experiments with much success for over 10 years. Thanks also go to undergraduates Dan Custar and Justin Struble, who tested the amine titration and worked out the details for the visible spectroscopy component. Chapter 3 is an accumulation of several years of work, contributed by Dr Gordon Johnson and his talented inorganic laboratory undergraduates: Michael Asam, Bryan Bonder, Chad Braun, Elizabeth Drotleff, Samantha Hudgins, and Sophia Sarifidou. We are also indebted to Dr Chris Fowler for his contribution to the visible spectroscopy experiments of Chapter 3. Much gratitude goes to undergraduate Chris Wager, who is responsible for the planning and testing of experiments in Chapter 4. Many students participating in inorganic laboratory helped to make the experiments in Chapter 5 "undergraduate friendly"! We especially recognize the efforts of: Janice Pour and Mark and Erin (Wimmers) Wilson— $[Co(ox)_2en]^-$ synthesis; Ryan Parks and Janice Pour—CV; Julia M. (Boon) Hinckley— Marcus theory; Nora Colburn, Ansley Scott, and Alison Ogilvie—inert atmosphere electron transfer kinetics. Thanks go once again to Dan Custar, who researched and developed Chapter 6. Chapter 7 was inspired by the work of Honors student Kimberly Levin. Many thanks to Sarah Collins, who researched and developed the solid-phase peptide synthesis experiment with advice from Dr. Elizabeth Ottinger, and to both Sarah and Emily Ross, who under the guidance of Dr Tania Gonzalez, worked out the biosynthesis of phytochelatins. Mary Kloc was instrumental in helping with the stability complex determination laboratory. We acknowledge Megan Chambers for her work on the chromium chemistry of Chapter 8. Very special thanks go to Drs Anthony Watson and Paul Fischer for their help with experiments in Chapter 9. We would like to thank Drs A. Graham Lappin, Michael Johnson, and Jose Alberto Portela Bonapace for their careful reading of the manuscript, corrections and helpful suggestions. Lastly, special thanks also go to science medical illustrator, Mr Kenneth Eward (Biografx), for his help with sketches of select figures, as well as careful reading of the manuscript, corrections, and helpful suggestions. Freelance art directors Jen Christiansen and Melissa Thomas have our deep appreciation for their insightful contributions to development of the cover design. Our thanks also go to Professor Heinz Berke of the University of Zurich, who kindly provided photographs of Alfred Werner's original coordination compounds for the book cover, as well as the University of Zurich's office of public affairs, who made available his portrait from their archives.

Instructor Notes are available for most chapters online. These include select answers to post-lab questions, tips on procedure, sample data, and literature results.

CHAPTER 1

Experimental Inorganic Chemistry: A History of Dazzling Color!

The starting point is the study of color and its effects on men [Wassily Kandinsky (1912), *Concerning the Spiritual in Art*]

EARLY DEVELOPMENTS IN INORGANIC CHEMISTRY[1-4]

Ask any inorganic chemist just why he or she entered the field and more often than not, the answer will be "color!" People have long been captivated by the magnificent colors of inorganic compounds. As early as 15,000 BC, cave painters used iron oxides for sources of yellows and reds, aluminum silicates for greens, and manganese oxide for browns. By 3000 BC, Egyptian and Syrian artisans and jewelers were staining metal surfaces with inorganic salts derived from ground minerals and alloys. The colors obtained depended mostly on the types of transition metals contained in the minerals and on their combinations. Soon synthetic pigments were developed as well. One of the first, Egyptian Blue, $CaCuSi_4O_{10}$, was prepared by heating malachite/sand (quartz or silica) mixtures to temperatures of 800–900 °C. Later in the fifteenth to eighteenth centuries, metal-containing pigments such as copper carbonate and the brilliant ferric ferrocyanide coordination complex, Prussian Blue, $Fe_4[Fe(CN)_6]_3$, were synthesized for use in the textile industry (Fig. 1.1). Iron oxides, lead chromates, and the ubiquitous white pigment, titanium dioxide, are just a few of the many inorganic compounds that are still important in this industry today.

The beauty of inorganic chemistry lies in the fact that minute changes in a metal ion environment induce dramatic changes in color. The fact that easily noted changes in color signify chemical and sometimes physical change was critical to the development of the field of chemistry. Even before the establishment of the modern science of chemistry, early metallurgists and alchemists relied on color change as a positive step toward what they believed was the transformation of base metals, primarily into gold. They documented procedures that included the manipulation of mixtures of lead, tin, copper, and iron through a series of black, white, yellow, and purple stages. During the sixteenth and seventeenth centuries, careful quantitative studies of color change brought forth new views on the nature of matter that gave birth to modern chemistry. The heating of mercury metal in the presence of air, for example, forms a red calx (mercury

Integrated Approach to Coordination Chemistry: An Inorganic Laboratory Guide. By Rosemary A. Marusak, Kate Doan, and Scott D. Cummings

Figure 1.1   Vials of different metal-containing compounds and complexes.

TABLE 1.1   Some Inorganic Compounds and Their Colors

Compound	Color	Compound	Color
$Co(OH)_3$	Black	PbI_2	Yellow
$Cu(OH)$	Yellow	PbO_2	Brown
Cu_2O	Red	PbS	Black
Cu_2S	Black	HgO	Yellow or red
$Au(OH)_3$	Yellow-brown	Hg_2O	Brown-black
AuOH	Dark violet	Hg_2I_2	Yellow
$Fe(OH)_3$	Red-brown	HgICl	Red
Fe_2S_3	Dark green	Ag_3AsO_4	Dark red
FeS	Black	Ag_3AsO_3	Yellow

oxide), which can subsequently be returned to the original metal by heating. In 1778, Antoine Lavoisier showed that this color change was caused by the addition and removal of oxygen at the metal center. In 1788, French chemist Joseph-Louis Proust argued that colors in alloys were a result of constituents in fixed and definite proportions, leading to his law of definite proportions. Between 1790 and 1830 geologists discovered a vast number of naturally occuring inorganic mineral types; and Swedish chemist Jöns Jakob Berzelius alone prepared, purified and analyzed over 2000 inorganic compounds in just 10 years. A few representative inorganic compounds and their colors are given in Table 1.1. Notice how switching the metal from copper to gold in the metal hydroxide [M(OH)] compounds changes the observed color from yellow to dark violet. Likewise, replacing iodides in the yellow PbI_2 with oxygen or sulfur changes the colors to brown or black, respectively. The colors of these common naturally occurring minerals are earth tones. Chemists have produced their most vibrant colors, however, by manipulating the metal ion environment in compounds known as complex ions.

COMPLEX IONS[3,7,8]

Although Prussian Blue, synthesized in 1704, was the first officially recognized *metal coordination complex* to be made, discovery of this group of transition metal complex ions is often credited to Taessert, who in 1798 prepared the first known cobalt ammonia salts. His work inspired a revolution in inorganic chemistry. At the turn of the nineteenth century, amidst the flourishing developments of organic chemistry, the striking colors

TABLE 1.2   Color Names Given by Edmond Frémy (Adapted from Brock[3])

Compound	Color	Original Name	Formula
$Co(NO_2)_3 \cdot 4NH_3$	Brown	*Flavo* complex	*cis*-$[Co(NH_3)_4(NO_2)_2]NO_2$
$Co(NO_2)_3 \cdot 4NH_3$	Yellow	*Croceo* complex	*trans*-$[Co(NH_3)_4(NO_2)_2]NO_2$
$CoCl_3 \cdot 6NH_3$	Yellow	*Luteo* complex	$[Co(NH_3)_6]Cl_3$
$CoCl(H_2O) \cdot 5NH_3$	Rose-red	*Roseo* complex	$[Co(NH_3)_5(H_2O)]Cl_3$
$CoCl_3 \cdot 5NH_3$	Purple	*Purpureo* complex	$[Co(NH_3)_5Cl]Cl_2$
$CoCl_3 \cdot 4NH_3$	Green	*Praseo* complex	*trans*-$[Co(NH_3)_4Cl_2]Cl$
$CoCl_3 \cdot 4NH_3$	Violet	*Violeto* complex	*cis*-$[Co(NH_3)_4Cl_2]Cl$

of the metal ammines and their unusual characteristics[†] piqued the interest of chemists, resulting in the synthesis of a plethora of metal ammine complexes over the next 50 years. In 1852, French chemist Edmond Frémy put forth a color-based naming scheme for these complexes, shown in Table 1.2. Unfortunately, this scheme has its limitations; for example, the cobalt complexes, $Co(NO_2)_3 \cdot 4NH_3$ and $CoCl_3 \cdot 6NH_3$, though both are yellow, bear unrelated names. To further this scheme of confusion, $Co(NO_2)_3 \cdot 4NH_3$ can be either yellow or brown. One can quickly see the shortcomings of Frémy's color-based nomenclature.

As new complex ions were synthesized, several bonding theories were postulated and rejected. The two most convincing theories, "the Blomstrand–Jorgensen chain theory" and "coordination theory" proposed by Alfred Werner, were debated extensively (a subject taken up in detail in Chapter 2) and it was coordination theory that eventually proved to be correct, winning Werner the Nobel Prize in 1913.

Thanks to Werner, the nature of complex ions is no longer as complex as it used to seem. A complex ion is a species that contains a central metal ion (M), surrounded by ions or molecules, called ligands (L) (Fig. 1.2). Although partial substitution may take place at the metal center, the complex tends to retain its identity in solution. Werner pointed out that complex ions, now termed *metal complexes*, have two valences: the primary valence is the charge of the metal ion itself (the oxidation state of the metal, n^+) and the secondary valence is the number of ligands bound to the metal. Werner noted that, unlike carbon, metal complexes can possess a maximum number of bound ligands beyond their primary valency. The total charge on the metal complex is the sum of the metal ion charge and the ligand charges. If the overall charge is not zero, it is balanced by *counterions* to give an overall neutral species. Although the early ambiguous formulations for the cobalt ammines in Table 1.2 could be found in texts as late as the mid-1950s, they have since been replaced by modern formulas to reflect the nature of bonding. As an example, consider the yellow *croceo* complex, $Co(NO_2)_3 \cdot 4NH_3$. The modern formula, *trans*-$[Co(NH_3)_4(NO_2)_2]NO_2$, gives the spatial relationship of the atoms in this octahedral complex: the complex cation consists of a cobalt ion surrounded by four NH_3 molecules in one plane and two NO_2^- ions situated $180°$ apart from each other (*trans* terminology). Because the counterion balancing the charge on the cation is NO_2^-, the overall charge on the complex cation is $+1$; thus the cobalt center contributes $+3$ charge. The structure of this complex is shown in Figure 1.3(*a*). The *cis* form of the complex

---

[†]Chemists found it odd that two stable compounds, $CoCl_3$ and NH_3, with seemingly saturated valences, combined to form a new stable compound. This reactivity was very different from that of carbon.

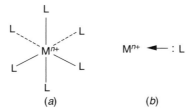

Figure 1.2   (*a*) Schematic of the octahedral complex and (*b*) the M–L coordinate bond.

[Fig. 1.3(*b*)] places the coordinating NO_2^- ligands orthogonal (90° angles) to each other. This seemingly simple change in ligand arrangement results in a color change from yellow to brown. Color change resulting from such *geometric isomerism* was critical to Werner's first predictions of metal complex molecular geometry (Chapter 2).

Ligands bind to the metal center through electron pair donation [Fig. 1.3(*b*)]. This type of *coordinate bonding* was first put forth in 1922 by Maurice L. Huggins, an undergraduate student of Gilbert N. Lewis, and was expanded upon by Thomas M. Lowry. In 1927, Nevil V. Sidgwick applied coordinate bonding to metal complexes. Ligands, then, are Lewis bases (electron pair donors) and metals are Lewis acids (electron pair acceptors). This bond type [Fig. 1.3(*b*)] gives metal complexes their name, *coordination complexes*, which was a term actually first used by Werner.

The ligand number—or *coordination number*—varies depending on the nature of the metal ion, as well as on the nature of the ligand. With a given number of ligands, each metal complex adopts one of a number of energetically favorable geometries. A metal complex of coordination number 6, for example, adopts an octahedral shape. The octahedral configuration for a series of hexammonium compounds, first predicted by Werner, was confirmed by X-ray analysis by the early 1920s. The topic of metal complex geometry is explored in Chapter 3 using the geometrically versatile nickel(II) ion.

The robust nature of Werner's original cobalt(III) ammine complexes enabled him to purify, isolate and study the solution chemistry of these complexes. (*Robust* complexes have a complex ion—metal and the primary coordination sphere ligands—that remains as one entity in solution.) The ability of a metal complex to resist decomposition by water or dilute acids was recognized early on; it is still a qualitative measure of stability today. One of the first chemists to quantitatively measure metal complex stabilities (*stability constants*) was Danish chemist Jannik Bjerrum, who published his Ph.D. dissertation on the formation of metal ammines in aqueous solution in 1940. His work suggested that important factors governing stability and coordination number include ligand polarizability and structure, and metal ion electron configuration and size. It was understood that electronegativity differences between the electropositive metal and its electron donating ligand led to an unequal sharing of electrons and a polar covalent bond, yet in his day, the extent of electron sharing was still unknown.

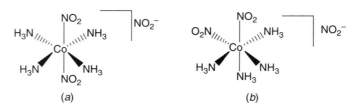

Figure 1.3   Yellow *trans*- (*a*) and brown *cis*-(*b*) [Co(NH₃)₄(NO₂)₂]NO₂.

ON THE NATURE OF THE COORDINATE BOND—BONDING MODELS[7,9–11]

Valence Bond Theory

Soon after the development of the quantum mechanical model of the atom, physicists such as John H. van Vleck (1928) began to investigate a wave-mechanical concept of the chemical bond. The electronic theories of valency, polarity, quantum numbers, and electron distributions in atoms were described, and the valence bond approximation, which depicts covalent bonding in molecules, was built upon these principles. In 1939, Linus Pauling's "Nature of the Chemical Bond" offered valence bond theory (VBT) as a plausible explanation for bonding in transition metal complexes. His application of VBT to transition metal complexes was supported by Bjerrum's work on stability that suggested electrostatics alone could not account for all bonding characteristics.

Although VBT is no longer a popular way to describe bonding in transition metal complexes, its historical importance requires discussion. In this model low-lying, empty d orbitals of transition metals (as with nontransition elements) are combined with appropriate s and p orbitals to form new hybrid orbitals that accept ligand electron pairs to form a coordinate covalent bond. Such a hybridization scheme for Werner's octahedral, $[Co(NH_3)_6]^{3+}$ complex is shown below. Magnetic studies revealed $[Co(NH_3)_6]^{3+}$ to be a diamagnetic (having no unpaired electrons) complex and, therefore, the six valence d electrons of cobalt(III) are paired in three d orbitals, leaving the other two d orbitals to form six d^2sp^3 hybrids (Fig. 1.4).

The paramagnetic (having unpaired electrons) octahedral cobalt(III) complex, CoF_6^{3-}, however, required hybridization of the $4s$, $4p$, and $4d$ orbitals. VBT could not explain this inconsistency. Further, although VBT accounted for the observed geometry, it could not predict geometry.

Magnetism and geometry were not the only two properties VBT could not predict: metal complex color also eluded this theory. Postulates put forth by Caven and Lander in 1939, which attributed the relationship of color and complex structure to "a looseness or an unsaturation in the electronic structures," or as Fajans stated, "a constraint or deformation of the electron systems of the co-ordinated molecules," could not be adequately explained, but because of Pauling's influence in the field, VBT was used by coordination chemists through the mid-1950s.

Crystal Field Theory

In 1951, chemists trying to make sense of metal complex optical spectra and color returned to an emphasis on the ionic nature of the coordinate covalent bond. Coordination chemists rediscovered physicists Hans Bethe's and John van Vleck's crystal field theory (CFT),

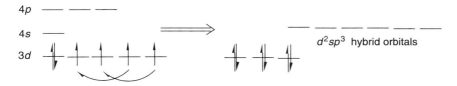

Figure 1.4   Valence bond theory: using orbital hybridization to describe bonding in transition metal complexes.

which accounted nicely for color and magnetic properties of crystalline solids. In contrast to VBT, CFT is an electrostatic model that ignores covalency in bonding and considers the metal ligand bond to be completely ionic. CFT thus takes a purely electrostatic approach to bonding and—though this is not fully realistic—this model is relatively successful at explaining many transition metal complex properties including color. CFT assumes:

- d electrons do not participate in bonding;
- ligands are point charges;
- metal–ligand bonds are purely electrostatic.

Although the following discussion focuses on CFT as applied to octahedral complexes exclusively, other geometries can be treated similarly. Recall the spatial representations of the five d-orbitals (Fig. 1.5). In an isolated gaseous ion (no ligands), these orbitals all have the same energy (i.e., are degenerate) and, when combined, form a spherical field of electron density about the metal ion. Placing ligands (represented as point charges ·) along the x, y, and z axes forms an octahedral ligand field around the metal ion. This ligand interaction splits the orbitals into two with higher energy (d_{x2-y2}, d_{z2}) and three of lower energy (d_{xy}, d_{xz}, d_{yz}). From group theory these are labeled e_g and t_{2g}, respectively, as seen in Figure 1.6. The orbitals are separated by an amount of energy, Δ_o (10 Dq), the crystal field splitting energy. The two e_g orbitals are repelled (+) by $0.6\Delta_o$ while the three t_{2g} orbitals are stabilized (−) by $0.4\Delta_o$ relative to the gas phase isolated ion (see Fig. 1.6). This preservation of the barycenter is common to all metal complexes despite geometry. These values are used to calculate the *crystal field stabilization energy* (CFSE) for a particular configuration.

For a d^4 $t_{2g}{}^3e_g{}^1$ configuration in an octahedral field for example, the CFSE is:

$$(3)(-0.4\Delta_o) + (1)(+0.6\Delta_o) = -0.6\Delta_o$$

A d^4 metal ion is therefore stabilized by $0.6\Delta_o$ in this configuration.

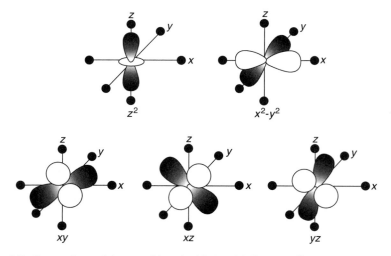

Figure 1.5   Interactions of the metal ion d orbitals with ligands, ●, in an octahedral field.

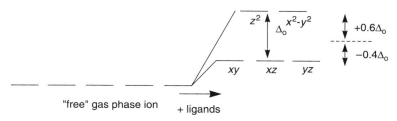

Figure 1.6 Crystal field splitting of metal ion d orbitals in an octahedral field of ligands.

Electrons fill the metal ion orbitals according to Hund's rule, but for d^4, d^5, d^6, and d^7 electronic configurations, electrons may fill the orbitals two different ways, resulting in high and low spin states. For example, in the case of d^6 cobalt(III), three electrons are immediately placed in the lower three t_{2g} orbitals. The fourth electron can either enter the higher energy e_g orbital, or pair with a t_{2g} orbital electron. Whether the former or latter occurs depends on the relative expense (in terms of overall energy) to pair the electrons (strong field, "low spin" case) vs putting them in the unfavorable e_g orbitals (weak field, "high spin" case). For a d^6 ion, the high spin ($t_{2g}{}^4 e_g{}^2$) case yields:

$$\text{CFSE} = (4)(-0.4\Delta_o) + (2)(+0.6\Delta_o) = -0.4\Delta_o$$

Note that there is *no* pairing energy, P, since the CFSE is the pairing *difference* between the ground state ion (no ligands) and the ion in the metal complex. Calculating the CFSE for a low spin ($t_{2g}{}^6$) case yields a stabilization energy of $-2.4\Delta_o + 2P$. When the Δ_o term is greater than the P term, the low spin state case is very favorable.

One of the factors governing the magnitude of Δ_o is the nature of the bound ligands. For a metal ion, the field strength increases according to the *spectrochemical series*:

weak

$$I^- < Br^- < SCN^- < Cl^- < N_3^-, F^-, H_2NC(O)NH_2, OH^- < ox^{2-}, O^{2-} < H_2O$$
$$< NCS^- < py, NH_3 < en < bpy, \, phen < NO_2^- < CH_3^- < C_6H_5^- < CN^-, CO$$

strong

Here CO and CN^- are strong field ligands and I^- is the weakest. The spectrochemical series explains why $Co(NH_3)_6{}^{3+}$ is low spin and diamagnetic while $CoF_6{}^{3-}$ is high spin and paramagnetic. Further, CFT as applied to coordination number 4 also explains nicely why many d^7 and d^8 complexes adopt a square planar over a tetrahedral geometry. Other factors affecting the crystal field splitting energy are oxidation state and row placement (principle quantum number) in the periodic table.

CFT is powerful in that it can be used to explain thermodynamic properties (Chapter 3), reactivity (Chapters 4 and 5) and color of metal complexes, as we will see below. However, as a model for bonding, it is not fully realistic. The spectrochemical series, for example, indicates that ligands such as CO cause a greater Δ_o (because of greater repulsion) than the small, negatively charged F^- and suggests that both ionic and covalent forces are important in transition metal bonding.

Molecular Orbital Theory and Ligand Field Theory

Molecular orbital theory (MOT), a quantum mechanical treatment of molecules that considers valence orbitals in chemical bonds for the analysis of electronic structural properties, when applied to transition metal complexes, is referred to as ligand field theory (LFT). LFT, unlike CFT, emphasizes the interaction of metal d valence orbitals with appropriate ligand orbitals, accounting for both covalency and ionicity in the bonding interactions and providing a more accurate description of the metal–ligand bond. For example, LFT indicates that pi-bonding capabilities of ligands at the high end of the spectrochemical series (CO) are responsible for the large increase in Δ_o. Discussion using LFT is similar to CFT; however, molecular orbital diagrams replace the simplified CFT diagram and the energy of stabilization is then referred to as ligand field stabilization energy (LFSE) instead of CFSE. By the mid-1960s coordination chemists had fully adopted the molecular orbital approach and our present computational methods are based on MOT. Some of these methods will be explored in Chapters 3 and 4 and in select advanced chapters.

ORIGIN OF COLOR IN TRANSITION METAL COORDINATION COMPLEXES[4]

Almost every metal ion forms coordination compounds with an accompanying color change (Table 1.3), and one of the great accomplishments of CFT is that it explains the origin of colors in transition metal complexes. We now know that changes in covalency, ligating atoms/molecules, ion packing (structure and impurities), and oxidation state give

TABLE 1.3 Colors of Select Aqueous (Uncomplexed) and Other Simple Monodentate[a] Ligand-Complexed Inorganic Ions. Note How Color Changes with Ligand

Aqueous Ion	Color	Aqueous Ion	Color
$Cr(H_2O)_6^{2+}$	Blue	$Fe(CN)_6^{4-}$	Yellow
$Cr(H_2O)_6^{3+}$	Blue-violet	$Fe(CN)_6^{3-}$	Red
$Cr_2O_7^{2-}$	Orange	$FeNCS^{2+}$	Red
$Fe(H_2O)_6^{2+}$	Pale blue-green	$FeOH^{2+}$	Yellow
$Fe(H_2O)_6^{3+}$	Pale violet	$VO_2Cl_2^-$	Yellow
FeO_4^{2-}	Purple	VS_3^{2-}	Black
$Cr(CN)_6^{3-}$	Yellow	$Ti(H_2O)_6^{3+}$	Purple
$Co(NH_3)_6^{3+}$	Orange-yellow	$Co(H_2O)_6^{2+}$	Red
$Co(NH_3)_5H_2O^{3+}$	Brick-red	$V(H_2O)_6^{2+}$	Violet
$Co(NH_3)_5Cl^{2+}$	Violet-red	$V(H_2O)_6^{3+}$	Green
$Ni(NH_3)_6^{2+}$	Violet-blue	VO^{2+}	Blue
$Ni(NO_2)_4^{2-}$	Yellow	VO_2^+	Yellow
$Mn(H_2O)_6^{2+}$	Pale rose	$Mn(CN)_6^{5-}$	Colorless
MnO_4^{3-}	Blue	$Mn(CN)_6^{4-}$	Blue
MnO_4^{2-}	Green	$Mn(CN)_6^{3-}$	Red
MnO_4^-	Purple	$PtCl_6^{2-}$	Yellow
$Ni(H_2O)_6^{2+}$	Green	PtI_6^{2-}	Black
$Pt(H_2O)_4^{2+}$	Yellow	$Pt(NO_2)_4^{2-}$	Colorless

[a]*Denticity* refers to the number of atoms that can simultaneously bind to one metal center.

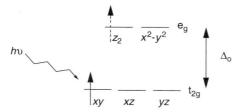

Figure 1.7 Absorption of light by a metal complex promotes a $d-d$ transition responsible for the color absorbed.

rise to different colors in transition metal systems because such changes alter the energy gap, Δ, between the highest occupied molecular orbital (HOMO) and the lowest unoccupied molecular orbital (LUMO) in a transition metal complex. The energy gap, Δ, is related to the energy, E, in Planck's equation (1.1):

$$\Delta \approx E = hc/\lambda \qquad (1.1)$$

Absorption of light of visible wavelength promotes an electronic transition as shown for the d^1 case in Figure 1.7. The light that is not absorbed is that which is transmitted and what we observe, giving a metal complex its characteristic color. Using CFT, we can understand just how changes in the environment of a metal ion (including ligands and geometry) can modify the orbital energies and therefore its color. Chapters 2–5 cover aspects of visible spectroscopy more fully.

COORDINATION COMPLEX REACTIVITY[7,12–17]

Substitution Reactions

Even back in 1912, Werner was keenly aware of the need to understand metal complex substitution. Stereochemical changes [e.g., *cis–trans* conversions, equation (1.2)] in cobalt(III) complexes were critical to the development of his coordination theory and he sought to understand how these occurred.

$$\text{trans-}[Co(en)_2Cl_2]^+ \longrightarrow \text{cis-}[Co(en)_2NH_3Cl]^{2+} \qquad (1.2)$$

Werner, in 1912, concluded: "when such a molecule (in the second sphere) gets included in the first sphere an acid group becomes transferred from the first sphere into the second." Werner's proposed mechanism for this type of rearrangement is shown in Figure 1.8. The ethylenediamine ligand in the cobalt complex is known as a *chelating* ligand, which is a molecule that binds the metal using two or more donating atoms.

There are many other reasons why coordination chemists study substitution kinetics, one of the most common being that substitution provides a route for the synthesis of new coordination complexes. The synthesis of $[Co(en)_3]^{3+}$ from Werner's purpureo complex, $[Co(NH_3)_5Cl]^+$, by ethylenediamine substitution, equation (1.3), and the synthesis of $[Cu(NH_3)_4]^{2+}$ by NH_3 substitution of the aqua complex, $[Cu(H_2O)_4]^{2+}$, equation (1.4), are two notable examples.

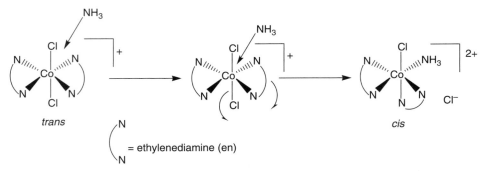

Figure 1.8   Werner's proposed mechanism for *trans* to *cis* rearrangement upon substitution.

$$[Co(NH_3)_5Cl]^{2+} + 3en \longrightarrow [Co(en)_3]^{3+} + 5NH_3 + Cl^-$$
$$\text{purple} \qquad\qquad\qquad \text{yellow}$$
$$(1.3)$$

$$[Cu(H_2O)_4]^{2+} + 4NH_3 \longrightarrow [Cu(NH_3)_4]^{2+} + 4H_2O$$
$$\text{blue} \qquad\qquad\qquad\qquad \text{dark blue}$$
$$(1.4)$$

The reaction in equation (1.3) involving cobalt(III) is very slow at room temperature (the reaction must be heated to take place), while that in equation (1.4) involving copper(II) is rapid at room temperature, despite similar NH_3 group ligation. The substitution rates depend markedly on the nature of the metal ion.

Nobel prize-winning coordination chemist Henry Taube defined substitution in 1951 as *a net process in which one group coordinated to the central ion is replaced by another, without change in oxidation state for any of the atoms participating*, and systematically compiled all known rate data. Taube noted how, along with the nature of the central metal ion, variations in the properties of the entering group, the group being replaced, and the remaining groups bound to the central atom affect the rate of reaction.

Years of experience led to a general classification of substitution reactions. The mechanism nomenclature of the 1950s and early 1960s, which was consistent with that of organic chemistry (S_N1, S_N2, etc.), was replaced by *stochiometric mechanisms*— associative (A), dissociative (D), associative interchange (I_a), and dissociative interchange (I_d)—by Harry Gray and Cooper H. Langford in their 1966 publication, *Ligand Substitution Processes*. General schemes for these mechanisms are noted below in equations (1.5)–(1.7).

$$MX_n \underset{+X}{\overset{-X}{\rightleftharpoons}} MX_{n-1} \underset{-Y}{\overset{+Y}{\rightleftharpoons}} MX_{n-1}Y$$
$$(1.5)$$

$$MX_n \underset{-Y}{\overset{+Y}{\rightleftharpoons}} MX_nY \underset{+X}{\overset{-X}{\rightleftharpoons}} MX_{n-1}Y$$
$$(1.6)$$

$$Y \dashrightarrow MX_{n-1} \dashrightarrow X$$
$$\text{ion pair}$$
$$(1.7)$$

The D and A pathways proceed through intermediates of reduced and increased coordination numbers, respectively. The I mechanisms are characterized by the lack of an intermediate with a modified metal ion coordination number in the reaction. When bond breaking is more important than bond making the mechanism is I_d; the transition state has a reduced coordination number. In an I_a mechanism bond making is more important than bond breaking; the transition state has an increased coordination number.

The importance of substitution reactions cannot be overstated. Systematic investigations of coordination compound substitution kinetics, and mechanisms shed light on the electronic structure of compounds and on their interactions. Although formally taken up in Chapter 4, substitution is encountered in all chapters of this book.

Electron Transfer Reactions

Valence d electrons of transition metals impart special properties (e.g., color and substitution reactivity) to coordination complexes. These valence electrons can also be removed completely from (oxidation) or added to (reduction) metal d orbitals with relative facility. Such oxidation–reduction (redox) reactions, like substitution reactions, are integral to metal complex reactivity. Consider the role of redox chemistry in the synthesis of $[Co(NH_3)_5Cl]^+$, equation (1.8). In general, the preparation of cobalt(III) complexes (Chapters 2 and 5) starts with substitutionally labile cobalt(II) salts that are combined with appropriate ligands with subsequent oxidation of the metal by H_2O_2 or O_2 to the substitutionally inert (robust) $+3$ state.

$$[Co(H_2O)_6]^{2+}_{(aq)} \xrightarrow[\text{brown solution}]{NH_3, NH_4Cl} \xrightarrow{H_2O_2 \text{ or } O_2} \xrightarrow{HCl} [Co(NH_3)_5Cl]^{2+}_{(aq)} \quad (1.8)$$
$$\text{pink} \qquad\qquad\qquad\qquad\qquad\qquad\qquad\qquad\qquad \text{purple}$$

Equation (1.8) also shows how redox reactions are intricately linked to substitution. As Taube stated in his Nobel lecture: "While substitution reactions can be discussed without concern for oxidation reduction reactions, the reverse is not true." The substitution properties of both cobalt(III) and cobalt(II) metal ions provides the rationale for this synthetic methodology.

The study of electron transfer reactions began in earnest when radioactive isotopes, produced for nuclear research and the atom bomb program during World War II, became accessible. Glen Seaborg, in a 1940 review of artificial radioactivity, noted the first attempt to measure the *self-exchange* reaction between aqueous iron(III) and iron(II), equation (1.9).[†]

$$[^*Fe(aq)]^{2+} + [Fe(aq)]^{3+} \longrightarrow [^*Fe(aq)]^{3+} + [Fe(aq)]^{2+} \quad (1.9)$$
$$\text{colorless} \qquad \text{yellow} \qquad\qquad \text{yellow} \qquad \text{colorless}$$

By 1950, such isotopic tracer methods began to revolutionize the study of redox reactions as color alone could not always be used to distinguish product formation; see equation (1.9). The importance of H^+ and other ions on electron transfer rates was soon discovered. A symposium on electron transfer took place in 1951 at the University of Notre Dame, during which a distinction between outer- and inner-sphere electron transfer was made.

[†]Equation (1.9) is an example of a "self-exchange" electron transfer reaction for which $\Delta G° = 0$. One cannot tell the difference between reactant and product by color alone, hence the need for isotopic labeling indicated by the asterisk.

In 1954, King and Taube published the 1980 Nobel Prize winning work that defined these two different types of electron transfer reactions. In an inner-sphere mechanism, the atoms undergoing redox form bonds to a common atom (or small group of atoms), which then serves as a bridge for electron transfer (ISPC = inner-sphere precursor complex and k_{et} = electron transfer rate constant).

$$
\begin{aligned}
A^{red} - X + B^{ox} &\rightleftharpoons A^{red} - X - B^{ox} \quad \text{ISPC} \\
A^{red} - X - B^{ox} &\rightarrow A^{ox} - X - B^{red} \quad k_{et} \\
A^{ox} - X - B^{red} &\rightarrow A^{ox} + X - B^{red} \\
&\quad\ \ \text{or } A^{ox} + B^{red}
\end{aligned}
$$

The bridging group, X, does not necessarily transfer from A to B; however, if this happens, it is strong evidence that an inner-sphere reaction has taken place. Indeed, this sort of evidence was how Taube first solved the puzzle. In the reduction of $[Co(NH_3)_5Cl]^{2+}$ by $[Cr(H_2O)_6]^{2+}$, the Cl^- on the inert cobalt(III) complex readily displaces a H_2O molecule at the labile chromium(II) center to form the bridged species:

$$[(H_3N)_5Co-Cl-Cr(OH_2)_5]^{4+}$$

Electron transfer from the chromium(II) center to the cobalt(III) center within this inner-sphere precursor complex results in a labile cobalt(II) center and inert chromium(III) center. Concurrent with electron transfer, atom transfer (Cl^-) also takes place as Cl^- remains bound to the inert chromium(III) center to yield $[ClCr(OH_2)_5]^{2+}$ and "$[(H_3N)_5Co]^{2+}$" decomposes to $[Co(H_2O)_6]^{2+}$. In 1961, Norman Sutin was able to provide direct mechanistic proof of atom transfer in the inner-sphere process by his introduction of the stopped-flow (rapid mixing) techniqe. The transition state in an inner-sphere process is relatively well defined, and because the kinetics are complex, the overall mechanism is also well defined.

The outer-sphere pathway is obtained when electron transfer is more rapid than substitution, and electron transfer takes place through the intact coordination spheres of the reaction partners (OSPC = outer-sphere precursor complex):

$$
\begin{aligned}
A^{ox} + B^{red} &\rightleftharpoons [A^{ox}, B^{red}] \quad \textit{OSPC} \text{ or ion pair} \\
[A^{ox}, B^{red}] &\rightarrow [A^{red}, B^{ox}] \quad k_{et} \\
[A^{red}, B^{ox}] &\rightarrow A^{red} + B^{ox}
\end{aligned}
$$

Unlike the inner-sphere reaction, the kinetic simplicity of outer-sphere reactions makes it difficult to obtain mechanistic details about the reactants in the electron transfer process. Experimental investigations using optically active metal complexes, along with theoretical calculations pioneered by Rudolph Marcus, have shed light on factors governing outer-sphere electron transfer processes.

The rates of many electron transfer reactions have now been measured and, as new coordination complexes are prepared, characterized, and their solution properties studied, our understanding of fundamental structure (both geometric and electronic)–reactivity relationships continues to grow. Both inner- and outer-sphere reactions will be explored in Chapter 5.

APPLICATIONS OF COORDINATION COMPLEXES[18]

As the fundamentals of coordination complex structure and reactivity were realized, new and exciting applications were discovered. In 1955, Ziegler and Natta and their coworkers developed a titanium-based catalyst, Figure 1.9, which catalyzes the polymerization of alkenes at atmospheric pressure and ambient temperature. The open coordination site allows an alkene ligand such as ethylene or propylene to bind to the Ti center, initiating the polymerization. Their work sparked tremendous interest in the field and garnered them the Nobel Prize in 1963.

In 1962 Max Perutz and John Kendrew earned the Nobel Prize for solving the crystal structures of metalloproteins myoglobin and hemoglobin. Central to function of these proteins is their iron heme coordination complex shown in Figure 1.10. The iron center serves to bind dioxygen, transporting and delivering this important life-sustaining molecule to all parts of the body.

Barnett Rosenberg, in 1969, serendipitously discovered the cytotoxic effects of *cis*-platin, Figure 1.11, a complex studied in Chapter 6. The area of metals in medicine has grown so rapidly that new journals and conferences dedicated solely to this topic now flourish.

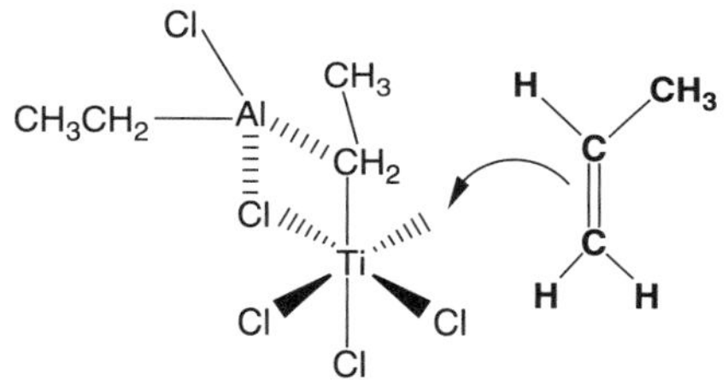

Figure 1.9 The Ziegler–Natta TiCl$_x$ catalyst.

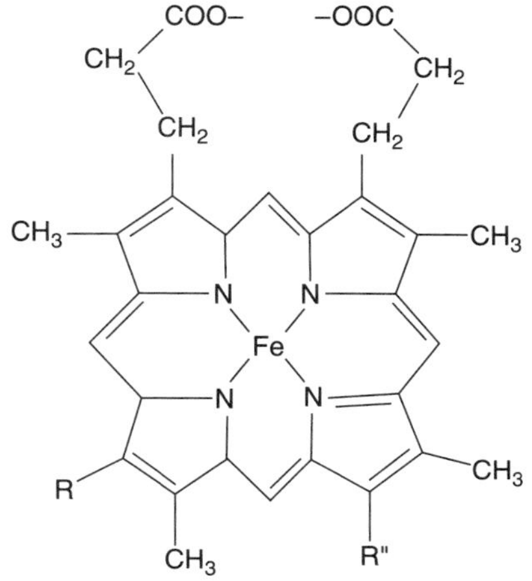

Figure 1.10 An iron heme complex.

Figure 1.11 The structure of *cis*-platin.

Applications and roles for new metal coordination complexes continue to be discovered daily. Coordination complexes in the development of new architectural materials such as nanostructures and in environmental applications (e.g., "green" catalysts and bioremediation; Chapter 7) are on the frontiers of inorganic chemistry in the twenty-first century; however, even these exciting and splashy new systems remain governed by the fundamentals of metal ion chemistry. Advancements in applied fields cannot be fully realized without the knowledge imparted by pure and basic research, especially as metal ions are placed in novel environments.

ASPECTS OF INORGANIC SYNTHESIS—GOVERNING FACTORS[2,19-25]

Why do inorganic chemists prepare new compounds? One reason is simple curiosity! The synthesis of the many fascinating borohydrides (B_xH_y), for example, came about when, in 1912, Alfred Stock speculated whether boron could form a variety of compounds similar to carbon, its neighbor on the periodic table. New syntheses also emerge by accident or chance coupled with astute observation. The anticancer agent, *cis*-platin (Chapter 6), discovered by Rosenberg in 1969, is such an example. Other complexes are designed and synthesized to fulfill a particular need. Many metal coordination complex catalysts have been designed to carry out specific organic reactions (Chapter 9), and other metal systems have been designed to recognize and bind specific biological targets (Chapter 8). Chemists sometimes synthesize new compounds to test theories. Alfred Werner synthesized many new metal complexes in support of his coordination complex bonding theory, including an optically active complex free of carbon. In fact, a combination of curiosity and desire to test theory is what inspired both Alfred Werner and Danish chemist, Sophus Jørgensen, the chemist who laid down the groundwork for Werner's contributions, to study metal ammine complexes (Chapter 2).

Synthesis: An Interplay Between Thermodynamics and Kinetics

Just like any other chemical reactions, synthetic reactions adhere to the principles of thermodynamics: the study of energy and its interconversions. Is your desired product thermodynamically stable (can it be made and will it exist at room temperature and atmosphere)? Will it be stable under your given synthetic conditions, including during and after isolation? Gibb's free energy, ΔG, is our measure of the driving force for reactions, equation (1.10).

$$\Delta G = -RT \ln Q = \Delta H - T\Delta S = -nFE \qquad (1.10)$$

When the reactants and products are at unit activity, equation (1.10) is expressed with standard state functions, equation (1.11):

$$\Delta G° = -RT \ln K = \Delta H° - T\Delta S° = -nFE° \tag{1.11}$$

A general expression for metal complex formation is given in equation (1.12). The formation constant, K_f, reflects the thermodynamic feasibility of complex formation.

$$M + L \rightleftharpoons ML \quad K_f = [ML]/[M][L] \tag{1.12}$$

For example, the reaction of aqueous iron(III) with the hexadentate ligand, $EDTA^{4-}$, equation (1.13), has a log $K_f = 25.0$ (25 °C, 0.1 M ionic strength). The large equilibrium constant tells us that the reaction as written is expected to favor the $FeEDTA^-$ product (i.e., the product should be very stable under these conditions in solution)—and it does!

$$Fe^{3+}{}_{(aq)} + EDTA^{4-}{}_{(aq)} \longrightarrow FeEDTA^- \quad K_f = 10^{25} \tag{1.13}$$

Formation constants for many metal complexes have been compiled by Ramunas Motekaitis and Art Martell, and these as well as techniques for measuring them in the laboratory will be covered in Chapters 3 and 8. One can, however, predict the relative stability of a desired complex based on simple bonding theories. Crystal field theory, as well as the Irving–Williams series and Pearson's hard–soft–acid–base theory (see the next section) enable us to predict what might happen in solution.

Knowing and understanding standard reduction potentials, E^0, also helps determine whether a particular oxidation state will be stable and appropriate oxidants and reductants to use in a synthetic scheme. In reviewing the cobalt complex syntheses in Chapters 2 and 6, for example, complex ions are formed by oxidizing cobalt(II) salts to the more stable +3 state.

Thermodynamic data only suggest whether a reaction is able to take place (i.e., if $\Delta G < 0$): They do not reveal how long it will take for the reaction to occur. Consequently, both thermodynamic and kinetic factors must be considered when devising a synthesis. Consider an example pertinent to the synthesis in Chapter 2. Thermodynamics predicts that $[Co(NH_3)_5Cl]^{2+}$ should be thermodynamically favorable compared with $[Co(NH_3)_6]^{3+}$, equation (1.14):

$$[Co(NH_3)_6]^{3+} + Cl^- \rightleftharpoons [Co(NH_3)_5Cl]^{2+} + NH_3 \quad K = 2 \times 10^4 \tag{1.14}$$

Although this reaction is thermodynamically favorable (large K), sluggish substitution at the cobalt(III) center of $[Co(NH_3)_6]^{3+}$ makes this reaction almost nonexistent under the given laboratory conditions. Vigorous conditions must be used, reinforcing the rule that simply knowing a reaction is thermodynamically favorable does not ensure that the reaction will be complete in a reasonable time frame. Alternatively, consider a metal complex that is thermodynamically stable but kinetically labile with respect to substitution at the metal center. Such a situation may hinder product isolation. The formation and rapid breakdown of the *violeto* complex, cis-$[Co(NH_3)_4Cl_2]^+$, for example, caused Werner considerable problems during his early defense of coordination theory, equation (1.15).

$$cis\text{-}[Co(NH_3)_4Cl_2]^+ \xrightarrow[\text{solvent}]{H_2O} cis\text{-}[Co(NH_3)_4(H_2O)_2]^{3+} \qquad (1.15)$$

A successful synthesis may require careful manipulation of reaction temperature. The rate of a reaction depends on the activation energy, E_a (the energy to reach the transition state) and this energy barrier, in turn, depends on temperature. The Arrhenius expression, equation (1.16), shows that the rate constant and, therefore, for most instances the rate, for a reaction will increase with an increase in temperature.

$$k = Ae^{-E_a/RT} \qquad (1.16)$$
$$\ln k = -E_a/RT + \ln(A)$$

However, if a reaction is exothermic ($\Delta H < 0$), Le Chatelier's principle states that increasing the temperature during the reaction will push the thermodynamic equilibrium towards reactants. The classic example of this dilemma is seen in the Haber synthesis of ammonia, equation (1.17).

$$N_2(g) + 3H_2(g) \longrightarrow 2NH_3(g) \quad \Delta H^\circ = -92\,kJ \qquad (1.17)$$

Although the equilibrium lies far to the right ($K \approx 10^6$), a large amount of energy (E_a) is required to break the N_2 triple bond. Because of the exothermicity, however, K decreases significantly when the temperature is increased. The Haber process requires the use of an iron-based catalyst at 250 atm and $\sim$400 °C.

Acid–Base Chemistry (Hard–Soft–Acid–Base Theory)

One of the simplest ways to predict the formation and stability of metal–ligand bonds is by using hard–soft–acid–base theory.[†] In 1968, Ralph Pearson noted that some Lewis acids prefer first row donors (bases). These "class a" acids (or hard acids) include all ions from the s block, other cations with charge $\geq +3$, and lighter transition metals with charges $> +1$ (Table 1.4). Other Lewis acids, "class b" acids, were shown to prefer second and third row donors. These soft acids include lower oxidation state metals (di- and monovalent metals), cations from the p block with a charge $< +3$ and heavier transition metals with a charge $< +3$. The soft bases are characterized by high polarizability and low charge density. In general, hard acids prefer hard bases and are stabilized by ionic-type interactions while and soft acids prefer soft bases, stabilized by covalent-like bonds. As seen in Table 1.4, metal ions are potent Lewis acids.

Because most coordination complex syntheses are carried out in aqueous solution, the formation of metal hydroxides, equation (1.18), which are less substitutionally labile and are often insoluble, can pose substantial problems.

$$M(H_2O)^{n+} \longrightarrow M(OH)^{(n-1)+} + H^+ \qquad (1.18)$$

[†]R. Bruce Martin (*Inorg. Chim. Acta* **1998**, *283*, 30–36) points out the limitations of HSAB theory, showing that metal ion stabilities correlate better with electron affinities than with hardness and softness. Still, this is a good qualitative general rule.

TABLE 1.4 Examples of Hard, Intermediate and Soft Acids and Bases

	Hard	Intermediate	Soft
Acids	H^+, Li^+, Na^+, K^+, Be^{2+}, Mg^{2+}, Ca^{2+}, Sr^{2+}, Mn^{2+}, Al^{3+}, Sc^{3+}, Ga^{3+}, In^{3+}, La^{3+}, Gd^{3+}, Cr^{3+}, Co^{3+}, Fe^{3+}, N^{3+}, As^{3+}, Si^{4+}, Ti^{4+}, U^{4+}, Ce^{3+}, Sn^{4+}, VO^{2+}	Fe^{2+}, Co^{2+}, Ni^{2+}, Cu^{2+}, Zn^{2+}, Pb^{2+}, Sn^{2+}, Sb^{3+}, Bi^{3+}, Rh^{3+}, Ir^{3+}, Ru^{2+}, Os^{2+}	Cu^+, Ag^+, Au^+, Tl^+, Hg^+, Pd^{2+}, Cd^{2+}, Pt^{2+}, Hg^{2+}, $Co(CN)_5{}^{2-}$, Pt^{4+}, Te^{4+}
Bases	H_2O, OH^-, F^-, CH_3COO^-, $PO_4{}^{3-}$, $SO_4{}^{2-}$, Cl^-, $ClO_4{}^-$, $NO_3{}^-$, NH_3, RNH_2	$C_6H_5NH_2$, C_5H_5N, $N_3{}^-$, Br^-, $NO_2{}^-$, $SO_3{}^{2-}$, N_2	R_2S, RSH, RS^-, I^-, SCN^-, R_3P, CN^-, RNC, CO, C_2H_4, C_6H_6, H^-, R^-

TABLE 1.5 Acid Hydrolysis Constants (pK_as) for Some Aquo Ions and Complex Metal Ions at 25 °C (Selected from Basolo and Pearson,[22] and pH Precipitation Ranges of Hydroxides or Hydrous Oxides (Adapted from Walton[23])

Ion	pK_a	Ppt Range	Ion	pK_a	Ppt Range
Tl^+	13.2		Ti^{3+}	2.2	0–2
$Hg_2{}^{2+}$	5.0		Cr^{3+}	3.8	4–6
Mg^{2+}	11.4	>10	Co^{3+}	0.7^a	
Ca^{2+}	12.6		Fe^{3+}	2.2	2–4
Ba^{2+}	13.2		($pK_{a2} = 3.3$)		
Mn^{2+}	10.6	8–10	$Co(NH_3)_5H_2O^{3+}$	6.6^b	
Fe^{2+}	9.5	6–8	cis-$Co(NH_3)_4(H_2O)_2$	6.0^b	
Co^{2+}	8.9	6–8	$Ru(NH_3)_5H_2O^{3+}$	4.2	
Ni^{2+}	10.6	6–8	$Rh(NH_3)_5H_2O^{3+}$	5.9	
Cu^{2+}	6.8	4–6	$Cr(NH_3)_5H_2O^{3+}$	5.3^b	
Zn^{2+}	8.8	6–8	cis-$Co(en)_2(H_2O)_2$	6.1	
Cd^{2+}	9.0			($pK_{a2} = 8.2$)^b	
Hg^{2+}	3.7		$trans$-$Co(en)_2(H_2O)_2$	4.5	
	($pK_{a2} = 2.6$)			($pK_{a2} = 7.9$)^b	
Sn^{2+}	3.9		cis-$Cr(en)_2(H_2O)_2$	4.8	
Pb^{2+}	7.8			($pK_{a2} = 7.2$)^b	
Al^{3+}	5.1	2–4	$trans$-$Cr(en)_2(H_2O)_2$	4.1	
Sc^{3+}	5.1			($pK_{a2} = 7.5$)^b	
In^{3+}	4.4		cis-$Co(en)_2(H_2O)(NO_2)^{2+}$	6.3	
	($pK_{a2} = 3.9$)		$trans$-$Co(en)_2(H_2O)(NO_2)^{2+}$	6.4	
Tl^{3+}	1.1		cis-$Pt(NH_3)_2(H_2O)_2{}^{2+}$	5.6	
	($pK_{a2} = 1.5$)			($pK_{a2} = 7.3$)^c	
Bi^{3+}	1.6		$trans$-$Pt(NH_3)_2(H_2O)_2{}^{2+}$	4.3	
V^{3+}	2.8			($pK_{a2} = 7.4$)^c	
			$Pt(NH_3)_5H_2O^{4+}$	4	
			$Pt(en)_2(H_2O)_2{}^{4+}$	Strong acid	

Systems are dilute unless indicated with ^{a}3 M $NaClO_4$ or ^{b}1 M $NaNO_3$. ^{c}Reaction at 20 °C.

Most metal ions effectively reduce the pK_a ($-\log$ of the acid dissociation constant, K_a) of a solvent water molecule, so knowing the pK_a values of metal aquo complexes can be useful. Some values are given in Table 1.5.

As with syntheses that occur at relatively high pH, acid–base reactions must be considered for those taking place at low pH. Being such strong Lewis acids, metal ions compete effectively for hydrogen ions bound to a ligand, LH. However, depending on the acid–base chemistry of the ligand and of the metal ion, placing a metal complex under very acidic conditions may result in a loss of the ligand through protonation, equation (1.19).

$$ML + H^+ \longrightarrow M^+ + LH \tag{1.19}$$

As you become familiar with the metal ion with which you are working, designing and manipulating experimental conditions for synthesis and further application will become more straightforward.

Stoichiometry

Once a synthetic route for preparing a complex has been designed, the stoichiometry of the reaction will help to determine the minimum amount of reagents needed. Often, excess reagents are used to maximize the yield of a desired product, in compliance with Le Chatelier's principle. A move towards "greener" approaches, however, asks chemists to think of a new definition of maximizing yield, one in terms of total atoms rather than just the desired product. This new approach asks synthetic chemists to adhere as closely as possible to the stoichiometric amounts.

The Art of Precipitation: Properties of Precipitates

Most often, synthetic products are obtained by precipitation from the reactant solution and purified by recrystallization. This section introduces some aspects of these processes.

Product Solubility   Precipitates vary widely in solubility. Metal complexes in particular are often polar and charged. This fact taken with the consideration that most syntheses are carried out in polar aqueous media, can lead to high solubility losses in the desired material. Careful control of precipitation and washing is necessary.

Simple general chemistry principles can be used to help maximize precipitation and product yield. The first is the *common ion effect*, formally defined as making a weak acid or weak base weaker by the addition of a salt that contains a common ion, a direct consequence of Le Chatlier's principle. The common ion effect can also be used to understand metal complex precipitation. Consider a saturated solution of $[Co(NH_3)_6]Cl_3$, equation (1.20).

$$[Co(NH_3)_6]Cl_{3(S)} \rightleftharpoons [Co(NH_3)_6]^{3+}_{(aq)} + 3Cl^-_{(aq)} \tag{1.20}$$

According to the solubility product relationship, equation (1.21), if the concentration of one ion is raised, the other must drop. Doubling the chloride ion concentration by addition of NaCl(s) for example, will divide the concentration of $[Co(NH_3)_6]^{3+}$ in equilibrium with the solid by a factor of 8, maximizing precipitation.

$$[Co(NH_3)_6]^{3+}[Cl^-]^3 = K_{sp} = 1.2 \times 10^{-1} \tag{1.21}$$

Another important factor for coordination chemists is the general principle stated simply by Basolo: "solid salts separate from aqueous solution easiest for combinations of either small cation–small anion or large cation–large anion, preferably with systems having the same but opposite charges on the counter ions." Other factors governing solubility of salts are pH, interionic attraction, and the diverse ion effect, which will not be covered in detail here.

Particle Size of Precipitates and Product Isolation   Three types of precipitates you may encounter are: crystalline, $0.1–1\,\mu m$ (e.g., $BaSO_4$), curdy, approximately $0.02 \geq x \leq 0.1\,\mu m$ (e.g., AgCl), and gelatinous, $\leq 0.02\,\mu m$ (e.g., hydrous Fe_2O_3). The more slowly a precipitate forms, the more effective the crystal packing and the larger and purer are the particles. Thus controlling the speed of precipitation (or particle size control) is often important in inorganic synthesis. The rate of crystal growth is influenced by several factors, including temperature, metal complex character and concentration, solvent, and nature of the counter ion.

The more concentrated a solution is, the greater the speed of precipitation. This is known as von Weimarn's rule: *the rate of precipitation can be slowed down by lowering the concentrations of the reactants and raising the solubility of the precipitate.* The solubility of the precipitate can be altered using a cosolvent as well as by controlling temperature and taking into consideration the factors mentioned above and by the nature of the counter ion. If the smallest, curdy or gelatinous precipitates form, they can pass through normal filters and special procedures (e.g., multiple filter papers and/or multiple filtering) must be taken to trap these precipitates. These precipitates should be washed with an electrolyte solution to avoid dispersion of aggregates.

Crystal packing of a precipitate (particle size) also influences the color of a precipitate. For example, large crystals of solid NaFeEDTA are dark brown in color, while powder precipitate is yellow. Do not be fooled by comparing the color of your product with a lab-mate's color!

Impurities in Precipitates   Coprecipitation is the process by which soluble impurities become incorporated into precipitates during their formation by either incorporation into the crystal lattice or by surface adsorption. In the former, less common process, the coprecipitates interlace. Once this occurs the impurity cannot be simply washed out of a desired product. An undesired coprecipitate may be converted into another form (e.g., a more soluble salt) prior to precipitation. Impurities due to surface adsorption are more common. For an impure crystalline substance, again the contaminant cannot just be washed out. The impure product should be redissolved and reprecipitated in a process called "Ostwald ripening," the dissolving of small particles followed by a redeposition of the material on the larger particles. If coprecipitated impurities are sensitive to pH, recrystallization in alkali or acid can be used to remove these. For curdy and gelatinous precipitates, because of the difference in precipitate size, coprecipitate impurities can often be washed out using a dilute electrolyte (e.g., HNO_3) that can then be removed by volatilization upon drying.

Postprecipitation is the precipitation of an impurity after the main product precipitate has been formed. With postprecipitation, contamination increases the longer the precipitate is in contact with its mother liquor. Contamination is faster at higher temperatures and can amount to even 100% by weight of the desired precipitate. Timing, temperature, and pH are often critical factors in minimizing postprecipitation problems. A challenging example of this process is encountered with the synthesis of $Ca[Co(ox)_2en]$ in Chapter 5.

Purification by Recrystallization Most impurities can be removed by the process of recrystallization. Recrystallization involves several steps that you should commit to memory:

- dissolving crude solid product in a minimum amount of solvent in which the product is slightly soluble, preferentially at a mild but elevated temperature;
- filtering the saturated solution by vacuum filtration to remove any insoluble particulates;
- allowing the product to crystallize from solution at room or reduced temperature (scratching the sides of the beaker often helps by providing a surface nucleation site for crystallization);
- filtering the purified product.

This section alerts you to only a few important considerations that should be kept in mind during synthesis. Keep a log in your notebook of others you may discover along the way.

One of the most exciting and challenging aspects for the experimental inorganic coordination chemist today is the need to draw on knowledge from all areas of chemistry: physical, organic, analytical, instrumental, and biochemistry. We hope this text convinces you of this. Although synthetic routes are outlined for you throughout the text, you should pay close attention to fundamental chemical principles governing a successful synthesis. Thinking scientifically about a synthesis will help you remain calm should mistakes occur during your procedure, as well as provide flexibility in modifying conditions during more advanced or first-time syntheses.

REFERENCES

1. webexhibits.org/pigments/indiv/technical/egyptblue.html (last accessed June 2004).
2. Zumdahl, S. S. *Chemical Principles*, 4th edn. Houghten Mifflin: Boston, MA, 2002.
3. Brock, W. H. *The Norton History of Chemistry*. Norton: New York, 1993.
4. Clifford, A. F. *Inorganic Chemistry of Qualitative Analysis*. Prentice-Hall: Englewood Cliffs, NJ, 1961.
5. Ball, P. *Bright Earth: Art and the Invention of Color*. Farrar, Straus and Giroux: New York, 2002.
6. Mason, S. F. *A History of the Sciences*. Macmillan: New York, 1962.
7. Basolo, F., Johnson, R. *Coordination Chemistry—the Chemistry of Metal Complexes*. Benjamin: New York, 1964.
8. Caven, R. M., Lander, G. D. *Systematic Inorganic Chemistry*. Blackie: London, 1948.
9. Lagowksi, J. J. *The Chemical Bond*. Houghten Mifflin: Boston, MA, 1966.
10. Pauling, L. *Nature of the Chemical Bond—and the Structure of Molecules and Crystals*, 2nd edn. Cornell University Press: Ithaca, NY, 1948.
11. Gray, H. B. *Chemical Bonds*. Benjamin: Menlo Park, CA, 1972.

12. Bailar, J. C. *Chemistry of the Coordination Compounds*. Monograph series no. 131. Reinhold: Baltimore, MD, 1956 (see p. 130 for Fajan's comments on color).

13. Kauffman, G. B., symposium chairman, *Werner Centennial*. Advances in Chemistry Series no. 62. American Chemical Society: Washington, DC, 1967.

14. Taube, H. *Chem. Rev.* **1952**, *50*, 69–126.

15. Langford, C. H., Gray, H. B. *Ligand Substitution Processes*. Benjamin: New York, 1966.

16. Taube, H., King, E. L. *J. Am. Chem. Soc.* **1954**, *76*, 4053.

17. Taube, H. *Electron Transfer Reactions of Complex Ions in Solutions*, Loebl, E. M. (ed.). Academic Press: New York, 1970.

18. Miessler, G. L., Tarr, D. A. *Inorganic Chemistry*, 3rd edn. Pearson Prentice Hall: Upper Saddle River, NJ, 2004.

19. Jolly, W. L. *Synthetic Inorganic Chemistry*. Prentice-Hall: Englewood Cliffs, NJ, 1960.

20. Bjerrum, J. *Metal Ammine Formation in Aqueous Solution*. Haase: Copenhagen, 1941.

21. Martell, A. E., Motekaitis, R. J. *Determination and Use of Stability Constants*, 2nd edn. Wiley-VCH: New York, 1992.

22. Basolo, F., Pearson, R. G. *Mechanisms of Inorganic Reactions: A Study of Metal Complexes in Solution*, 2nd edn. J Wiley: New York, 1967.

23. Walton, H. F. *Principles and Methods of Chemical Analysis*. Prentice Hall: New York, 1952.

24. Pearson, R. G. *J. Chem. Educ.* **1968**, *45*, 581–587.

25. Basolo, F. *Coord. Chem. Rev.* **1968**, *3*, 213–223.

Werner's Notion—Creating the Field: Synthesis and Analysis of Cobalt Ammine Coordination Compounds[1-15]

Nothing whatever is hidden; From of old, all is clear as daylight (Zen Verse)

PROJECT OVERVIEW

In 1893, at the age of 26, Alfred Werner "stepped outside the box" and proposed his coordination theory that revolutionized the field of inorganic chemistry. Werner's work, in part, made use of colorful cobalt compounds prepared by the reaction of ammonia with cobalt salts. Fortunately, these cobalt salts are quite stable ("robust") in aqueous solution. This stability, along with the rapid development of qualitative and quantitative analytical techniques, enabled Werner to carry out the rigorous and in-depth analyses that led to the determination of molecular formulas and prediction of structures and geometries. His work opened the door to understanding bonding and stability (Chapter 3) and reactivity (Chapters 4 and 5) in metal coordination complex systems.

This multiweek project begins with a dry lab exercise that introduces you to the historical development of Werner's coordination bonding theory. You will then prepare one of three different Werner cobalt complexes, and like the early inorganic chemists, you will analyze your compound for halide, ammonia, and cobalt content and carry out titration and ion exchange studies to determine the molecular formula. In the final experiment, you will confirm your product using visible spectroscopy. The focus of this chapter is on mastering standard classical analytical techniques used to determine percentage composition of atoms or groups in a molecule. Careful data and error analysis along with notebook keeping and scientific writing are critical to your development as an experimental chemist.

Integrated Approach to Coordination Chemistry: An Inorganic Laboratory Guide. By Rosemary A. Marusak, Kate Doan, and Scott D. Cummings
Copyright © 2007 John Wiley & Sons, Inc.

EXPERIMENT 2.1: EXPLORING THE WERNER–JØRGENSEN DEBATE ON METAL COMPLEX STRUCTURE[3-5,7-13]

Level 1

For this exercise, imagine yourself a chemist in the late nineteenth century: Tassaert's striking cobalt ammine complexes (Table 2.1) have recently been rediscovered and, given the color-less nature of most organic compounds, there is much curiosity about how these beautiful colors arise. Influenced greatly by *The Radical Theory of Structure* for organic compounds (1830), Danish chemist Sophus Jørgensen and others contributed to and proposed the "chain theory" of bonding for metal ammine complexes, called *The Blomstrand–Jørgensen chain theory*. At the same time, Swiss-German chemist Alfred Werner also worked on deciphering the bonding nature of metal ammines. In the following exercise, you will actively participate in the Werner–Jørgensen debate, providing the background for your own synthesis and analysis of cobalt ammine compounds in upcoming experiments.

Jørgensen's Chain Theory

Jørgensen's chain theory links ammonia molecules in metal compounds similar to the linking of carbon units in hydrocarbons. The five-carbon pentane (C_5H_{10}) molecule is shown in **I** as an example (Fig. 2.1).

Like carbon, each metal center is thought to have a fixed valence (valency being defined as the number of bonds formed by the atom of interest), with each metal stabilizing different chain lengths. For example, platinum is divalent in structure **II**, stabilizing an ammine chain length of two while trivalent cobalt in structure **V** favors an ammine chain length of four. Water linkages in metal hydrates (**V**) were treated similarly. Further, Jørgensen astutely noticed that anions such as chloride (Cl^-) and nitrite (NO_2^-) found in the complex show two distinct reactivities. Chain theory suggests that these anions occupy two different positions, proximal to (or bound directly to) and distal to (far from) the metal ion. Anions far from the metal can be readily removed during a chemical reaction, for example, Cl^- removal with $AgNO_3$, equation (2.1), whereas those bound directly to the metal do not react.

$$AgNO_{3(aq)} + Cl^-_{(aq)} \longrightarrow AgCl_{(s)} + NO^-_{3(aq)} \tag{2.1}$$

Q2.1 Given a fixed valence of 3 for cobalt and the number of precipitated Cl^- ions in reaction with $AgNO_3$, predict according to chain theory the structure for the metal complexes corresponding to the formulas in Table 2.2. Although chain theory proposes differences in reactivity for each $Co-NH_3$ linkage, these would be spatially equivalent if we consider these molecules to be, for example, trigonal planar arrangements. Jørgensen also noticed that compounds richest in ammonia always contained six ammonia molecules for one metal atom, and, upon replacing two ammonias with two anions (X^-), the new residues

TABLE 2.1 Taessert's Cobalt Amine Complexes

Compound	Color	Original Name	Formula
CoCl3 · 6NH3	Yellow	Luteo complex	$[Co(NH_3)_6]Cl_3$
CoCl3 · 5NH3	Purple	Purpureo complex	$[Co(NH_3)_5Cl]Cl_2$
CoCl3 · 4NH3	Green	Praseo complex	$[Co(NH_3)_4Cl_2]Cl$
CoCl3 · 4NH3	Violet	Violeto complex	$[Co(NH_3)_4Cl_2]Cl$

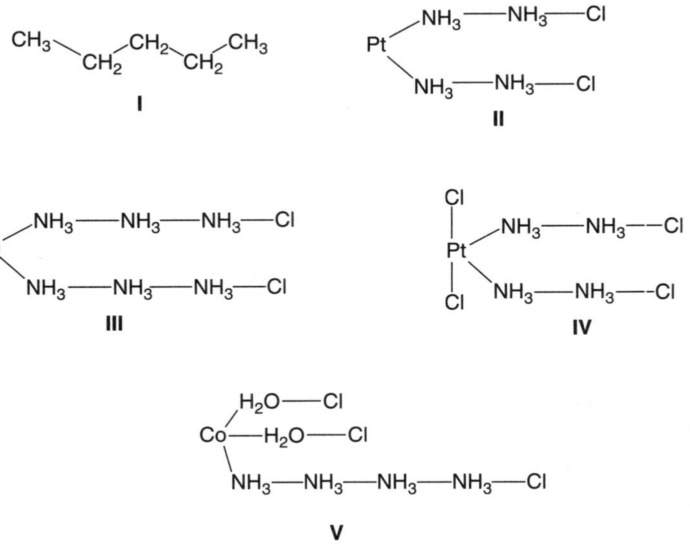

Figure 2.1 Examples of molecular formulas according to Jorgensen's Chain Theory.

TABLE 2.2 Chain Theory Structure Predictions for a Series of Cobalt Ammine Complexes Based on the Number of Reactive Cl⁻

Formula	Chain Theory Structure	Number of reactive Cl^-
$CoCl_3 \cdot 6NH_3$		3
$CoCl_3 \cdot 5NH_3$		2
$CoCl_3 \cdot 4NH_3$		1

showed direct bonding to the metal center, equation (2.2). It was this latter observation which Alfred Werner's work expanded on and which led him to propose new structures for metal ammine complexes.

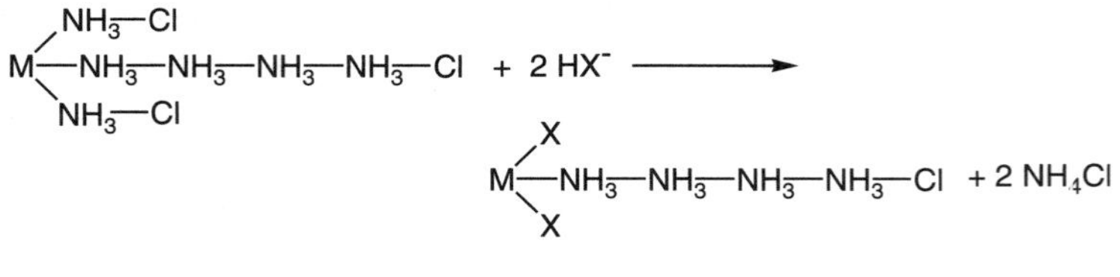

$$(2.2)$$

Werner's Coordination Theory

Jørgensen's work on amine displacement constituted the groundwork for Alfred Werner, who showed that, upon further replacement of amines with anions X⁻, these new substituents also exhibited properties as if bound to the metal atom. Werner concluded that ammonia molecules could not exist in the chains suggested by the Blomstrand–Jørgensen theory and proposed an alternative way of viewing metal ammine bonding. Pfeiffer in 1928 described this transforming event:

According to his own statement, the inspiration (for Coordination Theory) came to him like a flash. One morning at two o'clock he awoke with a start; the long-sought solution of this problem had lodged in his brain. He arose from his bed and by five o'clock in the afternoon the essential points of the coordination theory were achieved.

Unlike Jørgensen's fixed valence for metals, Werner proposed two valences for ligand-bound metals: a primary valence and a secondary valence. The primary valence, originally considered the "ionizable valence," is what we now term the *oxidation state* of the metal. The secondary valence is defined as the number of groups attached (or coordinated) directly to the metal (now termed the *coordination number*). Werner proposed a directional arrangement of the secondary valence and, from his replacement experiments, that this valence must always be satisfied. Recall that Jørgensen had already noted that most metals preferred six ammonias, and for complexes such as the cobalt ammines discussed above, Werner proposed that there must be a directional arrangement of metal coordinating groups, **VI**.

$$H_3N\bullet \quad \bullet NH_3$$
$$H_3N\bullet \; M \bullet NH_3$$
$$H_3N\bullet \quad \bullet NH_3$$
$$\textbf{VI}$$

Q2.2 Werner noted that six ammonia ligands could be arranged around a central metal ion (M) with four different possible geometries. Draw and name these below.

To distinguish between these choices consider the following thought experiment. If the formulas for each of these complexes contained mixed ligand types, A and B (where A is ammonia and B is a different ligand), then for MA_4B_2 and MA_3B_3, different numbers of geometrical isomers should exist.

Q2.3 Determine the possible number of isomers for each of your geometries given MA_4B_2 and MA_3B_3 formulas and record these below (show your work).

Geometry   Predicted no. of geometric isomers

(MA_4B_2) (MA_3B_3)

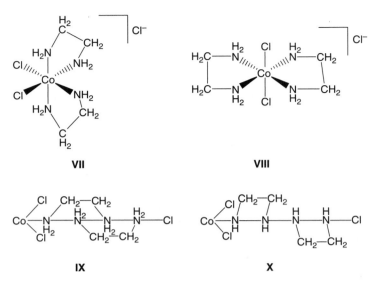

Figure 2.2 Isomeric structures of $CoCl_3 \cdot (en)_2$ as predicted by Werner's Coordination Theory (**VII,VIII**) and Jorgensen's Chain Theory (**IX,X**).

Q2.4 Assuming that two geometric isomers for both formulas exist, Werner predicted that the spatial arrangement about the metal center is: _____

Werner's experiments with the MA_3B_3 showed that indeed two geometrical isomers existed, but although his theory predicted two isomers for the MA_4B_2 formula as well, he had difficulty isolating the second isomer. The compound he was working with was $CoCl_3 \cdot (NH_3)_4$. Despite this, Werner was certain that a second isomer of $CoCl_3 \cdot (NH_3)_4$ should exist since both isomers of the analogous ethylenediamine (en) complexes were known, **VII** and **VIII**. Recall that en is a ligand with two NH_2 metal binding sites, Figure 1.8. Two en ligands can be considered equivalent to four NH_3 ligands. Jorgensen argued that $[Co(NH_3)_4Cl_2]^+$ was, however, different from $[Co(en)_2Cl_2]^+$ as his chain theory predicted. In fact, chain theory accounted for the two isomers of $[Co(en)_2Cl_2]^+$ (**IX** and **X**) and predicted only one for $[Co(NH_3)_4Cl_2]^+$. For these reasons Jørgensen contested Werner's theory stating that the formation of isomers was inherent to ethylenediamine as a ligand (Fig. 2.2).

Q2.5 Revisit the table in question 2.1. Given the number of reactive chlorides in the following table, predict the coordination structure for each of the following formulas according to Werner's coordination theory.

Testing the Theories

Werner's inability to isolate the second isomer for the MA_4B_2 formula cast doubts about his theory in the minds of not only Jørgensen, but the rest of the chemistry community as well. For several years, Werner and Jørgensen set out to prove their own theories to the scientific community. Experiments 1 and 2 lead you through a small sampling of their debate.

Experiment 1: Chloride Precipitation of $IrCl_3 \cdot 3 NH_3$

Q2.6 Draw the structures of $IrCl_3 \cdot 3 NH_3$ expected from both chain and coordination theories and predict the number of reactive chlorides expected.

Chain theory   Coordination theory

No. reactive Cl⁻: _____   No. reactive Cl⁻: _____

The laboratory experiment revealed that no Cl⁻ could be precipitated from $IrCl_3 \cdot 3 NH_3$ with $AgNO_3$, in support of _____ theory.

Experiment 2: Molar Conductivity Measurements   Considering Arrhenius's *electrolytic theory of dissociation*, Werner noted that evidence for his coordination theory may be obtained by determining the electrolytic conductivity of the metal complexes in solution. Werner and Jørgensen assumed that acid (ionic) residues bound directly to the metal would not dissociate and would thus behave as nonconductors, while those loosely held would be conductors. Molar conductivities of 0.1 molar percent aqueous solutions of some tetravalent platinum and trivalent cobalt ammines are given in Table 2.3.

Q2.7   Given that the expected conductances (Ω^{-1} mol^{-1} cm^2) for zero, one, two, and three ionizable groups are approximately 0, 100, 250, and 400, respectively, draw the coordination and chain structures expected for these formulas and predict the conductance values. Assume octahedral geometry for all Werner coordination complexes.

The theory that best accounts for the data is _____.

As you have just determined, Werner's coordination structures comply with most of the data. Only for $CoCl_3 \cdot (NH_3)_4$ and $CoCl_3 \cdot (NH_3)_3H_2O$ do his structures fail to predict appropriate conductance values. Werner explained that solvent water could replace ammonia and acid residues that are directly attached to the metal in some complexes; the anomalous conductivity results as well as the inability to isolate the second isomer of $CoCl_3 \cdot (NH_3)_4$ were, Werner argued, due to the instability of these complexes in

TABLE 2.3   Molar conductivities (Ω^{-1} mol^{-1} cm^2) for Several Platinum and Cobalt Complexes. Predict Values for WCT (Werner's Coordination Theory) and JCT (Jorgensen's Chain Theory) Given the Information Below. Measured Values are from the Literature[3–5,9]

Formula	WCT	JCT	Measured
$PtCl_4 \cdot (NH_3)_6$			523
$PtCl_4 \cdot (NH_3)_5$			404
$PtCl_4 \cdot (NH_3)_4$			228
$PtCl_4 \cdot (NH_3)_3$			97
$PtCl_4 \cdot (NH_3)_2$			~0
$K_2 \cdot PtCl_6$			256
$CoCl_3 \cdot (NH_3)_6$			432
$CoCl_3 \cdot (NH_3)_5$			261
$CoCl_3 \cdot (NH_3)_4$			365
$CoCl_3 \cdot (NH_3)_3H_2O$			393
$CoCl_3 \cdot (NH_3)_5H_2O$			413
$Co(NO_2)Cl_2 \cdot (NH_3)_5$			246
$Co(NO_2)_2Cl \cdot (NH_3)_4$			98
$Co(NO_2)_3 \cdot (NH_3)_5$			7
$K \cdot Co(NO_2)_4 \cdot (NH_3)_2$			99

aqueous solution towards further reactivity, equations (2.3) and (2.4).

$$CoCl_3 \cdot (NH_3)_3H_2O + 2H_2O \longrightarrow CoCl_3 \cdot (NH_3)_3(H_2O)_3 \qquad (2.3)$$

$$CoCl_3 \cdot (NH_3)_4 + 2H_2O \rightleftharpoons CoCl_3 \cdot (NH_3)_4(H_2O)_2 \qquad (2.4)$$

Further, the Jørgensen chain structures can account for only one of the platinum complexes, $PtCl_4 \cdot (NH_3)_4$, and fails with $Co(NO_2)_3 \cdot (NH_3)_5$.

Q2.8   Chain theory cannot provide structures for $K_2 \cdot PtCl_6$ and $K \cdot Co(NO_2)_4 \cdot (NH_3)_2$. Why not?

Q2.9   Werner correctly predicted the conductance of another platinum complex, $K \cdot PtCl_5(NH_3)$ to be _____. Give the primary and secondary valences and draw the structure Werner proposed for this complex.

Primary valence: _____

Secondary valence: _____

Structure:

Werner's Triumph!

The most convincing evidence for Werner's theory came in 1907, when his student, Czech chemist, J. V. Dubsky, successfully synthesized the elusive *violeo* (*cis*) isomer of $CoCl_3 \cdot (NH_3)_4$, predicted to exist by coordination theory but not by chain theory. As Werner had predicted, this complex underwent rapid solvolysis in aqueous solution, making isolation difficult. After Dubsky proved the existence of *cis*-$CoCl_3 \cdot (NH_3)_4$, Jørgensen conceded to coordination theory. Jørgensen died in 1908, having never met Werner face to face.

In 1913, Werner received the Nobel Prize for his work on coordination chemistry. Werner in his Nobel address praised the early work of Blomstrand and Cleve who "carefully tended (the metal ammonia compounds and)... recognized their theoretical importance (during) the great era of development of organic chemistry" and further expressed debt to Jørgensen, who, Werner acknowledged, "extended and deepened the field of metal ammonias by researches which have become classic." This generous acknowledgement of a rival's work set the collegial spirit of the coordination chemistry field that still exists in many circles today. Alfred Werner died shortly thereafter in 1919 from Alzheimer's disease, the onset of which was already taking place at the time of his Nobel address.

EXPERIMENT 2.2: SYNTHESIS OF A COBALT AMMINE COORDINATION COMPOUND[3-5,16]

Level 1

Pre-lab Questions

Q2.10   What should you do if you spill a small amount of aqueous ammonia? Concentrated hydrochloric acid? Hydrogen peroxide?

Q2.11   How accurate (to what decimal place) is a top-loading balance?

Q2.12   What creates the vacuum necessary to use the Büchner funnel?

Q2.13   What is a meniscus?

Q2.14   What is the difference between a *filtrate* and a *precipitate*?

Procedure 2.2.a: Synthesis of Cobalt Complex I

Safety warning! Operations set off between the lines must be performed in a fume hood.

Caution: concentrated ammonia solution (15 M) is a skin, eye, and nasal passage irritant. Avoid breathing the ammonia fumes.

Caution: concentrated HCl solution (12 M) is a skin, eye, and nasal passage irritant. Avoid breathing the HCl fumes.

Caution: 30% hydrogen peroxide solution (H_2O_2) can cause painful burns. Avoid contact with skin and eyes.

Reasonable places for procedure interruption are marked by asterisks. Masses of starting reagents can be measured using a top-loading balance. Volumes can be measured using graduated cylinders.

---

1. Dissolve 8.0 g of NH_4Cl and 12.0 g of $CoCl_2 \cdot 6H_2O$ in 20 ml of de-mineralized water in a 250 ml Erlenmeyer flask and bring the solution to a boil. *Carefully* add 1 g ($\sim$1 teaspoon) of decolorizing charcoal (Norit).

---

2. Add 25 ml of concentrated aqueous ammonia (**caution!**).

3. Cool the mixture and *keep the temperature between 5 and 10 °C* (using an ice/salt bath) while adding 32 ml of 30% H_2O_2, 1 ml at a time, with a medicine dropper (1 ml = 20 drops or 1 dropper-full = $\sim$1 ml), with continuous stirring. When all the H_2O_2 has been added, let the reaction mixture stand at room temperature until decomposition of H_2O_2 has subsided (about 5 min).

4. Warm the mixture on a hot plate, maintaining the temperature at 55 or 65 °C for 90 min.* Remove the stir bar.

---

5. Cool the mixture in an ice bath.*

6. Filter the precipitate using a Büchner funnel (clamp the filter flask to a ring stand) and Whatman no. 1 filter paper. The precipitate can be flushed out of the flask using a few milliliters of the filtrate from the filter flask. Discard the filtrate in a bottle marked for filtrate.*

7. Carefully transfer the precipitate to a 250 ml beaker containing a hot mixture (70 °C) composed of 4 ml concentrated HCl (**caution!**) and 100 ml de-mineralized water.

8. Heat to 90 °C. When all but the charcoal has dissolved, filter the hot suspension using a Büchner funnel (preheat the funnel in a beaker of very hot water) to remove the charcoal.

9. Carefully re-precipitate the product by adding 17 ml concentrated HCl (**caution!**) to the filtrate and cooling in an ice bath.*

10. Filter the product using a Büchner funnel and wash with four 5 ml portions of *cold* 95% ethanol; then wash with four 5 ml portions of *cold* acetone.

Recrystallization of Product

11. In a 250 ml beaker, dissolve the compound in a *minimum volume* of de-mineralized water at 85–90 °C. Start with approximately 50 ml, if this is not enough to

completely dissolve the compound after heating, add 10 ml of water and bring back up to 85 °C; repeat until all of the compound is dissolved. The hot solution may be filtered (*with preheated funnel*) if necessary. Cool the filtrate in an ice bath to 5–10 °C; then slowly add 5–10 ml concentrated HCl (**caution!**) to re-precipitate the product. Filter and wash as before.

12. The product should be dried in an oven at 100–110 °C for at least 1 h and then stored in a bottle with a tight fitting cap. Record the color and yield in grams.

Procedure 2.2.b: Synthesis of Cobalt Complex II

Safety warning! Operations set off between the lines must be performed in a fume hood.

Caution: concentrated ammonia solution (15 M) is a skin, eye, and nasal passage irritant. Avoid breathing the ammonia fumes.

Caution: concentrated HCl solution (12 M) is a skin, eye, and nasal passage irritant. Avoid breathing the HCl fumes.

Caution: 30% hydrogen peroxide solution (H_2O_2) can cause painful burns. Avoid contact with skin and eyes.

Reasonable places for procedure interruption are marked by asterisks. Masses of starting reagents can be measured using a top-loading balance. Volumes can be measured using graduated cylinders.

---

1. In a 500 ml Erlenmeyer flask, dissolve 7.5 g of NH_4Cl in 45 ml of concentrated aqueous ammonia. (**caution**).
2. While continuously stirring the solution with a magnetic stirrer, add 15.0 g of $CoCl_2 \cdot 6H_2O$.
3. When the cobalt salt has dissolved and formed a brown slurry, slowly (over 1–2 min) add 12 ml of 30% H_2O_2 in 1 ml portions, using a medicine dropper (1 ml = 20 drops or one dropper-full = ~1 ml) with continuous stirring (**caution**).*
4. When the effervescence has ceased, slowly and carefully add 45 ml of concentrated HCl (**caution.**) (Large amounts of vapor will be produced.)
5. Continue to stir and heat the solution on a hot plate in the hood to 80–85 °C for 20–30 min.* Remove the stir bar and cool the mixture to room temperature using an ice bath.

---

6. Filter the product by suction using a Büchner funnel (clamp the funnel to a ring stand) and Whatman no. 1 filter paper.
7. Wash the product with one 10 ml portion of ice-cold de-mineralized water, then with three 10 ml portions of cold 95% ethanol, and finally with three 10 ml portions of cold acetone. Let the vacuum suck the precipitate dry after the final portion of wash liquid.*

Recrystallization of Product

8. In a 250 ml beaker, *dissolve* the product in about 125–150 ml of hot (85 °C) 6 M aqueous ammonia.

9. Cool the flask to about room temperature in an ice bath. Then, *using a fume hood* (vapor is produced), slowly pour the contents into a 400 ml beaker containing 125 ml of concentrated HCl (**caution!**).*

10. Heat the resulting mixture on a hot plate at 80–85 °C for 20–30 min. Cool the solution to room temperature.

11. Filter and wash the product as above. Dry the product in an oven at 100–110 °C for at least 1 h and store in a bottle with a tight fitting cap. Record the color and yield in grams.

Procedure 2.2.c: Synthesis of Cobalt Complex III

Safety warning! Operations set off between the lines must be performed in a fume hood.

Caution: concentrated ammonia solution (15 M) is a skin, eye, and nasal passage irritant. Avoid breathing the ammonia fumes.

Caution: concentrated HBr solution (48%, ~6 M) is a skin, eye, and nasal passage irritant. Avoid breathing the HBr fumes.

Caution: 30% hydrogen peroxide solution (H_2O_2) can cause painful burns. Avoid contact with skin and eyes.

Reasonable places for procedure interruption are marked by asterisks. Masses of starting reagents can be measured using a top-loading balance. Volumes can be measured using graduated cylinders.

1. To 10.0 g of $CoCO_3$ in a 500 ml beaker, *slowly* add 26 ml of 48% HBr. The mixture will effervesce and turn purple.

2. Filter the solution by suction using a fritted-glass filter and a 250 ml side arm flask, to remove undissolved cobalt(II) oxide.

3. Next add 20 g NH_4Br followed by 100 ml of concentrated NH_4OH solution (approximately 1.4 mol).

4. Equip your flask with a stir bar and *carefully*, with stirring, add 16 ml of 30% H_2O_2. The solution turns dark red.

5. When effervescence ceases (completing oxidation), remove the excess ammonia by drawing a stream of air through the mixture for 2 h (Fig. 2.3).*

6. Neutralize the solution (~pH 7.5) with approximately 25 ml 48% HBr, and filter the scarlet-colored precipitate. Wash with 50 ml of cold dH_2O followed by four 5 ml portions of cold 95% ethanol and four 10 ml portions of cold acetone.

7. *Remove half the precipitate and allow it to air-dry over the week. Label and save this precipitate.*

8. Return the other half of the precipitate to the filtrate. Add 18 ml of 48% HBr to the solution and reduce the volume using a steam bath for 2 h (until precipitate crust forms along top of solution).

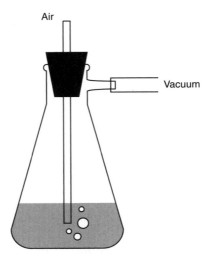

Figure 2.3   Diagram of apparatus for solution aeration.

9. Cool the solution and filter the product using a Buchner funnel, and wash with ~60 ml of cold dH$_2$O followed by four 5 ml portions of cold 95% ethanol and four 10 ml portions of cold acetone.

10. Allow product to air dry over the week. Record your color and yield in grams.

Optional Recrystallization

---

11. Add a 10 g portion of product to a solution of 25 ml of NH$_4$OH in 200 ml H$_2$O that has been heated to 90 °C in a 500 ml beaker.

12. Filter the solution and add 16 ml of 48% HBr.

13. Reduce the volume using a steam bath (2 h). Filter the product and wash and dry as done previously.

Name _____   Section _____

Results Summary for Synthesis of a Coordination Compound (I, II, or III)

1. Write a balanced chemical equation describing the reaction to form this product, basedon your procedure. Hint: consider all the reagents you used (HCl does not need to be considered as a reagent and assume CoCO$_3$ + 2 HBr → CoBr$_2$ + H$_2$O + CO$_2$), and balancing both mass and charge.

2. Calculate the moles of the following reagents used in your procedure. (The concentrated aqueous NH_3 solution is 15 M and the concentrated HCl is 12 M. The 48% HBr contains 48 g of HBr per 100 ml of solution and the 30% H_2O_2 contains 30 g of H_2O_2 per 100 ml of solution. Assume the density of both solutions to be approximately 1 ml^{-1}.)

   $NH_4Cl(Br)$ _____

   $NH_3(aq)$ _____

   HCl (Br) (aq) _____

   $CoCl_2 \cdot 6H_2O$ _____
   (*or* $CoCO_3$)

   H_2O_2 _____

3. Determine the limiting reagent: _____

4. What was your product yield? _____ g
   a. Calculate the theoretical yield _____ g
   b. Calculate your percentage yield _____ g
   c. Note the colors of all three cobalt complexes prepared in the lab. What do you think accounts for the differences in color?

Post-lab Questions

Q2.15 Why did you start with a cobalt(II) salt instead of a cobalt(III) salt for your synthesis?

Q2.16 Approximately what percentage of the ammonia ends up in your compound? Why do you think an excess of ammonia is used?

Q2.17 What happens to the *excess* hydrogen peroxide? Write a balanced chemical equation to describe this process. Why is the peroxide added slowly?

Q2.18 It has been shown that, under the conditions of all three syntheses, an equilibrium is established between $[Co(NH_3)_6]^{3+}$ and $[Co(NH_3)_5(H_2O)]^{3+}$ ($pK_a \approx 6.6$), which rapidly becomes $[Co(NH_3)_5(OH)]^{2+}$. The resulting equilibrium is given below:

$$Co(NH_3)_6^{3+} + H_2O \rightleftharpoons Co(NH_3)_5OH^{2+} + NH_4^+$$

The equilibrium position for this reaction is independent of NH_3 concentration. What added reagent is used to shift the equilibrium to obtain the desired product? (*Hint*: compare all three syntheses.)

Q2.19 The precipitate is washed with (cold water), cold ethanol and cold acetone.
   a. What impurities are removed by the washing process?
   b. Why are the wash liquids cold?
   c. Why is complex I *not* washed with water?

Q2.20 In the preparation of the *bromopurpureo* complex, a red intermediate was isolated. Make an educated identification of this complex. Explain.

EXPERIMENT 2.3: ANALYSIS OF COBALT COORDINATION COMPOUNDS I: SPECTROPHOTOMETRIC DETERMINATION OF COBALT[1-5]

Level 1

Pre-lab Questions

Q2.21 3.50 ml of a 0.0850 M solution of Co^{2+}(aq) is diluted to 10.00 ml. Calculate the concentration of the resulting solution.

Q2.22 How many significant figures are in the following values?

    a. 3.50 ml

    b. 0.0850 M

    c. 10.00 ml

Q2.23 Calculate the %Co in $CoSO_4 \cdot 7H_2O$

Q2.24 Why should the mass of the Co(III) complex be measured immediately after removing it from the dessicator?

Q2.25 What does a "quantitative" transfer mean?

Pre-lab 2.3.a Introduction to Visible Spectroscopy and Beer's Law[15]

Color changes have always been important identifiers of reaction for chemists. Colorimetry—the use of light absorption of colored solutions for the quantitative analysis of dissolved substances—quickly became an important tool in inorganic chemical analysis. Although strictly speaking colorimetry refers to analysis by direct color comparison, today this refers to any technique that involves absorption by visible light. In Chapter 1 we saw that the wavelength or color of light absorbed is what gives a transition metal complex its particular color. For this reason, visible spectrophotometry is often used for both qualitative and quantitative analysis of inorganic compounds.

The visible region of the electromagnetic spectrum includes wavelengths from about 380 to 750 nm. The familiar color wheel, Figure 2.4, shows that a solution that is red transmits red light and absorbs primarily green light (~500 nm). The absorption of light leads to a decrease in the *intensity* of light passing through the solution. The ratios of intensities going into and out of the solution are measured using a visible *spectrophotometer* and are used to calculate the *absorbance* of the solution.

A typical instrument used in visible analysis is the diode array spectrophotometer, Figure 2.5. This instrument uses a limited number of optical components in its construction and a single deuterium lamp source provides significant radiant throughput in the 200–820 nm range. A single collimated beam of light is formed after passing through the source lens, which enters the cuvette that holds the sample. The transmitted light reaches a grating that disperses the light into wavelengths. The light is collected and converted into measurable current using a photodiode array detector. [Photodiode arrays are composed of silicon diode elements (typically 1024) with reverse-biased on junctions.] The current is proportional to the amount of light reaching each diode element. The full spectrum is acquired almost simultaneously and allows multiple scans that can be averaged to enhance the signal to noise ratio. The resolution for a diode array instrument is typically 2 nm.

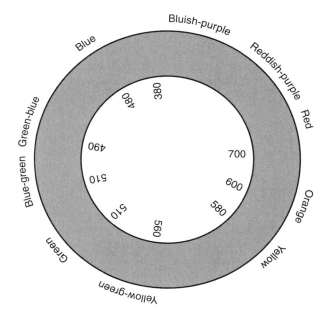

Figure 2.4   Color wheel.

The Beer–Lambert equation predicts that the absorbance A of a solution at wavelength λ is linearly proportional to the concentration (c, in mol l^{-1}) of the absorbing solute and the pathlength of sample that the light passes through (b, in cm). The Beer–Lambert equation, equation (2.5) is:

$$A = \varepsilon \cdot b \cdot c \qquad (2.5)$$

where ε is a proportionality constant known as the molar absorptivity (or molar extinction coefficient) that depends on the nature of the absorbing solute; these values are typically

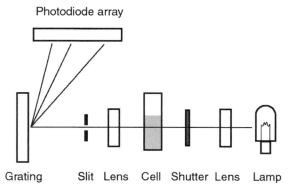

Grating   Slit Lens Cell Shutter Lens Lamp

Figure 2.5   Schematic of a diode array spectrophotometer.

small for transition metal complexes. The molar absorptivity has units of $l\,mol^{-1}\,cm^{-1}$, and is determined by the electronic structure of the metal complex.

A plot of A vs c for a solution is commonly known as a "Beer's law plot." Such a plot can be used to determine ε for a metal complex electronic transition, provided that the pathlength, b, is known. A Beer's law plot can also be used to determine the concentration of a solute in solution when ε for the solute is known simply by measuring the absorbance of the solution. In order to prepare a Beer's law plot, several solutions (at least four) of precisely known metal complex concentrations must be prepared and the absorbance of each measured.

Pre-lab 2.3.b: Weighing by Difference

Preparing solutions of precisely known concentration is a common technique in chemistry. Preparing dilute solutions often requires a more concentrated "standard" solution to be prepared and then diluted, because directly preparing the dilute solution would require the weighing of masses too small to measure precisely. The standard glassware used in such dilutions includes volumetric or transfer pipettes and volumetric flasks. Accurate measurement of solute mass for the "standard" solution should be done by a technique known as "weighing by difference." Weighing by difference involves first measuring the mass of a vial containing a solid sample. Some of this solid is then transferred to another piece of glassware, usually by directly pouring, and the vial is then re-weighed. The difference in mass is equal to the mass of solid transferred. This technique yields *accurate* mass measurement, because no solid is "lost" on weighing paper or spatulas.

Pre-lab 2.3.c: Visible Spectroscopy of the Hexaaquacobalt(II) Ion

This experiment begins a series of analytical investigations that will be used to confirm the product composition of your cobalt ammine coordination compound. We begin by determining the percentage composition by mass of cobalt in the coordination compound synthesized in Experiment 2.1. The first step involves quantitative decomposition and reduction of the cobalt(III) complex into the cobalt(II) ion. The decomposition of pentaammine-chlorocobalt(III)chloride $[Co(NH_3)_5Cl]Cl_2$ is shown as an example in equation (2.6).

$$6[Co(NH_3)_5Cl]Cl_2 \xrightarrow[\text{blue}]{} 6CoCl_{2(s)} + 22NH_{3(g)} + N_{2(g)} + 6NH_4Cl \qquad (2.6)$$

Upon dissolving the $CoCl_{2(s)}$ in acidic water, the hexaaquacobalt(II) ion, $[Co(H_2O)_6]^{2+}$ is formed, equation (2.6). The visible spectroscopy of the $[Co(H_2O)_6]^{2+}$ is well studied; the molar absorptivity coefficient, ε, for $[Co(H_2O)_6]^{2+}$ has been determined to be $4.84\,M^{-1}\,cm^{-1}$ at 512 nm.

$$CoCl_{2(s)} + 6H_2O \overset{H+}{\rightleftharpoons} [Co(H_2O)_6]^{2+} \qquad (2.7)$$

In this experiment, you will prepare a series of $[Co(H_2O)_6]^{2+}$ complexes of varying concentrations from a standardized stock solution of known concentration.

By constructing a plot of absorbance vs concentration, you will be able to determine the concentration of cobalt in your sample.

Pre-lab Preparation

1. A stock solution of aqueous cobalt(II), Co^{2+}(aq), of known concentration ($\sim$0.085 M) prepared from $CoSO_4 \cdot 7H_2O$–cobalt(II) sulfate heptahydrate will be provided by the instructor.

2. *Each student* must dry his/her sample of cobalt coordination compound in a weighing bottle (top ajar) for about 1 h at 120 °C and cool to room temperature in a desiccator. It is recommended that this be carried out during a pre-lab lecture or at the start of Procedure 2.2.a.

Procedure 2.3.a: Preparation of Co^{2+}(aq) Standard Solutions

1. Record the exact concentration of the Co^{2+}(aq) stock solution.

2. Prepare *five* standard solutions (10 ml each) from the Co^{2+}(aq) stock solution in *clean, dry* test tubes. The *most concentrated* will be the stock solution. The other four will be dilutions of the stock solution having concentrations between $\sim$0.040 and 0.085 M. Record your concentrations in the Results Summary section.

You can easily prepare these standard solutions using only 10 and 5 ml graduated pipettes. Keep in mind a useful equation for preparing solutions: $C_1V_1 = C_2V_2$, where C_1 is the concentration of your stock solution, V_1 is the volume of stock solution that you use, C_2 is the concentration of your final solution and V_2 is the volume of your final solution (10 ml, which is V_1 plus the volume of water added to dilute it).

Keep in mind the importance of accuracy and precision in your use of the volumetric measurements you are making. The concentration of each standard solution should be known to three significant figures. The graduated pipettes should be read to ± 0.02 ml.

3. Securely seal each tube with parafilm to prevent any evaporation of the solvent.

Procedure 2.3.b: Preparation of the Cobalt Coordination Compound Sample Solution—Decomposition and Reduction of Cobalt(III) to Cobalt(II).

Safety warning! Operations set off between the lines must be performed in a fume hood.

Caution: flasks heated on a hot plate will be hot to the touch. Use tongs to remove flasks.

Caution: use appropriate care when handling aqueous solutions of sulfuric acid (H_2SO_4).

Carry out the following procedure in duplicate (*each* student prepares two samples of their own cobalt product)

1. Remove your dried and cooled sample of cobalt coordination compound from the desiccator, and immediately weigh (*by difference*, using an analytical balance) 0.7–0.8 g of your coordination compound into a *clean* and *dry* 125 ml Erlenmeyer flask.
2. Record the mass to ± 0.1 mg. Evenly distribute the sample in the bottom of the flask.

---

3. In the hood, place the flask on a hot plate set to at least 400 °C until the sample liquefies, foams, and turns blue. Do not over- or under-heat.
4. Remove your sample from the hot plate and *cool to the touch*. Use your wash bottle *and less than 10 ml of water* to wash down any sample that may have spattered onto the sides of the beaker and watch glass.
5. Add 10 ml of a prepared 1:10 sulfuric acid solution (a 1 ml concentrated sulfuric acid and 10 ml water solution). Boil gently on a hot plate for a few seconds until all solids are dissolved. **Cool to room temperature!**
6. *Quantitatively* transfer (that is, transfer all of the solution) the contents of the beaker to a clean 50 ml volumetric flask. Fill flask "to the mark" with deionized water and mix thoroughly.

Procedure 2.3.c: Measurement of the Visible Spectrum of the Aquated Co(II) Ion: $[Co(H_2O)_6]^{2+}$

This may be done as a group exercise.

1. Run a "blank" spectrum of your cuvette and solvent. Obtain the absorption spectrum of the cobalt ion stock solution.
2. Determine the wavelength of maximum absorbance. Record in the Results Summary.

Procedure 2.3.d: Preparation of Beer's Law Calibration Plot and Sample Analysis

1. Select the wavelength of maximum absorbance (λ_{max}) in the spectrophotometer software.
2. Measure the absorbance at λ_{max} of the five standard solutions prepared in Procedure 2.3.a.
3. Using the same spectrophotometer sample cuvette, measure the absorbance at λ_{max} of your sample solutions prepared in Procedure 2.3.b.
4. Using a plotting program, plot absorbance vs [Co(II)]. This is your Beer's law calibration plot. Obtain a linear least-squares fit to the data points.
5. Use your calibration plot to determine the concentration of Co(II) in the sample solution, and then calculate the percent cobalt in the sample.

6. Complete the Results summary sheet.

Name _____   Section _____

Results Summary for Spectrophotometric Determination of Cobalt

1. λ_{max} for Co(II) stock solution   ____ nm
2.

Solution	ml Co(II) stock	ml water	[Co(II)] (M)	A @ λ_{max}
1	_____	_____	_____	_____
2	_____	_____	_____	_____
3	_____	_____	_____	_____
4	_____	_____	_____	_____
5	_____	_____	_____	_____
Blank	_____	_____	_____	_____

ε (with units) for Co^{2+}(aq) (*determined from linear fit to Beer's law plot*):_____

	Sample 1	Sample 2
3. Mass of sample (mg):	_____	_____
Proposed formula for Co(III) product	_____	
Molar mass of proposed product	_____ g	_____
4. Absorbance of sample solution:	_____	_____
[Co(II)] in sample solution (M):	_____	_____
Moles of Co in 50 ml sample solution:	_____	_____
Mass of Co in 50 ml sample solution (mg):	_____	_____
Mass of sample (mg):	_____	_____
Percentage of Co in sample:	_____	_____
Average percentage of Co in your sample:	_____	
Theoretical percentage of Co in your sample:	_____	
Percentage difference based upon theoretical percentage Co:	_____	
Theoretical percentage Co in $[Co(NH_3)_6]X_3$:	_____	
Theoretical percentage Co in $[Co(NH_3)_5X]X_2$:	_____	
Theoretical percentage Co in $[Co(NH_3)_4X_2]X$:	_____	
Theoretical percentage Co in $[Co(NH_3)_5(H_2O)]X_3$:	_____	

Let X be Cl^- or Br^- depending on your synthesis.

Post-lab Questions

Q2.26 Why does oxidation state variability [i.e., reducing the metal center from Co(III) to Co(II)] not affect the analysis of total cobalt in the compound?

Q2.27 What was the wavelength of light at which the Co(II) stock solution had the maximum absorbance? What color of light does this correspond to? What color was the Co(II) stock solution?

Q2.28 Draw the crystal field diagram for the $[Co(H_2O)_6]^{2+}$ complex.

Q2.29 Use your diagram in question 2.28 to discuss the absorption of visible light.

Q2.30 What would happen to your Beer's law calibration plot if you had forgotten to seal your test tubes containing your metal complex solution?

EXPERIMENT 2.4: ANALYSIS OF COBALT COORDINATION COMPOUNDS II—ACID–BASE TITRIMETRIC DETERMINATION OF AMMONIA[16]

Level 2

The careful attention required by the experimental set-up, along with difficulty in obtaining highly accurate results, increases the level of difficulty for this experiment.

Pre-lab Questions

Q2.31   Define the equivalence point.

Q2.32   What makes a good acid–base indicator?

Q2.33   Define pK_a.

Pre-lab 2.4.a: Acid–Base Titrations of Solutions Containing Two Acids[17]

In this experiment, we wish to accurately measure the amount of NH_3 that is released from a known amount of cobalt ammine complex. To do so, the released NH_3 is delivered into a beaker containing a *known excess* amount of strong acid, HCl. The reaction of NH_3 with HCl quantitatively introduces the second weaker acid, NH_4^+, into the solution, equation (2.8).

$$HCl_{(aq)} + NH_{3(aq)} \longrightarrow NH_4^+ \qquad (2.8)$$

To determine the amount of NH_4^+ present, a titration of a solution containing the two acids, one strong (HCl) and one weak ($NH_4^+ - pK_a = 9.26$, 25 °C), is carried out. Since $pK_a(NH_4^+) - pK_a(HCl) > 4$, the titration steps are distinct, Figure 2.6, with a

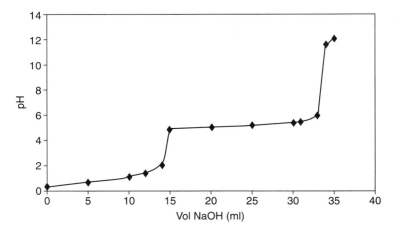

Figure 2.6   Simulation of the titration of HCl (0.44 M) and NH_3^+ (0.56 M) by NaOH (0.3 M) following the decomposition of the $[Co(NH_3)_6]Cl_3$ complex.

large $\Delta pH/\Delta V$ for the first equivalence point. The contribution of H^+ by NH_4^+ can be neglected since the excess H_3O^+ will suppress its dissociation; however, this assumption becomes less valid as the first equivalence point is approached.

By choosing an indicator suitable for the first HCl equivalence point, bromcresol green (pH range 3.8–5.4), one can determine by titration with a strong base the amount of HCl present in solution following the reaction with the NH_3 that was released from the metal complex. In summary, the total HCl added to the solution minus that determined to be in solution by titration is equal to the NH_4^+, equation (2.9)

$$H^+{}_{total} = H^+(Cl^-) + NH_4^+(Cl^-) \tag{2.9}$$

Procedure 2.4

Safety warning! Operations set off between the lines must be performed in a fume hood.

Caution: pay close attention to the experimental set-up in Figure 2.7. The test tube should be secured in place with a clamp (not shown in figure). The glass tubing must be *above* the level of liquid in the test tube at all times. When heating the test tube with a Bunsen burner, gently pass the flame back and forth over the lower end of the test tube for slow heating. Alternatively, a hair dryer may be used for heating the tube; however, results tend to be less accurate.

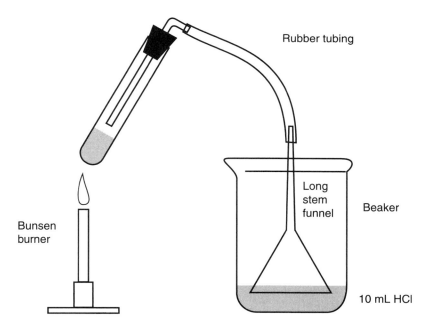

Figure 2.7   Schematic of experimental set-up. Be sure that the inserted glass tubing is *above* the surface of liquid in the test tube as shown. Use a clamp and ring stand to secure the test tube in place. Heed cautionary note above. Adapted from Steve Marsder's Chemistry Resources for Students and Teacher.[16]

1. Place 0.25 g of your cobalt(III) ammine complex into a test tube equipped with a rubber stopper through which a bent glass tube is placed (Fig. 2.7). Add 3 ml 3 M NaOH to your sample. Connect the glass tube to a long-stemmed glass funnel using a short piece of rubber tubing. Record the actual amount of metal complex and the HCl molarity.

2. Pipet 10 ml of 1.0 M HCl into a 125 ml beaker and add approximately three drops of bromcresol green indicator.

3. Insert the mouth of the funnel *into* the HCl solution, just forming a seal of liquid around the funnel.

4. *Carefully* and gradually heat your cobalt ammine sample to a gentle boil. Continue boiling, releasing NH_3 from the compound, until a solid black residue remains and there is no further change in the color of the HCl solution.

5. Rinse the tube and funnel in the HCl solution with a small amount of dH_2O.

---

6. Titrate the resulting HCl solution with 0.30 M NaOH to the bromcresol green end-point (yellow to blue) and calculate the percentage of NH_3.

7. Repeat steps 1–7 two more times and calculate the average percentage NH_3.

Name _____   Section _____

Results Summary for Titrimetric Determination of Ammonia

1. Compound color   _____
2. Record accurate values:

    NaOH titrant   _____ M

    HCl solution   _____ M

    Mass of cobalt complex sample   _____ g

	Trial 1	Trial 2	Trial 3
3. Volume of NaOH added to reach equivalence point[†]	_____	_____	_____
4. Moles NaOH added to reach equivalence point (=moles H^+ from HCl titrated)	_____	_____	_____
5. Moles NH_4^+ in solution (=0.01 − moles H^+ titrated)	_____	_____	_____
6. Grams NH_4^+ in solution	_____	_____	_____
7. Grams NH_3 released released from sample	_____	_____	_____	_____ average

8. Average percentage NH_3 in total sample ____ (include error analysis)
9. Theoretical percentage NH_3

[†]When using an acid–base indicator, the change in color refers to the *end point* of the titration. With proper choice of indicator, the end point should be very close to the *equivalence (or stoichiometric) point*.

$[Co(NH_3)_6]Cl_3$ _____

$[Co(NH_3)_5Cl]Cl_2$ _____

$[Co(NH_3)_4Cl_2]Cl$ _____

$[Co(NH_3)_5Br]Br_2$ _____

$[Co(NH_3)_5(H_2O)]Br_3$ _____

Identify the complex best supported by your titration _____

Discuss places for error in your procedure.

EXPERIMENT 2.5: ANALYSIS OF COBALT COORDINATION COMPOUNDS III: POTENTIOMETRIC ANALYSIS OF IONIC HALIDE (CHLORIDE/BROMIDE)[1-5]

Level 2

This experiment involves more challenging technique and theory.

Pre-lab 2.5

Potentiometric titration using a silver ion electrode.[18,19]

Using electrochemistry in this experiment, you will determine the percentage by mass of ionic halide (X^-) in your coordination compound, and calculate which possible cobalt(III) complex has a percentage X^- closest to this value.

A potentiometric titration will be run using a solution of your cobalt(III) complex and aqueous silver nitrate ($AgNO_3$) as a titrant. $AgNO_3$, which is soluble in aqueous solution, undergoes a double displacement reaction with the cobalt ammine complex according to equation (2.10).

$$[Co(NH_3)_nX_m]X_i + iAgNO_3 \rightleftharpoons [Co(NH_3)_nX_m](NO_3)_i + iAgX(\downarrow) \qquad (2.10)$$

The precipitation of AgX out of solution drives the equilibrium to the right. Similar to acid–base titrations, potentiometric titrations measure the volume of a solution of one reactant that is required to completely react with a measured amount of another reactant, or until the equivalence point (or stoichiometric point) is reached. At the equivalence point, the silver nitrate has reacted stoichiometrically with all of the *ionic* halide present in your sample solution. In our case, the numbers of moles of halide and silver ions are equal, equation (2.11).

$$\text{moles } X^- = \text{moles } Ag^+ \longrightarrow \text{equivalence point} \qquad (2.11)$$

Further, like acid–base titrations, a rapid change in potential indicates the endpoint of the titration. A plot of potential E vs volume titrant added enables the equivalence point to be found, Figure 2.8(a).

You will be using an Ag^+ ion-selective electrode (ISE) in this experiment. The earliest ISE was the familiar H^+ selective or pH glass electrode. Ion-selective liquid membrane electrodes are commonly used to measure the movement of free, unbound ions (e.g., H^+ or Ag^+) in a sample solution. When the electrode encounters a solution of Ag^+ whose concentration is different from that inside the electrode, an electric potential develops. The potential is

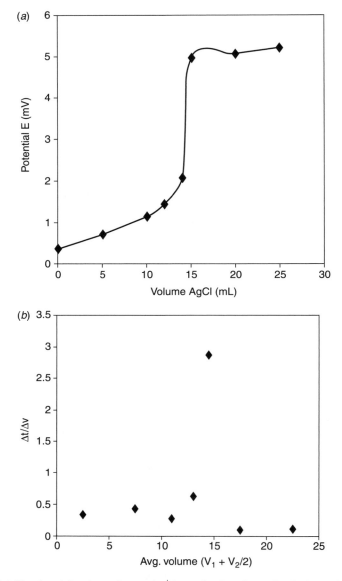

Figure 2.8   (*a*) Simulated titration using an Ag^+ ion-selective electrode. (*b*) Second derivative plot.

measured as ion exchange, a result of the potential difference, occurs between the aqueous phase and an ion exchanging hydrophobic membrane inside the electrode.

Procedure 2.5

Safety warning! Silver nitrate stains skin and clothing. Use gloves when working with this reagent, and clean all spills with water.

Caution: use appropriate care when handling aqueous solutions of nitric acid (HNO_3).

Carry out the following procedure in triplicate:

1. Equip a millivolt/pH meter with a silver-indicating electrode. The laboratory instructor will demonstrate the proper use of the instrument.

2. Rinse your burette with a small amount of the $AgNO_3$(aq) solution (this is called "poisoning" the burette), then fill with ~40 ml titrant.

3. Record the precise concentration of the $AgNO_3$(aq) solution, as found on the bottle.

4. Using weighing-by-difference technique as you did in the spectrophotometric experiment, measure a 0.10–0.15 g (to the nearest 0.0001 g) sample of your coordination compound into a 250 ml beaker.

5. Just prior to beginning your titration, dissolve the sample in 100 ml *cold* distilled water, sonicating briefly if needed. Add 1 ml of 6 M HNO_3. Use a magnetic stir plate and bar to slowly stir the solution. Immerse the electrodes (be careful that the stir bar does not hit the electrodes) and begin titrating with the $AgNO_3$(aq) solution.

6. Record the potential E as a function of volume of titrant added (to the nearest 0.05 ml). Titrate the sample somewhat *past* the equivalence point (signaled by a sudden and substantial change in E).

7. Using Excel or another plotting program, plot E vs volume of titrant added. Create a first derivative plot $[\Delta E/\Delta V (=S)]$ vs average volume, ∇, of titrant $(V_1 + V_2/2)$ of these data. Determine the equivalence point (ml of silver nitrate) graphically at the apex of the first derivative plot. If the maximum is large and hard to locate, a second derivative plot (ΔS vs ∇) can be constructed, Figure 2.8(*b*). For best results, E values and volumes should be obtained from smoothed titration curves rather than experimental values.

8. Based on reaction stoichiometry, calculate the percentage ionic halide in your samples, and the average value for your cobalt(III) coordination compound.

Name _____   Section _____

Result Summary for Potentiometric Analysis of Ionic Halide

Calculation of percentage halide:

	Sample I	Sample II	Sample III
Mass sample used			
Molarity of $AgNO_3$ solution (see label on bottle)	_____	_____	_____
Volume of $AgNO_3$ added to reach equilibrium point	_____	_____	_____
Moles of $AgNO_3$ used to reach equilibrium point	_____	_____	_____
Moles of ionic halide in sample	_____	_____	_____
Grams of ionic halide in sample	_____	_____	_____
Percentage of ionic halide in sample	_____	_____	_____
Average percentage ionic halide	_____		

Theoretical percentage ionic Cl in $[Co(NH_3)_6]Cl_3$:   _____

Theoretical percentage ionic Cl in $[Co(NH_3)_5Cl]Cl_2$:   _____

Theoretical percentage ionic Cl in $[Co(NH_3)_4Cl_2]Cl$:   _____

Theoretical percentage ionic Br in $[Co(NH_3)_5Br]Br_2$:   _____

Theoretical percentage ionic Br in $[Co(NH_3)_5(H_2O)]Br_3$:   _____

Write a balanced chemical equation to describe the reaction of silver nitrate with your (proposed) cobalt(III) coordination compound:

EXPERIMENT 2.6: ANALYSIS OF COBALT COORDINATION COMPOUNDS IV—DETERMINATION OF EQUIVALENT WEIGHT BY ION EXCHANGE CHROMATOGRAPHY[1−5]

Level 2

This experiment involves more challenging technique and theory.

Pre-lab Questions

Q2.34   At the equivalence point of a titration, what can be said about the moles of acid and base present?

Q2.35   Why should de-ionized water be used in this experiment?

Q2.36   What is an eluent?

Q2.37   Should your mass be measured using weighing-by-difference technique?

Q2.38   Calculate the theoretical equivalent weight of NaCl.

Q2.39   If 0.2000 g of NaCl(s) were dissolved in water and passed down an ion-exchange column, how many milliliters of 0.1000 M NaOH(aq) would be required to reach the equivalence point?

Hint: you may want to calculate the approximate volume of NaOH needed to reach the equilibrium point for your proposed Co(III) complex.

Pre-lab 2.6

Liquid column ion exchange chromatography and the determination of equivalent weight by ion displacement.

The equivalent weight of a compound is its molar mass to charge ratio, equation (2.12).

$$\text{equivalent weight} = (\text{molar mass})/(\text{charge of complex cation}) \qquad (2.12)$$

Therefore, the equivalent weight of a coordination compound is related to the charge of the complex cation. Cation exchange, liquid column chromatography can be used to determine the equivalent weight, and thus the charge of your compound.

A schematic of a typical chromatography set up is shown in Figure 2.9. In this experiment a cation exchange resin made of the high molecular weight polymer, polystyrene, with numerous functional groups (in this case the sulfonate group, SO_3H) on the surface, will be used. Your resin is strongly acidic and therefore, the H^+ on the sulfonate groups will dissociate completely upon displacement by your complex ion. The number of

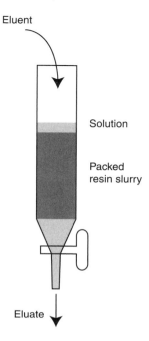

Eluent

Solution

Packed
resin slurry

Eluate

Figure 2.9   Schematic of ion-exchange column. Note the definitions of *eluent* and *eluate*.

H^+ ions that are displaced from the functional group depends upon the charge on the complex cation as illustrated by equations (2.13)–(2.15)

$$resin - (SO_3H)_3 + [complex]^+ \longrightarrow (HO_3S)_2 - resin - SO_3[complex] + H^+ \quad (2.13)$$

$$resin - (SO_3H)_3 + [complex]^{2+} \longrightarrow (HO_3S) - resin - (SO_3)_2[complex] + 2H^+ \quad (2.14)$$

$$resin - (SO_3H)_3 + [complex]^{3+} \longrightarrow resin - (SO_3)_3[complex] + 3H^+ \quad (2.15)$$

The number of H^+ displaced from the column by the sample will then be determined by titration with a standard base.

Procedure 2.6: Preparation of the Cation Exchange Column

1. A cation exchange column of Dowex 50W-X8, 20–50 mesh, hydrogen-form resin will be provided. *Always store the column with the barrel full of de-ionized water.* Flush the column with 20 ml portions of de-ionized water, draining each to just above the top of the resin bed before adding the next portion. When pouring solutions on the column, avoid stirring up the resin at the top of the bed. When draining the column, ***never*** *drain the solution below the top of the resin bed.* Continue this process until the eluate (solution that has passed through the column) is colorless and gives a neutral (yellow) test with methyl orange indicator.

Cation Exchange Experiments (Analyses should be done in Triplicate)

2. Precisely weigh (to the nearest 0.1 mg) 0.15–0.25 g samples of your cobalt compound into a 100 ml beaker.

3. Dissolve each sample in a minimum amount (~40–50 ml) of de-ionized water and record the volume used. Briefly sonicate the solution to dissolve the solid, if needed. (*Note:* for some compounds somewhat more than 50 ml may be required.) To guard against the effects of substitution reactions, dissolve the samples just prior to each analysis and do not heat the solutions.

4. Carefully pour the first sample on the column. Allow it to drain to the top of the resin bed, then flush it through with five or six 20 ml portions of de-ionized water and record the volume used. Collect the eluent (solution that comes off the column) in a 500 ml Erlenmeyer flask.

5. Titrate the eluate with a standard 0.1000 M NaOH solution (provided) using methyl orange indicator. The endpoint of the titration is reached at the color change of the indicator; *this will take place within one drop of titrant.* Record the volume of titrant added (to the nearest 0.05 ml) at the endpoint.

6. Through the column, in 20 ml portions, run a sample of de-ionized water equal in volume to the above eluent and titrate it as a "background" blank. Repeat the procedure with the other two samples. The column can be used numerous times without recharging. When the dark bands have moved about two-thirds of the way down the length of the resin bed, the resin should be discarded or recharged with HCl. (*Note:* consult the laboratory instructor.)

7. From the sample weight and the number of equivalents of base used calculate the equivalent weight for each sample and average. The formula weight of each complex is expected to be an integral multiple (1,2,3, ...) of the equivalent weight. Determine the charge on the complex cation.

Name _____   Section _____

Results Summary for Equivalent Weight Determination

1. Compound color   _____
2. Proposed cobalt (III) coordination compound formula: _____

	Trial:	1	2	3
3. Mass of sample used (g)		_____	_____	_____
4. Volume of 0.1000 M NaOH used (A)		_____	_____	_____
Data for background determination:				
Volume of 0.1000 M NaOH used (B)		_____	_____	_____
Corrected volume of NaOH				
(Volume A − B)		_____	_____	_____
Moles of NaOH used		_____	_____	_____
Moles of H^+ released		_____	_____	_____
Moles of complex cation in sample		_____	_____	_____
Calculated equivalent weight		_____	_____	_____

Equivalent weight = (mass of sample used)/(moles of H^+ released)

Average equivalent weight	_____	_____	_____
Charge on the cation based upon your determined equivalent weight	_____	_____	_____
Theoretical equivalent weight of $[Co(NH_3)_6]Cl_3$:			_____
Theoretical equivalent weight of $[Co(NH_3)_5Cl]Cl_2$:			_____
Theoretical equivalent weight of $[Co(NH_3)_4Cl_2]Cl$:			_____

Theoretical equivalent weight of $[Co(NH_3)_5Br]Br_2$:   _____

Theoretical equivalent weight of $[Co(NH_3)_5H_2O]Br_3$:   _____

Percentage difference between theoretical and actual equivalent weights: _____

Post-lab Question

Q2.40   Review your results obtained in Experiments 2.2–2.6. Identify your-complex based on your quantitative analysis results. What were some sources of error?

EXPERIMENT 2.7: CONFIRMING YOUR ANALYSIS—VISIBLE SPECTROSCOPY OF COBALT AMMINE COMPLEXES

Level 2

This experiment involves more sophisticated technique and theory.

Pre-lab 2.7: Visible Spectroscopy of Cobalt Ammine Complexes[20]

Review your answer to the Post-lab Question of Experiment 2.6. In this experiment, you will use visible spectroscopy to confirm your product identity. The strong ammine (NH_3) ligands of your octahedral cobalt(III) complexes promote a low spin d^6 electronic configuration. The energy gap, Δ_o, however is sensitive to NH_3 substitution; replacement of even one NH_3 is significant enough to change the maximum wavelength of absorption and therefore, complex color. Literature values for some pentaamminecobalt(III) complexes are given in Table 2.4.

You can see from Table 2.4 that several complexes can be described as having the same color. Further, color specification can be very subjective, especially if done without comparison. Although two complexes appear to have the same color, we see that we can distinguish them from one another by their actual wavelengths of maximum absorption. Given the information in Table 2.4, you can verify your complex identity determined from your quantitative analyses of Experiments 2.2–2.6, by visible spectroscopy.

Another important quantity is the proportionality constant in Beer's law, equation (2.5), the molar absorptivity, ε (M^{-1} cm^{-1}). The molar absorptivity is related to the probability of the absorption event for a substance in solution and, therefore, gives insight into the molecular and electronic structure of a metal complex. Precisely determined ε values

TABLE 2.4   Spectral Characteristics of some $[Co(NH_3)_5X]^{n+}$ Complexes

Metal Complex	Color	Maximum Wavelength of Absorption (nm)
$[Co(NH_3)_5(NO_2)]Cl_2$	Yellow	456
$[Co(NH_3)_5(NH_3)]Cl_3$	Yellow	472
$[Co(NH_3)_5(H_2O)]Cl_3$	Rose-red	487
$[Co(NH_3)_5(OH)]Cl_2$	Carmine	502
$[Co(NH_3)_5Cl]Cl_2$	Red-purple	522
$[Co(NH_3)_5Br]Br_2$	Red-purple	540

Adapted from reference 20.

are also necessary for Beer's law to serve as a basis for quantitative analysis by absorption measurements. The ε values for octahedral, d–d transitions are relatively small $(1-500 \text{ M}^{-1} \text{ cm}^{-1})$; the underlying reasons for this are covered in Chapter 3.

Procedure 2.7

1. Determine the approximate amount of pure, dry complex needed for a 50 ml ddH$_2$O solution that gives an absorbance reading at maximum wavelength in the visible region ($\sim$300–700 nm) of $\sim$0.8 for Cl$^-$ containing complexes or $\sim$0.4 for Br$^-$ containing complexes. Use the longest wavelength absorption to identify your complex.

2. Weigh out, by the *difference method*, twice the amount of cobalt complex determined above and quantitatively transfer to a 100 ml volumetric flask. Add ddH$_2$O to the mark. This is your most concentrated stock solution from which dilutions for a Beer's law plot will be prepared.

3. Prepare four dilutions of *total volume* 10 ml according to Table 2.5.

4. Obtain visible spectra from 300 to 700 nm for your solutions and record the absorbances at the maximum wavelengths in Table 2.6.

5. Plot absorbance vs [Co ammine]. The slope is equal to ε (M^{-1}cm^{-1})

Name _____   Section _____

Results Summary for Visible Spectroscopy of Cobalt Ammine Complexes

1. Compound color   _____
2. Wavelength of maximum
   absorbance in visible region   _____

TABLE 5   Solution Preparation for Visible Spectroscopy

Solution	ml Stock	ml H$_2$O
Stock A	10	0
B	8	2
C	6	4
D	4	6
E	2	8

TABLE 6   Results for Visible Spectroscopy

Solution	Concentration (M)	Absorbance @ ___ nm
Stock A		
B		
C		
D		
E		

3. Compound identity _____

4. Mass of compound in 100 ml
   Stock A _____

5. Fill in Table 2.6. _____

6. Molar absorptivity ($\pm$ error) _____ $\mathrm{M}^{-1}\,\mathrm{cm}^{-1}$

7. Obtain molar absorptivities for other complexes determined by your classmates. Comment.

8. Review your results obtained in Experiments 2.2–2.6. How well were you able to identify your complex? What were some sources of error?

REFERENCES

1. Nathan, L. C. *A Laboratory Project in Modern Coordination Chemistry*. Brookes Cole: Monterey, CA, 1981.

2. Loehlin, J. H., Kahl, S. B., Darlington, J. A. *J. Chem. Educ.* **1982**, *59(12)*, 1048–1051.

3. Wilson, L. R. *J. Chem. Educ.* **1969**, *46(7)*, 447–450.

4. Sebera, D. K. *J. Chem. Educ.* **1963**, *40(9)*, 476–478.

5. Creswell, C. J. *Preparation and analysis of a coordination compound,* Neidig, H. A., Spencer, J. N. (eds). Chemical Education Resources: Palmyra, PA, 1972.

6. Kauffman, G. B., symposium chairman, *Werner Centennial*. Advances in Chemistry Series no. 62. American Chemical Society: Washington, DC, 1967. [The Zen verse was used by Chemical Historian, Luigi Cerruti, to remember Alfred Werner.]

7. Rogers, G. E. *Introduction to Coordination, Solid State and Descriptive Inorganic Chemistry*. McGraw-Hill: New York, 1994.

8. Bailar, J. C. *Chemistry of the Coordination Compounds*. Monograph Series no. 131. Reinhold: Baltimore, MD, 1956.

9. Tassaert, *Ann. Chem. Phys. [1]* **1798**, *28*, 92.

10. P. Pfeiffer, *J. Chem. Educ.* **1928**, *5*, 1096.

11. Whisnant, D. M. J. *The Werner–Jorgensen Controversy*. J. Chemical Education Software: Spartanburg, SC (1993).

12. Kauffman, G. B. (ed.) *Coordination Chemistry—a Century of Progress*. ACS Symposium Series no. 565. American Chemical Society: Washington, DC, 1994.

13. Werner, A. On the constitution and configuration of higher-order compounds. Nobel Lecture, 11 December 1913. The Nobel e-Museum (http://www.nobel.se/).

14. Booth, H. S. (ed.) *Inorganic Syntheses*, Vol. 1. McGraw-Hill: New York, 1939, pp. 186–188.

15. Schwedt, G. *The Essential Guide to Analytical Chemistry*. Wiley: New York, 1997.

16. Steve Marsden's Chemistry Resources for Students and Teachers; http://www.chemtopics.com/aplab/cocomplex.pdf (accessed 22 June 2004).

17. Day, R. A. Jr, Underwood, A. L. *Quantitative Analysis*, 5th edn. Prentice-Hall: Englewood Cliffs, NJ, 1986, pp. 209–210.

18. Skoog, D. A., Holler, F. J., Nieman, T. A. *Principles of Instrumental Analysis*, 5th edn. Saunders College Publication: Philadelphia, 1998.

19. Schwedt, G. *The Essential Guide to Analytical Chemistry*, 2nd edn. Wiley: New York, 1996.

20. Bailar, J. C., Moeller, T., Kleinberg, J., Guss, C. O., Castellion, M. E., Metz, C. *Chemistry*, 3rd edn. Harcourt Brace Jovanovich: New York, 1989, pp. 820–821.

Molecular Geometry and Stability: Solid and Solution Phase Analysis of *N,N'*-disalicylaldehyde-1,3-propanediimine nickel(II)

There's no such thing as a bare naked metal ion ... except in the gas phase (Gordon L. Johnson, 1993)

PROJECT OVERVIEW[1-5]

Werner's work provided a solid basis for further investigation of coordination theory. To acquire a deeper understanding of the significance of Werner's coordination number and coordination theory in general, Jannik Bjerrum in the 1930s, studied the stability of metal ammine complexes. While determining consecutive formation constants and characteristic coordination numbers (2, 4, and 6 and sometimes greater) for a variety of metal ion systems, the influence of the metal's electron configuration and ionic radius on complex geometry and stability became apparent to him. With the discovery of new elements following World War II, along with advances in instrumentation, a rebirth in inorganic synthesis and characterization took place and trends in metal complex stability were rationalized. An understanding of structure and stability has brought forth many new applications for metal complexes. When we now think of metal complex structure, we think not only of the geometry about the metal center, but also about overall metal–ligand design. Chiral complexes for organic catalysis, structures that pack efficiently for new magnetic materials, macrocyclic ligands for medicinal targeting, and supramolecular structures for remediation efforts are just some examples of designs that have pushed forward the frontiers of inorganic synthesis.

The first row d^8 transition metal ion, nickel(II) provides an excellent system in which to investigate how the nature of the metal ion and its ligands affect coordination number, geometry, and stability. Nickel(II) offers a plethora of stereochemistries with various ligands and, further, complexes of the same chemical constituency can adopt different geometries with a range of stabilities. This behavior of nickel(II) complexes was first noted in 1939 by coordination chemist Lifschitz and his co-workers. Lifschitz reported that several

Integrated Approach to Coordination Chemistry: An Inorganic Laboratory Guide. By Rosemary A. Marusak, Kate Doan, and Scott D. Cummings

nickel(II) complexes, including [Ni(stien)$_2$] (Cl$_3$CCOO)$_2$, [Ni(phenen)$_2$](ClO$_4$)$_2$, and [Ni(phenen)$_2$](NO$_3$)$_2$ (stien = 1,2-diphenylethylendiamine and phenen = monophenyl ethylenediamine) occurred in two different forms: yellow diamagnetic and blue paramagnetic. After decades of work, the yellow and blue forms were shown to be square planar and octahedral complexes, respectively.

In this series of exercises, you will synthesize a nickel(II) Schiff base complex, *N,N'*-disalicylaldehyde-1,3-propanediimine nickel(II), [Ni(salpd)], which crystallizes in two different forms having characteristic colors, brown and green. Elemental analysis shows that these products differ by the presence of one water molecule. Using a variety of techniques such as infrared, spectroscopy, magnetic susceptibility, differential scanning calorimetry, and Hartree−Fock calculations, you will analyze and propose molecular structures for the two different complexes. The second half of this series examines the solution properties of [Ni(salpd)]. Using UV−visible and NMR spectroscopy, you will investigate molecular structure in solution.

EXPERIMENT 3.1: SYNTHESIS OF *N,N'*-DISALICYLALDEHYDE-1,3-PROPANEDIIMINENICKEL(II), [Ni(salpd)] AND ISOLATION OF THE BROWN AND GREEN PRODUCTS

Level 3

The ligand synthesis requires two consecutive days to complete. Isolation of a pure brown nickel(II) product can be challenging. Two methods are offered. However, enough pure product for further characterization is achievable.

Pre-lab 3.1.a: Schiff Base Ligands: *N,N'*-disalicylaldehyde-1,3-propanediimine, salpd

In this exercise you will synthesize and isolate two different products (green and brown) from the synthesis of the *N,N'*-disalicylaldehyde-1,3-propanediiminenickel(II) complex, [Ni(salpd)]. The preparation of the salpd ligand (**1**) involves a reaction between two salicylaldehyde species (**2**) and 1,3-propanediamine (**3**) in ethanol solvent to produce salpd. The reaction between an aldehyde and an amine, followed by a dehydration step, yields compounds known as *imines* or *Schiff bases*, equation (3.1).

$$(3.1)$$

In your reaction, the aldehyde R group is aromatic (**2**). Aromatic R groups produce more stable imine products than do alkyl R substituents. Salpd is a tetradentate ligand: it contains four atoms, two N and two O, that can readily bind to a metal ion. Metal complexes containing multidentate ligands show enhanced stability over those with analogous monodentate donors. Bjerrum first noted the enhanced stability of five-membered rings

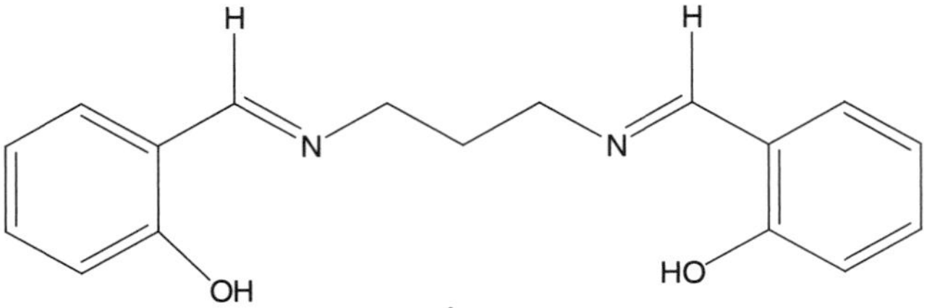

Figure 3.1 Structure of the salpd ligand, **1**, prepared from reactants salicylaldehyde, **2**, and 1,3-propanediamine, **3**.

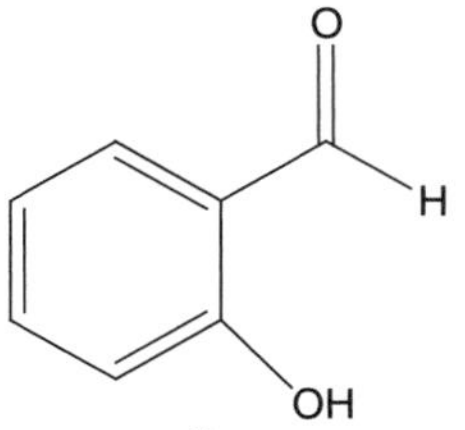

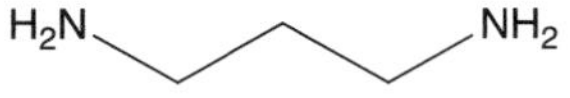

over six and indeed discovered that 10- and 18-membered rings would not form. Although Bjerrum was aware of this in the 1930s, it was Schwarzenbach, in 1952, who first coined the term *chelate* (meaning "claw") *effect*, referring to this enhanced complex stabilization. Upon reaction with a nickel(II) salt, the salpd ligand forms a chelate involving six-membered rings. When the synthesis is carried out in the presence of water, two products of different colors, green and brown, are formed. Your job will be to determine the compositions and structures of these products.

Pre-lab 3.1.b: Geometries of Ni(II) Coordination Complexes[6]

To begin postulating plausible structures for your two products, you must become familiar with the chemistry of nickel(II) and its complex ions. Nickel(II) is a d^8 metal ion that forms a wide variety of metal complexes with coordination numbers ranging from 3 to 6. Coordination number 3 is rare and will not be considered further. For coordination number 4, tetrahedral geometry is favored with bulky ligands while small ligands lead to the very stable square planar geometry. Because distortions from ideal symmetry are prevalent, magnetic properties rather than strict symmetry are used to characterize complexes as tetrahedral (paramagnetic) or planar (diamagnetic). Many 5 coordinate nickel(II) complexes are also known. The majority of these complexes are trigonal bipyramidal (tbp), low spin, and diamagnetic. However, high-spin complexes of tbp also exist. Five coordinate nickel(II) can also adopt the square pyramidal geometry, of which the majority of complexes are also high spin and paramagnetic. Lastly, many octahedral nickel(II) complexes containing two unpaired electrons are known. These often involve relatively small, neutral ligands as in the $[Ni(H_2O)_6]^{2+}$ and $[Ni(en)_3]^{2+}$ complexes. Because different geometries give different electronic structures and colors, both magnetism and visible spectroscopy are valuable tools for deciphering molecular structure of nickel(II) complexes.

Procedure 3.1.a: Synthesis of *N,N'*-disalicylaldehyde-1,3-propanediimine

All procedures are from Asam *et al.* (unpublished).

> **Safety warning!** Operations set off between the lines must be performed in a fume hood. *Note:* all procedures in this experiment are carried out in the hood.
>
> **Caution:** obtain proper training before using a rotoevaporator. Always wear goggles and never leave it unattended.

---

1. Dissolve 5.0 g (5.7 ml, 0.068 mol) 1,3-propanediamine in 20 ml ethanol.
2. In a separate beaker, dissolve 16.5 g (14.2 ml, 0.135 mol) salicylaldehyde in 20 ml ethanol.
3. Combine both solutions in a beaker. This will produce a yellowish solution composed initially of two layers.
4. Stir the mixture for approximately 15 min then cool in an ice bath. Remove the product and allow it to sit for 5 min. Yellow product crystals will begin forming. Formation of the crystals can be accelerated by gently swirling the round-bottom flask.

---

5. Vacuum filter the crystals and wash them several times with a small amounts of cold ethanol. Air dry over one week. Record your yield.

Procedure 3.1.b: Synthesis of *N,N'*-disalicylaldehyde-1,3-propanediiminenickel(II)

---

1. Dissolve 5 g (0.018 mol) of the *N,N'*-disalicylaldehyde-1,3-propanediimineligand in 175 ml ethanol in a round-bottom flask equipped with a magnetic stir bar. Add two equivalents of NaOH (35.5 ml of 1 M). Difficulty in getting the ligand to dissolve can be remedied through gently heating the solution and stirring.
2. Dissolve the 4.25 g (0.018 mol) $NiCl_2 \cdot 6H_2O$ in a solution of 75 ml ethanol and 25 ml water.
3. Add the ligand solution to a round-bottom flask, and add dropwise the $NiCl_2 \cdot 6H_2O$ solution. Over the course of the addition the color of the liquid will gradually turn from bright yellow to dark brown.
4. Attach the water condenser to the round-bottom flask and reflux for approximately 2 h. A green product should appear in the liquid. Cease heating and allow the reaction to stir overnight.
5. After 24 h the product solution will consist of two phases, a top murky brown phase, and a thick, light-green phase on the bottom. Vacuum filter and evaporate under reduced pressure to produce a green mixture of product containing both brown and green products.

Procedure 3.1.c: Isolation of the Brown Product

Method 1

1. Pour approximately 100 ml of *extra dry, freshly opened* acetone into a 250 ml round-bottom flask equipped with a drying tube and heat.
2. Add 1 g impure N,N'-disalicylaldehyde-1,3-propanediiminenickel(II) to the hot acetone, stir, and allow the mixture to sit for several minutes with a watch glass on top of the flask to prevent evaporation.
3. Vacuum filter the product. If the product is very green, the compound can be retreated with hot, dry acetone as above. Alternatively, small amounts of brown complex can be isolated by evaporating the brown filtrate.
4. Record your yield and store the brown product in a vacuum dessicator.

Method 2

1. Pour approximately 100 ml dry DMSO into a 250 ml round-bottom flask.
2. Add 1 g N,N'-disalicylaldehyde-1,3-propanediiminenickel(II) and evaporate under reduced pressure to dryness while heating the flask in an oil bath set to 120 °C.
3. Record your yield and store the brown product in a vacuum dessicator.

Procedure 3.1.d: Isolation of the Green Product

1. Remove approximately 5 g of the green and brown mixture from the round-bottom flask and place it in an Erlenmeyer flask along with a magnetic stir bar.
2. Prepare 125 ml of a hot solution of ethanol:water (4:1) in a separate flask.
3. Add the ethanol:water solution to the green brown mixture and heat the solution with stirring for 15 min.
4. Vacuum filter the product and wash with additional portions of the 4:1 ethanol:water mixture.
5. Allow the green product to dry. Record your yield.

Name _____   Section _____

Results Summary for the Synthesis of N,N'-disalicylaldehyde-1,3-propanediiminenickel(II)

1. Limiting reagent for ligand synthesis: _____
2. Limiting reagent for NiL synthesis: _____
3. Record product yields in Table 3.1.

Post-lab Questions

Q3.1   Elemental analysis gives the following percentages for your two products of the nickel(II) complex: *brown*—%C, 60.72; %H, 4.87; %N, 8.13; %Ni, 17.26; %O, 9.00; *green*—%C, 57.25; %H, 5.14; %N, 7.28; %Ni, 17.06; %O, 13.15. Given these

Table 3.1   Results Summary for the Synthesis of
N,N′-disalicylaldehyde-1,3-propanediiminenickel(II)

Compound	Molecular Mass[a]	Actual Yield (g)	Theoretical Yield (g)	Percentage Yield
N,N′-disalicylaldehyde-1,3-propanediimine	282			
N,N′-disalicylaldehyde-1,3-propanediiminenickel(II)— **brown product**	339			
N,N′-disalicylaldehyde-1,3-propanediiminenickel(II)— **green product**	357			

[a]To calculate the molecular masses of the brown and green products, complete Question 3.1.

percentages, and assuming one nickel per complex, predict the molecular formulas for the brown and green products.

Q3.2   Propose plausible structures for the two different forms that could result from the reaction between nickel(II) and the Schiff base ligand, N,N′-disalicyl-aldehyde-1,3-propanediimine.

EXPERIMENT 3.2: MAGNETIC SUSCEPTIBILITY MEASUREMENTS OF THE BROWN AND GREEN PRODUCTS OF N,N′-DISALICYLALDEHYDE-1,3-PROPANEDIIMINENICKEL(II), [Ni(salpd)]

Level 3

Pre-lab 3.2.a: Introduction to Magnetic Properties of Transition Metal Complexes[6-11]

The valence d electrons of transition metal complexes influence properties such as color, reactivity, and magnetism. By investigating these properties for a particular complex, we can gain insight into metal ion electronic structure. This experiment investigates the magnetism of the brown and green products of [Ni(salpd)], giving valuable insight into oxidation state, electronic configuration, and bonding nature.

Metal complexes exist with either all of their spins paired (diamagnetism) or with unpaired electrons (paramagnetism). The spin associated with unpaired electrons of paramagnetic metal ions generates a magnetic field. Because of this, paramagnetic compounds are attracted to a magnetic field while diamagnetic substances are repelled. The vector quantity that is the measure of the torque exerted on a paramagnetic compound when in a magnetic field is called the magnetic moment, μ, measured in units of Bohr magnetons, BM. The magnitude of μ is dependent on the number of unpaired electrons. Although both electron spin and orbital motion contribute to the total μ for transition metals, for first row elements like nickel, only the spin component is significant. This spin only moment, μ_s, is calculated using equation (3.2), where g is the electronic g factor and S is the total spin quantum number.

$$\mu_s = g[S(S + 1)]^{1/2} \tag{3.2}$$

Values for μ_s for a given number of unpaired electrons, n, are given in Table 3.2.

Table 3.2 Theoretical Values of μ_s (BM) for Given Numbers of Unpaired Electrons, n

n	μ_s (BM)	n	μ_s (BM)
1	1.73	5	5.92
2	2.83	6	6.93
3	3.87	7	7.94
4	4.90		

Experimental magnetic moments are not measured directly, but rather are calculated from *magnetic susceptibility* measurements obtained using a magnetic susceptibility balance. The balance measures the change in current required to keep a set of suspended permanent magnets in balance after their magnetic fields interact with a given sample. The mass magnetic susceptibility, χ_g, for the sample is calculated using equation (3.3).

$$\chi_g = (L/m)\ \{[C(R-R_0)/(1\times 10^9)] + \chi_v'A\} \tag{3.3}$$

where L = sample length (cm), m = sample mass (g), C = balance calibration constant, R = reading when the sample (in the sample tube) is in place in the balance, R_0 = reading when the empty sample tube is in place in the balance, χ_v' = volume susceptibility of air (0.029×10^{-6} erg G^{-2} cm^{-3}), A = cross-sectional area of the sample (cm^2), and 1×10^9 converts $C(R-R_0)$ to cgs units.

The volume susceptibility of air can often be ignored for solid samples and the mass magnetic susceptibility simplifies to equation (3.4).

$$\chi_g(\text{erg } G^{-2}g^{-1}) = [CL(R-R_0)]/10^9 \tag{3.4}$$

From χ_g, a molar susceptibility, $\chi_M{}^{corr}$, corrected for diamagnetic contributions (χ_{dia}) from inner core electrons and other atoms and ions in the molecule, is calculated using equations (3.5) and (3.6).

$$\chi_M = \chi_g \cdot (\text{molecular mass}) \tag{3.5}$$

$$\chi_M{}^{corr} = \chi_M + \chi_{dia} \text{ (negative values)} \tag{3.6}$$

The diamagnetic contribution of the ligand can be obtained by measuring its susceptibility or by summing the contribution of contributing atoms. Diamagnetic susceptibilities for molecules, substituents and ions are found in Table 3.3.

Once $\chi_M{}^{corr}$ is obtained, the effective magnetic moment, μ_{eff} (BM), is calculated using equation (3.7):

$$\mu_{eff} \text{ (BM)} = [(3kT\chi_M{}^{corr})/(N\beta^2)]^{1/2} \tag{3.7}$$

where k is Boltzmann's constant, T is temperature (°C), N is Avogadro's number, and β is the Bohr magneton. With substitution, this becomes equation (3.8):

$$\mu_{eff} \text{ (BM)} = (2.828)(T\chi_M{}^{corr})^{1/2} \tag{3.8}$$

The μ_{eff} for a particular metal complex can then be compared to theoretical spin-only moments given in Table 3.2, as well as with values obtained previously for other complexes

Table 3.3   Diamagnetic Susceptibilities [$\times 10^6$ (cgs units) mol^{-1}]

Cations	Anions	Neutral Atoms	Common Ligands	Constitutive Corrections
Li$^+$(-1.0)	F$^-$ (-9.1)	H (-2.93)	H$_2$O (-13)	C=C (5.5)
Na$^+$ (-6.8)	Cl$^-$ (-23.4)	C (-6.00)	NH$_3$ (-18)	C=C—C=C (10.6)
K$^+$ (-14.9)	Br$^-$ (-34.6)	N-ring (-4.61)	C$_2$H$_4$ (-15)	C+C (0.8)
NH$_4$$^+$ (-13.3)	I$^-$ (-50.6)	N-open chain (-5.57)	CH$_3$COO$^-$ (-30)	C-benzene ring (0.24)
Mg^{2+} (-5.0)	NO$_3$$^-$ (-18.9)	N-imide (-2.11)	en (-46)	N=N (1.8)
Zn^{2+} (-15.0)	ClO$_4$$^-$ (-32.0)	O-ether, alcohol (-4.61)	ox^{2-} (-25)	C=N—R (8.2)
Ca^{2+} (-10.4)	CN$^-$ (-13.0)	O-aldehyde, ketone (-1.73)	acac (-52)	C—Cl (3.1)
1st row TM (approx -13)	OH$^-$ (-12.0)	F (-6.3)	py (-49)	
	O^{2-} (-12.0)	Cl (-20.1)	bpy (-105)	
		Br (-30.6)	phen (-128)	
		I (-44.6)		

(Table 3.4). Magnetic properties of nickel(II) complexes vary according to geometric structure. Magnetic moments for octahedral complexes are generally in the range 2.9–3.4 BM; tetrahedral are expected to be ~4.2 BM but are lower (3.5–4.0 BM) due to distortion; both trigonal bipyramidal and square planar are usually low-spin, diamagnetic.

Procedure 3.2[†]

According to equations (3.4), (3.6), and (3.8), you must experimentally obtain and record values for R, R_0, L, m, and T to obtain a value for μ_{eff} (BM) for your reference, salpd, and [Ni(salpd)]. Follow the instructions and record data in the Results Summary section.

   Note: *this procedure is written for a Johnson–Matthey X magnetic susceptibility balance and may vary according to your instrumentation.*

   For each sample:

1. Zero your instrument and place an empty tube of known weight into the instrument to obtain R_0.
2. Carefully fill and tightly pack your sample tube with either reference compound,[‡] your ligand, or your metal complex. Critical to a proper reading for R, your sample must be well-powdered and tightly packed. Sample lengths (L) are typically 2.5–3.5 cm.
3. Obtain the mass of your tube with sample. The mass of the sample, $m =$ (mass of tube plus sample − mass of tube).

[†]An inexpensive alternative to a commercial balance is given by Eaton, S., Eaton, G. *J. Chem. Educ.* **1979**, *56*, 170.
[‡]If instrument calibration is necessary, it is generally carried out using either Hg[Co(NCS)$_4$] or [Ni(en)$_3$]S$_2$O$_3$, which have χ_g values of 1.644×10^{-5} and 1.104×10^{-5} erg G^{-2} g^{-1}, respectively. We will use Hg[Co(NCS)$_4$] for calibration in this experiment.

Table 3.4   Experimental Magnetic Moments for Various Transition Metal Ions

Ion (S)	obsd Moments, μ_{eff} (BM)	Ion (S)	obsd Moments μ_{eff} (BM)	Ion (S)	obsd Moments μ_{eff} (BM)
Ti^{3+} (1/2)	1.73	Mn^{4+} (3/2)	3.80–4.0	Fe^{2+} (2)	5.10–5.70
V^{4+} (1/2)	1.7–1.8	Mn^{3+} (2)	4.90–5.00	Co^{3+} (2)	~5.4
V^{3+} (1)	2.6–2.8	Mn^{3+} (1)	3.18	Co^{2+} (3/2)	4.30–5.20
V^{2+} (3/2)	3.8–3.9	Mn^{2+} (5/2)	5.65–6.10	Co^{2+} (1/2)	1.8
Cr^{3+} (3/2)	3.70–3.90	Mn^{2+} (1/2)	1.80–2.10	Ni^{3+} (1/2)	1.8–2.0
Cr^{2+} (2)	4.75–4.90	Fe^{3+} (5/2)	5.70–6.0	Ni^{2+} (1)	2.8–4.0
Cr^{2+} (1)	3.20–3.30	Fe^{3+} (1/2)	2.0–2.5	Cu^{2+} (1/2)	1.7–2.2

4. Obtain three readings with gentle but firm packing in between readings. Calculate and record the average R reading.

5. Determine the temperature to 0.1°C.

Name _____   Section _____

Results Summary for the Magnetic Susceptibility Measurements of the Two Products of N,N′-disalicylaldehyde-1,3-propanediiminenickel(II)

Determination of the Calibration Constant (C)

1. χ_g of $Hg[Co(NCS)_4]$ 1.644×10^{-5} erg $G^{-2}g^{-1}$

2. Mass of $Hg[Co(NCS)_4]$ (m)   _____ g

3. Length of sample (L)   _____ cm

4. Reading of empty sample tube (R_0)   _____

5. Readings of sample (R)   Trial 1   Trial 2   Trial 3   Average

   _____   _____   _____   _____

6. Use equation (3.4) to find C _____

Diamagnetic Corrections

(A) Determine χ_M for your ligand using equations (3.4) and (3.5)—*include error analysis*

1. Mass of ligand sample (m)   _____ g

2. Length of sample (L)   _____ cm

3. Reading of empty sample tube (R_0)   _____

4. Readings of sample (R)   Trial 1   Trial 2   Trial 3   Average R_1

   _____   _____   _____   _____

5. Using equation (3.4) and your value for C, calculate χ_g _____ erg $G^{-2} g^{-1}$

6. Repeat (A) two more times   Average R_2 ____ χ_g ____   Average R_3 ____ χ_g ____

7. Use equation (3.5) to calculate your ligand χ_M for each trial.

   Trial 1 ____   Trial 2 ____   Trial 3 ____ χ_M (average) ____

(B) Use Table 3.8 to obtain molar susceptibilities for inner core metal ion electrons and other ions and molecules.

	Green	Brown
1. χ_M (metal ion electrons)	_____ erg G^{-2} mol^{-1}	_____ erg G^{-2} mol^{-1}
2. χ_M (other ions)	_____ erg G^{-2} mol^{-1}	_____ erg G^{-2} mol^{-1}
3. χ_M (other molecules)	_____ erg G^{-2} mol^{-1}	_____ erg G^{-2} mol^{-1}
(C) χ_{diam} [χ_M A(avg) + B]	_____ erg G^{-2} mol^{-1}	_____ erg G^{-2} mol^{-1}

Determination of the Magnetic Moments of the Brown and Green Products of N,N'-disalicylaldehyde-1,3-propanediiminenickel(II)

(A) Determine χ_M for your nickel(II) complex using equations (3.4) and (3.5)—*include error analysis.*

1. Nickel(II) complex (brown or green)   _____

2. Mass of sample (*m*)   _____ g

3. Length of sample (*L*)   _____ cm

4. Reading of empty sample tube (R_0)   _____

5. Readings of sample (*R*)   Trial 1 ____   Trial 2 ____   Trial 3 ____   Average R_1 ____

6. Using equation (3.4) and your value for *C*, calculate χ_g _____ erg G^{-2} g^{-1}

7. Repeat (A) two more times   Average R_2 ____ χ_g ____   Average R_3 ____ χ_g ____

8. Use equation (3.5) and the molecular mass of your complex form to calculate χ_M (erg G^{-2} mol^{-1}) for each trial.

   Trial 1 ____   Trial 2 ____   Trial 3 ____

9. $\chi_M{}^{corr} = \chi_M + \chi_{diam}$ (diamagnetic corrections for your form)

   Trial 1 ____   Trial 2 ____   Trial 3 ____

10. Use equation (3.8) to calculate
    μ_{eff} (BM) for your compound   _____   _____   _____

11. Number of unpaired electrons: _____   _____   _____

12. Proposed geometry: _____

13. Proposed structure:

EXPERIMENT 3.3: THERMAL ANALYSIS OF THE GREEN AND BROWN PRODUCTS OF N,N′-DISALICYLALDEHYDE-1,3-PROPANEDIIMINENICKEL(II)

Level 3

Pre-lab 3.3: Differential Scanning Calorimetry[1,2]

Differential scanning calorimetry (DSC) is a calorimetric method that finds widespread use in many fields, including protein dynamics, polymers, pharmaceuticals, and inorganic materials. DSC measures energy (heat) flow into a sample and a reference substance as a function of controlled increase or decrease of temperature. In a typical power-compensated DSC (Fig. 3.2), the sample and reference are placed on metal pans in identical furnaces each containing a platinum resistance thermometer (thermocouple) and heater. During a thermal transition (e.g., when a physical change in the sample occurs),

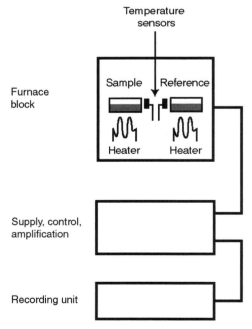

Figure 3.2   Schematic of DSC instrumentation.

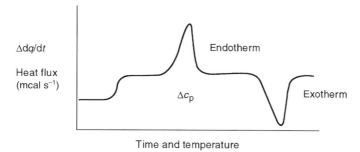

Figure 3.3 Heat flux ($\Delta dq/dt = \Delta dH/dt$), where Δ refers to dH/dt (sample) $- dH/dt$ (reference). Integration of the area under the curve (dH/dt) gives (ΔH). Further, since $C_p = (dH/dT)_p$, heat capacities and changes in heat capacities are also readily obtained. Adapted from reference 13.

both the reference and sample are maintained at the temperature of the user-specified program and the amount of energy needed to maintain a zero temperature differential between the reference and sample is recorded. The rate of the heat flow is proportional to the specific heat of the sample, which changes during a thermal transition. The thermal transition results in a peak, either positive (endothermic process) or negative (exothermic process; Fig. 3.3). Since DSC is carried out at constant pressure, the area of the peak is proportional to the total enthalpy change for the transition.

The green and brown [Ni(salpd)] products differ by only one water molecule. DSC will be used to investigate the release of the water molecule from the green product with subsequent transformation to the brown form. Important information obtained by DSC for the nickel(II) system includes:

1. the temperature of the transition, which gives insight into the process responsible for the transition;
2. the enthalpy change, ΔH, of the thermal transition (sum of all enthalpies contributing to the transition).

The ΔH for your transition will involve bond breaking and making as well as geometry change. Table 3.5 and crystal field theory can be used to estimate enthalpy changes associated with geometry changes.

Procedure 3.3: Thermal Analysis of Green and Brown *N,N'*-disalicylaldehyde-1,3-propanediiminenickel(II) Products

1. Calibrate your DSC instrument (if necessary).
2. Weigh out 4–5 mg sample (this may be modified depending on instrumentation).
3. Scan T from 0 to 150 °C at a rate of 10 °C/min.
4. For your green complex, complete two consecutive runs. Note any physical changes in your sample.
5. Complete the Post-lab Questions and Results Summary Worksheet.

Table 3.5 Energy Levels of d Orbitals in Crystal Fields of Different Symmetries. Energies are in Units of Dq where 10Dq = Δ. (Reproduced with permission from Huheey et al.[7])

Geometry	d_{z2}	d_{x2-y2}	d_{xy}	d_{xz}	d_{yz}
Tetrahedral	−2.67	−2.67	1.78	1.78	1.78
Square planar	−4.28	12.28	2.28	−5.14	−5.14
Trigonal bipyramidal	7.07	−0.82	−0.82	−2.72	−2.72
Square pyramidal	0.86	9.14	−0.86	−4.57	−4.57
Octahedral	6.00	6.00	−4.00	−4.00	−4.00
Pentagonal bipyramidal	4.93	2.82	2.82	−5.28	−5.28

Name _____   Section _____

Results Summary for the Thermal Analysis of *N,N'*-disalicylaldehyde-1,3-propanediiminenickel(II)

Product color

　　　　　　　　　　　　　　　Green　　　　　　　　　*Brown*

1. Thermal transition temperature (°C) _____   _____
2. Enthalpy change (ΔH) _____ (cal g^{-1}) _____ (cal g^{-1})
   _____ (kJ mol^{-1}) _____ (kJ mol^{-1})
3. Observations on two consecutive runs of the green complex:
4. Assignment of the thermal transition: _____

   ΔH (thermal transition) =

5. Analysis of observed thermal transition—fill in the table of enthalpy changes in Table 3.6 for the observed thermal transition (see help below). Use the following questions to help fill in the table.
   a. Propose a geometry and structure for the brown form.
   b. Predict at least *three* possible geometries for the green form. Draw your structures.
   c. Using your molecular structures (Table 3.5), crystal field splitting energies (to obtain $\Delta H_{geometry\ change}$), and outside sources (e.g., CRC or *Lange's Handbook*

Table 3.6 Results Summary

Product Color	Geometry	ΔH_{form} (kJ mol^{-1})	$\Delta H_{geometry\ change}$ (kJ mol^{-1})	Bond Changes	$\Delta H_{bond\ changes}$ (kJ mol^{-1})
Brown	1.		n/a		
Green	1.				
	2.				
	3.				
	4.				

of Chemistry[38]) for bond breaking and bond making processes (ΔH_{bond}), estimate all enthalpic contributions involved in the green to brown color transformation for each predicted green form.

The enthalpy changes associated with the geometry change, equation (3.9) can be calculated from CFSE and assuming negligible contribution from ΔS.

$$\Delta H_{geometry\ change} = \Delta H_{brown} - \Delta H_{green} \qquad (3.9)$$

For each geometry, determine ΔH_{form} (ΔH_{brown}, ΔH_{green}):

    i. Calculate the CFSE in units of Dq. Assume pairing energy contributions are negligible.

    ii. Assume $\Delta E = hc/\lambda_{max} = 10$ Dq and solve for Dq. The λ_{max} for the green and brown forms can be estimated from their colors (a more rigorous determination is given in Chapter 6).

6. ΔH_{calc} (thermal transition) $= [\Delta H_{geometry\ change} + \Delta H_{bond\ changes} + \Delta H_{vap}$ (H_2O)]

    Green form 1 to brown form   _____ (kJ mol^{-1})

    Green form 2 to brown form   _____ (kJ mol^{-1})

    Green form 3 to brown form   _____ (kJ mol^{-1})

7. Compare your measured ΔH with your calculated ΔH values for the thermal transition. Discuss assumptions and sources of error in your calculated ΔH values.

EXPERIMENT 3.4: INFRARED ANALYSIS OF THE GREEN AND BROWN PRODUCTS OF *N,N'*-DISALICYLALDEHYDE-1,3-PROPANEDIIMINENICKEL(II)

Level 3

Pre-lab 3.4: Introduction to IR Spectroscopy of Transition Metal Complexes[14-22]

The infrared region of the electromagnetic encompasses wavelengths from 0.78 to 1000 μm and is divided into three regions: near-, mid-, and far-IR, Table 3.7. Most

Table 3.7   EM Regions of IR Spectroscopy

Region	Wavelength (λ) Range, μm	Wavenumber (v) range, cm^{-1}
Near	0.78–2.5	12,800–4000
Mid	2.5–50	4000–200
Far	2.5–15	4000–670

frequently used is the $4000-670$ cm^{-1} range in which vibrations of organic compounds are typically measured. Table 3.8 lists some mid-IR transitions and their frequencies.

When donor atoms bind to Lewis acids such as metal ions, the frequency of the infrared absorption generally decreases. Carbonyl binding is a prototypical example, but similar frequency shifts are often seen for C–N, S–O, P–O, and N–O stretching vibrations as well. In the M–CO interaction, σ donation with concomitant π back donation into the carbonyl π^* orbital weakens and lengthens the C–O bond, decreasing the force constant, Figure 3.4. A shift to lower frequencies for this absorption is observed. In metal cyanide (C + N) and RC≡O systems, the CN or CO frequency will either increase or decrease depending on the predominant bonding (σ donation or π back bonding) interaction. Given these frequency shifts, IR spectroscopy can often be a useful tool for assessing the ligating atom and is especially useful in ambidentate cases such as the NCS$^-$ ligand. Pertinent to our present system, Schiff-base imines or azomethines

Table 3.8 Mid-IR Transitions and their Frequencies

Bond Type	Compound Type	Frequency Range, cm^{-1} [a]
C—H	Alkanes	2850–2970 (s), 1340–1470 (s)
	Alkenes	3010–3095 (m), 675–995(s)
	Alkynes	3300 (s)
	Aromatic rings	3010–3100 (m), 690–900 (s)
O—H	Monomeric alcohols and phenols	3590–3650 (v)
	H-bonded alcohols and phenols	3200–3600 (v, b)
	Monomeric carboxylic acids	3500–3650 (m)
	H-bonded carboxylic acids	2500–2700 (b)
	Water	3000–3600 (vb, s)[b]
N—H	Amines, amides	3300–3500 (m)
C=C	Alkenes	1610–1680 (v)
	Aromatic rings	1500–1600 (v)
C + C	Alkynes	2100–2260 (v)
C—N	Amines, amides	1180–1360 (s)
C + N	Nitriles	2210–2280 (s)
KCN	Cyano	2080
C—O	Alcohols, ethers, carboxylic acids, esters	1050–1300 (s)
C=O	Aldehydes, ketones, carboxylic acids, esters	1690–1760 (s)
R$_2$C=N—C	Substituted imines	1600–1675 (m)/1640–1690 (v)
	Azomethine	1614–1635
	Ortho-disubstituted benzene	400–500 (w), 700 (w), 750 (s)
	Aromatic alcohol	1150–1300 (s), 1300–1400 (m)
	Aromatic aldehyde	825–1025 (m), 1210–1240 (m), 1260–1325 (m), 1350–1400 (m), 1650–1775 (s)
R$_2$C=N—H	(NH)	3300–3400 (m)

[a](s) = strong, (m) = medium, (w) = weak, (v) = variable, (b) = broad, (vb) = very broad.
[b]A strong and broad absorption in the range 3200–3400 cm^{-1} for some metal complexes indicates the presence of H$_2$O.

π-back donation

σ-donation

Figure 3.4 The M–CO interaction.

(R—C=N—C) generally show a shift in frequency from approximately 1610–1635 to ~1600–1620 cm^{-1} upon metal ligation.

Along with determining metal ion ligation, mid-IR spectroscopy is also useful for assessing changes in symmetry of a ligand upon complexation. Normally infrared inactive vibrations of small molecules such as N_2, O_2, and H_2 can become active upon metal ion binding. Further, changes in the number of IR frequencies can indicate a change in symmetry of the molecule due to structural changes about the metal center. A decrease in symmetry results in a greater number of observed peaks while an increase in symmetry decreases the number of observed peaks. We will take advantage of this situation in assessing our two Ni(salpd) complexes.

Far-IR Spectroscopy

The far-IR region (<650 cm^{-1}) is very useful for inorganic chemists and will be used in both this chapter and in Chapter 4 in the analysis of metal coordination compounds. In this region, stretching and bending vibrations of M–L bonds occur. Some assignments useful for our study are found in Table 3.9.

Procedure 3.4: Infrared Analysis of the Green and Brown Forms of *N,N*'-Disalicylaldehyde-1,3-propanediiminenickel(II)

1. Prepare samples of your *N,N*'-ligand, and both brown and green nickel(II)salpd products for IR analysis by grinding to a fine powder a mixture of compound (~2 mg) and dry, IR-quality KBr (100 mg) and pressing into a thin, clear, pellet.
2. Obtain mid-IR spectra (400–4000 cm^{-1}) and far-IR spectra (200–475 cm^{-1}) for your samples.
3. Record results in Table 3.10 in the Results Summary.

Table 3.9 Stretches and Bends of M–L Bonds

Compound or Bond Type	Frequency Range, cm^{-1}	Compound or Bond Type	Frequency Range cm^{-1}
Phenol	510 (s), 535 (m)	Ni–O (azomethine)	~398 400–450
2-Methylphenol	450 (s), 535 (m), 590 (w)	Ni–N (azomethine)	276–295 430–480
Methylbenzene	345 (s), 470 (vs)		
Aromatic acids and esters	270–370		

Table 3.10   Mid- and Far-IR Assignments for [Ni(salpd)] and salpd

Compound	Functional Group	Frequency (cm^{-1})
N,N′-Disalicylaldehyde-1,3-propanediimine (salpd)		
[Ni(salpd)] (green form)		
[Ni(salpd)] (brown form)		

Name _____   Section _____

Results Summary for the Infrared Analysis of *N,N′*-disalicylaldehyde-1,3-propanediiminenickel(II)

1. List and assign all important transitions observed in the IR spectra for your ligand, green, and brown [Ni(salpd)] complexes in Table 3.10.

2. Compare your salpd mid-IR spectrum with those for your brown and green forms. Highlight any absorptions that have shifted upon binding to the nickel(II) ion.

3. Compare the mid-IR spectra for your brown and green complexes. What conclusions can you draw?

4. Compare the far-IR spectra for your brown and green complexes. Propose likely structures for your brown and green complexes.

Post-lab Question

Q3.3   Look up the reference Elmali, A., Zeyrek, C. T., Elerman, Y., Durlu, T. N. *J. Chem. Crystallogr.* **2000**, *30*, 167–171, on the two solid-state forms (green and brown) isolated for the [Ni(Cl₂salpd)] complex. Are your conclusions similar or different for your isolated forms? Explain.

EXPERIMENT 3.5: HARTREE–FOCK CALCULATIONS AND STRUCTURE PREDICTIONS USING SIMPLIFIED BROWN AND GREEN *N,N′*-DISALICYLALDEHYDE-1,3-PROPANEDIIMINENICKEL(II), [Ni(salpd)], COMPLEXES

Level 3

Pre-lab 3.5: Introduction to the Hartree–Fock Approximation[23,24]

In recent years, dramatic advances in computational power combined with the marketing of packaged computational chemistry codes have allowed quantum chemical calculations to become fairly routine in both prediction and verification of experimental observations. The 1998 Nobel Prize in Chemistry reflected this impact by awarding John A. Pople a shared prize for his "development of computational methods in quantum chemistry."

The Hartree–Fock approximation is a basic approach to the quantum chemical problem described by the Schrödinger equation, equation (3.10), where the Hamiltonian (H) operating on the wavefunction (Ψ) yields the energy (E) multiplied by the wavefunction.

$$H\Psi = E\Psi \tag{3.10}$$

Closed-shell (diamagnetic) systems can be investigated using a restricted Hartree–Fock (RHF) calculation, while unrestricted Hartree–Fock (UHF) calculations are able to accommodate open-shell (paramagnetic) systems as well. The Hartree–Fock approximation is also important in serving as a foundation for a variety of more accurate quantum chemical calculations that account for electron correlation.

In the Hartree–Fock approximation, the Hamiltonian operator consists of several mathematical terms corresponding to the kinetic energy, the potential energy for attraction of the electrons to the nuclei, and the electron–electron terms known as the Coulomb and exchange operators. In a Hartree–Fock calculation, a trial wavefunction is employed that is improved through an iterative process known as self-consistent field (SCF) cycles. This trial wavefunction generally is composed of several gaussian functions per atom, known as the basis set. Through the iterative process, the wavefunction and energy are refined until preset tolerance criteria are met between cycles. At this point, the calculation is converged for the SCF cycle.

If a geometry optimization is being performed, a complete SCF convergence will be performed for a fixed set of atomic coordinates; then the atomic coordinates will be updated and a new SCF cycle will be performed. This series of SCF cycles, each with updated atomic coordinates, continues until the present geometry tolerance criteria are met. Generally, post-calculation analyses are performed at this point, when the calculation is converged both for the final SCF cycle as well as the geometry optimization.

In this experiment we will use Hartree–Fock calculations to predict the structure of our [Ni(salpd)] complexes according to their displayed magnetism. Recall that the two nickel complexes $C_{17}H_{16}N_2O_2Ni$ (brown form) and $C_{17}H_{18}N_2O_3Ni$ (green form) differ by only one water molecule yet have very different properties. The brown form is diamagnetic and due to molecular orbital splitting, should be in the singlet state. The green form is paramagnetic and, due to different molecular orbital splitting, should be in the triplet state.

Because of time constraints, this lab will not deal with proposed structures in their entirety, but rather will focus on simplified nickel complexes. These simplified structures are given in Figure 3.5. The simplified structures all contain two ammonia ligands and two hydroxyl ligands. Two structures are 4 coordinate and the other two include a water ligand as a possible 5 coordinate structure.

As mentioned, Hartree–Fock calculations are important as a basic approach to the Schrödinger equation that does not rely on parameterization and in serving as a foundation for more accurate quantum chemical calculations. Accordingly, all commercially available quantum chemical codes likely are able to perform Hartree–Fock calculations. A higher-order method or density functional calculation should yield better results for geometry optimizations of transition metal complexes. If such methods are available, it would be interesting to perform the geometry optimizations in this experiment using alternative methods and comparing the final structures with those obtained using UHF.

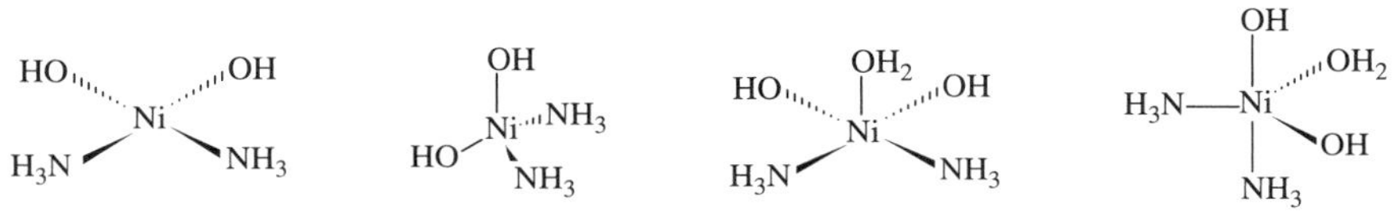

Figure 3.5 Simplified nickel complexes used in this lab. All contain two ammonia ligands and two hydroxyl ligands. Shown in this figure are the geometries used in this lab: square planar, tetrahedral, bipyramidal, and square pyramidal, respectively.

Procedure 3.5: Hartree–Fock Calculations and Structure Predictions Using Simplified Brown and Green *N,N'*-disalicylaldehyde-1,3-propanediimine nickel(II), [Ni(salpd)], Complexes[25]

In this lab we will perform Hartree–Fock calculations in Spartan '02[©] to determine the relative energies of simplified versions of the nickel(II) complexes in the triplet and singlet states. Since the green complex is known to be paramagnetic, this information will help in predicting the structure of the complex.

Splitting Anticipated from Crystal Field Theory

1. Prior to running any calculations, determine the *d*-orbital splitting patterns you would expect for each of the geometries to be investigated and the anticipated filling of these orbitals for a Ni(II) complex. Which do you predict to be diamagnetic and which do you anticipate to be paramagnetic? Record your answers in the Results Summary section.

Performing Single-Point UHF Calculations on the Simplified Ni Complexes

1. Open the Spartan program.
2. Go to **File**, **New** to open a new project.
3. Use the pallets on the right side of the screen to create one of the simplified nickel structures as shown in Figure 3.5. Note that Ni is under the **Exp.** pallet and that you may choose the geometry under the element choices. Place the appropriate Ni coordination geometry in the workspace by selecting Ni and the geometry in the pallet and then clicking in the workspace. You may add the ligands by clicking on the active yellow sites of the nickel. Ammonia and water ligands are available in the Exp. pallet under ligands.
4. Once your structure is created, save under **File**.
5. Now under **Setup**, select **Calculations**. Choose **Single Point Energy** with **Hartree–Fock** and leave the basis as the default [3-21G(*)]. Leave the **Symmetry** option checked. Under **Multiplicity**, choose triplet. In the **Options** box, type "SCFCYCLE = 10000, SCF = UNRESTRICTED, GUESS = CORE" without the quotes. Check the box for **Converge**; you may leave the **Global Calculations** box checked as well. Now click **Submit** (not OK).
6. The calculation may take a few minutes or more, depending on computer speed, etc. Under **Options**, **Monitor** you can watch the progress of the calculation. A pop-up box will inform you when the calculation is complete. View the calculation results under Display, Output. Record the final energy.
7. Rerun the calculation by going to **Setup** and leaving every option as before except change **Multiplicity** to singlet. Record the final energy for this as well.
8. Repeat the same procedure for each of the other structures. The multiplicity with the lower UHF energy should indicate the preferred magnetic state for that simplified complex. How do these compare with your original predictions based on *d*-orbital splitting patterns for the different geometries? (Record in Results Summary section.)

Optimizing the Geometry of the Preferred Triplet Structure(s)

1. Reopen the one structure that showed an energetic preference for the triplet state in your single point energy calculations.

2. Save the structure under a new name, such as "triplet" with **Save As** under **File**.

3. Under **Setup, Calculations**, change type of calculation to **Equilibrium Geometry**. In the **Options** line, type "SCFCYCLE = 10000, GEOMETRYCYCLE = 10000" without the quotes. All other settings given in step 5 of the previous section are fine, including the triplet **Multiplicity**. Now click **Submit**.

4. When the calculation is finished, note the final energy and geometry of the optimized structure. Save under **File**.

5. Reopen the original, single-point energy (non-optimized) structure that you opened in step 1.

6. Save the structure under a third name, such as "singlet," with **Save As** under **File**.

7. Again, under **Setup, Calculations**, change type of calculations to **Equilibrium Geometry**. Also, change the **Multiplicity** to singlet. Click **Submit**.

8. When the geometry optimization is finished, again note the final energy and the geometry of the optimized structure. Save under **File**.

Name _____   Section _____

Results Summary for the Hartree–Fock Calculations and Structure Predictions Using Simplified Brown and Green N,N'-disalicylaldehyde-1,3-propanediiminenickel(II), [Ni(salpd)], Complexes

Magnetism Expected from CF Splitting

1. Record the expected magnetism for the four investigated forms (Fig. 3.5) based on crystal field splitting patterns (see Table 3.5 for d orbital energies)

Square planar   _____

Tetrahedral   _____

Pentagonal bipyramidal   _____

Square pyramidal   _____

Single-Point UHF Calculations

2. Record the final energies (eV) obtained for each geometry in the triplet and singlet states obtained from your single-point UHF calculations. Check ($\sqrt{}$) the preferred spin state of each geometry.

	Triplet	Singlet
Square planar	_____	_____
Tetrahedral	_____	_____
Pentagonal bipyramidal	_____	_____
Square pyramidal	_____	_____

Preferred Geometry for the Triplet State Structure

3. Preferred triplet state structure   _____

	Triplet	Singlet
4. Final energy obtained from equilibrium geometry calculation	_____	_____
5. Geometry of optimized structure from equilibrium geometry calculation	_____	_____

6. Observations and conclusions from equilibrium geometry calculation:

Post-lab Questions

Q3.4 Describe the geometries for your final green and brown products. Are they ideal or do you find distortion? Read about the geometries of nickel(II) complexes in Cotton, F. A., Wilkinson, G. *Advanced Inorganic Chemistry*, 5th edn. Wiley Interscience: New York, 1988. Comment on your calculation results.

Q3.5 *Briefly* sum up the results from Experiments 3.2–3.4 and from the calculations in this experiment. Propose final structures for your green and brown [Ni(salpd)] complexes.

EXPERIMENT 3.6: VISIBLE SPECTROSCOPIC ANALYSIS AND JOB'S METHOD FOR DETERMINING THE BINDING OF PYRIDINE TO *N,N'*-DISALICYLALDEHYDE-1,3-PROPANEDIIMINENICKEL(II), [Ni(salpd)]

Level 3

Pre-lab 3.6.a: Behavior of Nickel(II) Coordination Complexes in Solution[6]

Nickel(II) complexes have long been known to undergo several types of structural and conformational changes in solution. These include:

1. The formation of 5 and 6-coordinate complexes by the addition of ligands to square planar complexes—Lifschitz's substituted ethylenediamine complexes are examples of complexes that exhibit such behavior, Figure 3.6.
2. Monomer–polymer equilibria driven by temperature and concentration, Figure 3.7.
3. Square planar–tetrahedral equilibria driven by both steric and electronic factors of the ligands, Figure 3.8.

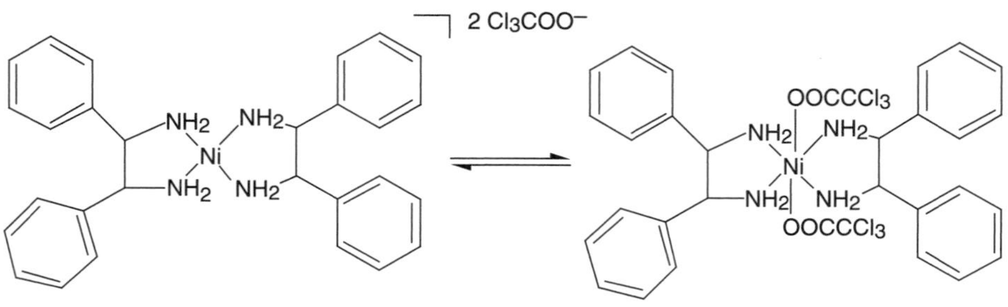

Figure 3.6 Conversion of yellow square planar [Ni(stien)$_2$] (Cl$_3$CCOO)$_2$ to the blue octahedral form.

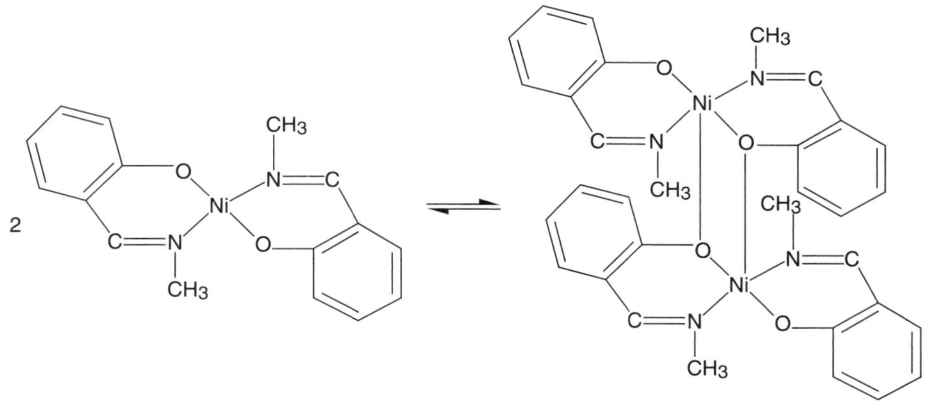

Figure 3.7   Dimerization of diamagnetic *N*-(methyl)salicylaldiminato nickel(II).

Each of these behaviors should be kept in mind when examining the solution chemistry of [Ni(salpd)].

Pre-lab 3.6.b: Electronic Spectra of Metal Complexes[7]

Three types of transitions that give rise to UV–vis spectra in transition metal complexes are:

1. $d \leftrightarrow d$ (visible);
2. charge transfer (M $\leftrightarrow$ L) (UV with possible tailing in the visible);
3. intraligand (UV).

For most transiton metal complexes the $d \leftrightarrow d$ transitions are responsible for their colors, but in some cases, M $\leftrightarrow$ L transitions are the source. In order for a transition to be allowed, the following rules must be met:

1. $\Delta S = 0$ (no changes in spin state are allowed);
2. $\Delta l = \pm 1$ (e.g., $p \rightarrow d$ are "allowed" but $d \rightarrow d$ are "forbidden");
3. $g \leftrightarrow u$ (a change in parity or orbital sign must take place).

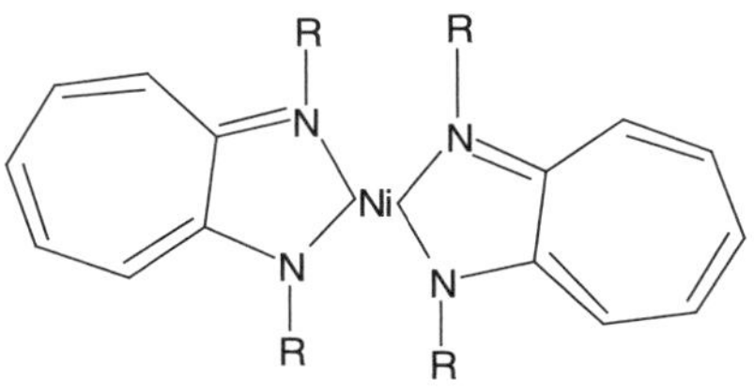

Figure 3.8   An example of a nickel(II) complex that undergoes square-tetrahedral equilibrium.

Electronic transitions in octahedral and square planar transition metal complexes disobey rules (2) and (3). Consider the promotion of a $3d^1$ electron from a t_{2g} to an e_g orbital in a complex of octahedral symmetry. The parity ($g \leftrightarrow g$) remains the same during the excitation of the electron from one $3d$ orbital to another. Such transitions are, therefore, formally forbidden. For various reasons including molecular vibration, however, the above rules can be "relaxed" and the transitions become partially allowed. Because they are formally forbidden, $d-d$ transitions are weak. This is reflected in the small molar absorptivities, ε (1–500 M^{-1} cm^{-1}) associated with many transition metal ion complexes.

For tetrahedral molecules, which do not have a center of symmetry, rule (3) is relaxed due to $d-p$ mixing of the t_2 orbitals. This is reflected in the higher ε values of tetrahedral complexes, 10^2–10^3 M^{-1} cm^{-1}.

Three-spin allowed transitions are observed for octahedral nickel(II) complexes (see Pre-lab 6.2 for theory on state assignment) with wavelength ranges 769–1428, 500–900, and 370–525 nm, and with ε values typically 10 M^{-1}cm^{-1}. Tetrahedral complexes generally show one $d-d$ transition in the visible region at $\sim$650 nm ($\varepsilon \sim$ 100 M^{-1}cm^{-1}). Square planar complexes show a band in the 450–600 nm ($\varepsilon \approx$ 60–100 M^{-1} cm^{-1}) range and the absorption responsible for their characteristic color is at $\sim$400 nm ($\varepsilon \approx$ 1000 M^{-1} cm^{-1}).

Pre-lab 3.6.c: Job's Method[26–28]

Job's method, or the method of continuous variation, was first published in 1928 as a means for determining the nature of metal complexes in solution. This spectrophotometric method is particularly useful for identifying metal complexes that, for stability reasons, may be difficult to isolate as a solid from solution. We will use this method to investigate the solution structure of the [Ni(salpd)] complex.

The formation of metal complexes can be viewed as a series of equilibria, equations (3.11)–(3.13), characterized by successive formation or equilibrium constants, K_n.

$$M + L \rightleftharpoons ML \quad K_1 \rightleftharpoons [ML]/[M][L] \tag{3.11}$$

$$ML + L \rightleftharpoons ML_2 \quad K_2 \rightleftharpoons [ML_2]/[ML][L] \tag{3.12}$$

$$ML_2 + L \rightleftharpoons ML_3 \quad K_3 \rightleftharpoons [ML_3]/[ML_2][L] \tag{3.13}$$

To decipher the number of ligands, L that bind to a metal, M, Job's method relies on the additive property of absorbance of species in solution. For a multicomponent system, absorbances, A, are additive for y non-interacting species[†] with concentrations c_y and molar absorptivities, ε_y, equation (3.14).

$$A_{\text{total}} = A_1 + A_2 +, \ldots, + A_y = \varepsilon_1 b c_1 + \varepsilon_2 b c_2 + \ldots + \varepsilon_y b c_y \tag{3.14}$$

Solutions of M and L are prepared such that the mole fractions of each component are varied, but the total number of moles is kept constant. If the mole fraction of ligand is x,

[†]Nickel(II) Schiff base complexes can associate in solution (Fig. 3.7), showing a deviation from Beer's law in a plot of A vs [NiL]. Concentrations must be chosen carefully to avoid interactions.

then the mole fraction of metal is $(1 - x)$. A plot of absorbance of the ML_n species, A_{ML}, where $A_{ML} = (A_\lambda{}^{measured}) - [(A_\lambda{}^{metal})(1 - x) + (A_\lambda{}^{ligand})(x)]$, at a particular wavelength λ, vs the mole fraction, x, of one component, is then constructed. If this plot gives a straight line, no metal complex formation has occurred. If, however, the plot deviates from linearity, the deviation will reach a maximum at the mole fraction corresponding to the composition of the complex. The mole fraction x corresponding to the apex of the plot is used to determine the number of ligands, n, equation (3.15).

$$n = x/1 - x .\qquad\qquad\qquad (3.15)$$

The plot can be repeated at a variety of wavelengths where changes in absorbance are greatest as the determination is independent of wavelength. In cases where multiple forms exist, the different forms should be determined at wavelengths corresponding to the different species.

Procedure 3.6.a: UV–Visible Spectroscopic Analysis of [Ni(salpd)]

Safety warning! Operations set off between the lines must be performed in a fume hood.
Caution: pyridine (py) should be used in the hood. Avoid contact with skin. Py can cause sterility in males. Wear goggles and gloves. Clean up spills immediately.

---

1. Quantitatively prepare 5 ml solutions of 0.015 M brown [Ni(salpd)] and pyridine (py) in $CHCl_3$.

---

2. Obtain visible spectra from 450–900 nm. Determine the molar absorptivities at 600 nm.
3. Add a small amount of green [Ni(salpd)] to $CHCl_3$ and to DMSO. Record observations.

Procedure 3.6.b: Job's Method

---

1. Prepare 25 ml of 15 mM [Ni(salpd)] (MM $= 339$ g mol^{-1}) in $CHCl_3$—label Stock 1.
2. Prepare 25 ml of 15 mM pyridine (MM $= 71.9$ g mol^{-1}, d. 0.978 g ml^{-1}) in $CHCl_3$—label Stock 2.
3. Using Stocks 1 and 2, prepare a series of 10 solutions of differing mole ratios as outlined in Table 3.11.

---

4. Obtain a UV–vis spectrum of each solution and record the absorbances at 600 nm.
5. For each series of wavelength measurements, plot the mole ratio, x, of pyridine vs A_{ML}. Using either software or a ruler, draw two straight lines following the increasing slopes to an apex.
6. Determine where the two lines intersect and record the x coordinate value. From this value of x, and equation (3.15) determine n or the number of equivalents of pyridine ligand bound to [Ni(salpd)].

Table 3.11 Preparation of Job's Method Solutions in 10 ml Total Volume

Solution No.	Volume [Ni(salpd)] Stock 1 (ml)	Volume Pyridine Stock 2 (ml)	Mole Fraction of Pyridine, (x_{py})	Mole Fraction of [Ni(salpd)], $(1 - x_{py})$
1	10	0	0	1.0
2	9	1	0.1	0.9
3	8	2	0.2	0.8
4	7	3	0.3	0.7
5	6	4	0.4	0.6
6	5	5	0.5	0.5
7	4	6	0.6	0.4
8	3	7	0.7	0.3
9	2	8	0.8	0.2
10	1	9	0.9	0.1
11	0	10	1.0	0

Name _____   Section _____

Results Summary for the UV–Visible/Job's Plot Analysis of Pyridine Binding to N,N'-disalicylaldehyde-1,3-propanediiminenickel(II)

UV–Visible Spectroscopic Analysis

1. Complete Table 3.12.
2. Record observations on dissolving the green form in DMSO and $CHCl_3$.

Job's Plot Analysis   For the calculation of A_{ML},

$$A_{ML} = (A_\lambda^{measured}) - [(A_\lambda^{metal})(1 - x) + (A_\lambda^{ligand})(x)]$$

1. Complete Table 3.13.
2. Prepare your plot and use equation (3.15) to calculate n: _____

Post-lab Questions

Q3.6 Discuss your UV–visible spectrum for [Ni(salpd)] in light of what is known of other nickel(II) complexes. What geometry(ies) does your spectrum support? Can you use optical spectroscopy alone to determine the nature of your complex?

Table 3.12   Results Summary for UV–Visible Spectroscopic Analysis

Compound	A_{600}	ε (M^{-1} cm^{-1})
[Ni(salpd)] 15 mM (solution 1)		
Pyridine 15 mM (solution 11)		

Table 3.13   Results Summary for Job's Plot Analysis

Solution No.	Mole Fraction of L (x)	Mole Fraction of Ni(II), $(1 - x)$	A_{600}	A_{ML}^a (600 nm)
$1 = A_L$	1	0		
2	0.9	0.1		
3	0.8	0.2		
4	0.7	0.3		
5	0.6	0.4		
6	0.5	0.5		
7	0.4	0.6		
8	0.3	0.7		
9	0.2	0.8		
10	0.1	0.9		
$11 = A_{Ni}$	0	1.0		

Q3.7 Using simple CFT (or LFT), suggest assignments for your transitions. Use the literature to support your arguments.

Q3.8 Discuss your observations upon dissolving the green form in DMSO.

Q3.9 What did your Job's plot reveal? Draw a complex supported by your results.

EXPERIMENT 3.7: DETERMINATION OF THE FORMATION CONSTANT FOR PYRIDINE BINDING TO *N,N'*-DISALICYLALDEHYDE-1,3-PROPANEDI IMINENICKEL(II), [Ni(salpd)], USING THE DOUBLE RECIPROCAL (BENESI–HILDEBRAND) METHOD

Level 3

Pre-lab 3.7.a: Metal Complex Formation[29–31]

The important pioneering studies on metal complex equilibria in solution by Bjerrum and Schwarzenbach laid the groundwork for rationalizing stability trends. Other important contributors to the understanding of periodic trends in complex stability include Ahrland, Chatt and Davies, and Pearson in formulating hard–soft–acid–base theory (see Chapter 1), and Martell and Smith, who have compiled an extensive database of critical stability constants. Motekeitis and Martell have also made very accessible to the chemist a computer program for calculating formation constants from aqueous potentiometric titrations (see Chapter 8).

The formation of metal complexes can be viewed as a series of equilibria, equations (3.16)–(3.18), characterized by successive formation or equilibrium constants, K_n.

$$M + L \rightleftharpoons ML \quad K_1 = [ML]/[M][L] \tag{3.16}$$

$$ML + L \rightleftharpoons ML_2 \quad K_2 = [ML_2]/[ML][L] \tag{3.17}$$

$$ML_2 + L \rightleftharpoons ML_3 \quad K_3 = [ML_3]/[ML_2][L] \tag{3.18}$$

Alternatively, the *overall* stability constants, β_n, can be expressed, equations (3.19)–(3.21).

$$M + L \rightleftharpoons ML \qquad \beta_1 = [ML]/[M][L] \qquad (3.19)$$

$$M + 2L \rightleftharpoons ML_2 \qquad \beta_2 = [ML_2]/[M][L]^2 \qquad (3.20)$$

$$M + 3L \rightleftharpoons ML_3 \qquad \beta_3 = [ML_3]/[M][L]^3 \qquad (3.21)$$

The successive equilibrium constants are related to the overall stability constants by: $\beta_n = K_1 K_2 K_3, \ldots, K_n$.

Both steric and electronic effects contribute to the magnitude of the formation constant. The effect of sterics are often assessed with the use of molecular mechanics calculations (Chapter 6) and electronic effects, by studying linear free energy relationships—the relationship between free energies and rates of complex formation. HSAB and CFT (or LFT) are used regularly to predict and rationalize metal complex stability.

Pre-lab 3.7.b: Double Reciprocal (Benesi–Hildebrand) Method[32,33]

Many methods including potentiometry, spectrophotometry, NMR spectroscopy, and reaction kinetics can be used to obtain K_n values in solution. Because ligands are often Arrhenius bases and metal–ligand complexes tend to be soluble in aqueous solution, potentiometric (pH) titration is one of the most widely used procedures. However, for complexes like [Ni(salpd)], where a non-aqueous medium is required, an alternative, spectrophotometric method is preferred. As you will see when reading through the derivation, this method requires several criteria to be met. One of the most important is that one component, either [Ni(salpd)] or pyridine, must be in excess; in our case this is pyridine. Another factor that will simplify the math is that pyridine does not absorb in the region of analysis.

To obtain the formation constant, the number of metal-bound ligands, L or pyridine, must be independently determined. The Job's plot analysis in Experiment 3.6 revealed that a 1:1 Ni:salpd–pyridine complex is formed; we will assume this is the case.

The absorbance, A, of a solution containing M, L, and ML is given by equation (3.22),

$$A = \varepsilon_M b[M]_f + \varepsilon_L b[L]_f + \varepsilon_{ML} b[ML] \qquad (3.22)$$

where $[M]_f$ and $[L]_f$ are the unbound "free" metal [Ni(salpd)] and ligand (pyridine) concentrations, respectively. If pathlength, b, is 1 cm, equation (3.22) reduces to equation (3.23).

$$A = \varepsilon_M [M]_f + \varepsilon_L [L]_f + \varepsilon_{ML} [ML] \qquad (3.23)$$

Our goal is to obtain an expression in terms of *total* M and *total* L concentrations, $[M_T]$ and $[L_T]$, respectively, because these are measurable quantities. Under conditions where $[L]_T \gg [M]_T$, we can assume that $[L]_f = [L]_T$. Further, we can solve for $[M]_f$ from conservation of mass, equation (3.24). The mass balance expression then becomes equation (3.25) and upon rearrangement, we obtain equation (3.26):

$$[M]_f = [M]_T - [ML] \tag{3.24}$$

$$K = [ML]/\{[M]_T - [ML]\}[L]_T \tag{3.25}$$

$$K[M]_T[L]_T = [ML]\{K[L]_T + 1\} \tag{3.26}$$

We are now only left with having to put [ML] into a measurable quantity. If we carry out our experiment at a wavelength where the absorbance of $[L]_T$ is negligible, equation (3.23) is dependent only on the M and ML concentrations, equation (3.27).

$$A = \varepsilon_M[M]_f + \varepsilon_{ML}[ML] \tag{3.27}$$

Rearranging and combining with equation (3.24),

$$[ML] = \{A - \varepsilon_M[M]_T\}/(\varepsilon_{ML} - \varepsilon_M) \tag{3.28}$$

Equation (3.28) is then substituted into equation (3.25); the reciprocal is taken of each side and then both sides of the equation are multiplied by $(1 + K[L]_T)$ to yield the $y = mx + b$, equation (3.29). A plot of $1/(A - \varepsilon_M[M]_T)$ vs $1/[L]_T$ yields a slope of the form $1/Kq$ and intercept, $1/q$.

$$\frac{1}{[M]_T(\varepsilon_{ML} - \varepsilon_M)} + \frac{1}{[L]_T[M]_T(\varepsilon_{ML} - \varepsilon_M)} = \frac{1}{\{A - \varepsilon_M[M]_T\}} \tag{3.29}$$

Procedure 3.7: K_f Determination by Visible Spectroscopy

Safety warning! Operations set off between the lines must be performed in a fume hood.

Table 3.14   **Preparation of 7.5 mM [Ni(salpd)] with Varying [Pyridine] Solutions in 1.5 ml (1500 μl) Total Volume**

Solution No.	Volume [Ni(salpd)] Stock 1 (μl)	Volume [Ni(salpd)]/Pyridine (py) Stock 2 (μl)	[py] (mM)
1	0	1500	7500
2	500	1000	5000
3	700	800	4000
4	900	600	3000
5	1100	400	2000
6	1300	200	1000
7	1350	150	750
8	1400	100	500
9	1440	60	300
10	1460	40	200
11	1480	20	100
12	1485	15	75

Caution: pyridine (py) should be used in the hood. Avoid contact with skin. Py can cause sterility in males. Wear goggles and gloves. Clean up spills immediately.

1. Prepare 50 ml of 7.5 mM [Ni(salpd)] (MM = 339 g mol^{-1}) in CHCl$_3$—label Stock 1.
2. Prepare 25 ml of 7500 mM pyridine (MM = 71.9 g mol^{-1}, d. 0.978 g mL^{-1}) that also contains 7.5 mM [Ni(salpd)] in CHCl$_3$—label Stock 2.
3. Using Stocks 1 and 2, prepare a series of 12 solutions of constant [Ni(salpd)] and varying [pyridine] concentration as outlined in Table 3.14.

4. Obtain a visible spectrum (400 nm–900 nm) of Stock 1 and of each solution and record the absorbances at 600 nm. You should overlay all of your spectra to observe absorbance behavior.
5. Construct a *binding curve*: for each series of wavelength measurements, plot the change in absorbance, ΔA_{600} vs total pyridine concentration (ΔA_{600} = measured absorbance at 600 nm for each solution minus A_{600} for 7.5 mM [Ni(salpd)], i.e., Stock 1. Note that this is equal to $\{A - \varepsilon_M[M]_T\}$ in equation (3.28).
6. Construct a *double reciprocal* (*Benesi–Hildebrand plot*): plot $1/\Delta A_{600}$ vs $1/[py]_T$. Do a linear regression and determine the *y*-intercept ($1/q$). Determine K from the slope ($1/Kq$), see equation (3.28).

Name _____   Section _____

Results Summary for the UV–Visible Determination of Formation Constant for Pyridine Binding to N,N'-disalicylaldehyde-1,3-propanediiminenickel(II)

K_f *Determination*

1. Complete Table 3.15.
2. K_f: _____

Table 3.15   Results Summary for Formation Constant Determination

Solution No.	[py], mM	$1/[py]$, mM^{-1}	A_{600}	ΔA_{600}	$1/\Delta A_{600}$
Stock 1	0				
1	7500				
2	5000				
3	4000				
4	3000				
5	2000				
6	1000				
7	750				
8	500				
9	300				
10	200				
11	100				
12	75				

Post-lab Questions

Q3.10 Describe how the absorbance spectrum changes upon addition of pyridine. Discuss.

Q3.11 Discuss your K_f value.

EXPERIMENT 3.8: NMR SPECTROSCOPIC ANALYSIS OF N,N'-DISALICYLALDEHYDE-1,3-PROPANEDIIMINENICKEL(II), [Ni(salpd)]

Level 4

This experiment incorporates advanced theory and data analysis. In this experiment, you will use one- and two-dimensional ^{1}H NMR spectroscopy to investigate the solution structure of the brown [Ni(salpd)] complex. A brief introduction to pulsed NMR spectroscopy is given Appendix 1.

Pre-lab 3.8: Paramagnetic NMR Spectroscopy[33−37]

The magnetic moment of an unpaired electron is $\sim 10^3$ times that of a nuclear spin. As a consequence, the presence of a paramagnetic center significantly affects a resonating nucleus. Two of the most important effects are isotropic hyperfine shifts and increased nuclear relaxation. Many metal ions cause significantly enhanced spin relaxation (the unpaired electron provides a mechanism $\sim 500,000$ times more efficient than a ^{1}H nucleus), which results in the broadening and obscuring of ligand ^{1}H resonances. Paramagnetic metal ions with long electron spin relaxation times (τ_s) are line-broadening agents and include Mn(II) and Cr(III). Paramagnetic nickel(II) and cobalt(II) have short τ_s, however, and promote sharp ligand resonances with hyperfine shifts that can be studied. These ions are called shift reagents. Hyperfine shifts arise by two mechanisms: contact and pseudo-contact magnetic interactions between a nucleus and an unpaired electron. Contact shifts (δ_c), which are transmitted through bonds, occur by delocalization of the unpaired electron and overlap with an s orbital at the nucleus. Pseudo contact shifts (δ_p), occur through space, mediated by spin−orbital and dipolar interactions. Contact shifts are proportional to the unpaired electron density at the nucleus and thus give information about electronic structure. Pseudo-contact shifts provide information on molecular geometry.

For systems that exhibit contact shifts in solution, the total spin density, ρ, centered on the s orbital of the shifted nucleus, i, is equal to the number of unpaired electrons on the paramagnetic center, equation (3.30). The value for ρ_i for each resonance is

$$\Sigma_i \rho_i = 2S \tag{3.30}$$

obtained from the measured contact shift, A_i, for each nucleus, equation (3.31), where $\hbar$ is reduced Planck's constant, g is the electronic g factor, and β is the Bohr magneton.

$$A_i = (8\pi/3)\gamma_N \hbar g\beta\rho_i \tag{3.31}$$

The value for A_i is the difference in chemical shift of the ligand resonance bound to nickel(II) and bound to an analogous diamagnetic probe [e.g., zinc(II)].

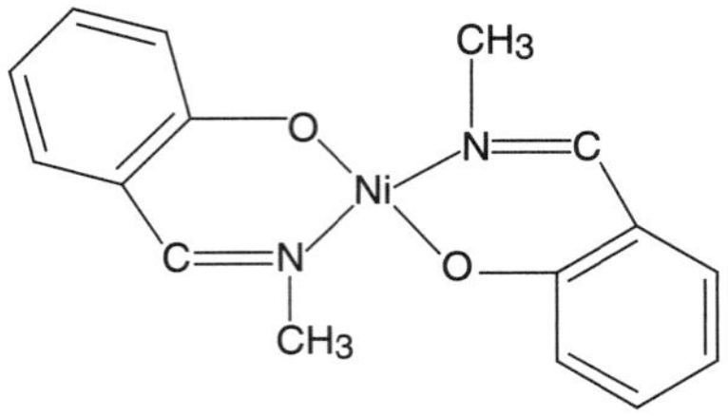

Figure 3.9 Structure of *bis*(N-methylsalicylaldimine)nickel(II).

Another means for attaining the electronic spin state of a paramagnetic shift reagent is to use the Evans method for determining the magnetic moment. The frequency shift of a solvent peak due to the presence of a paramagnetic center, Δv (Hz), is related to the magnetic susceptibility, χ_M (cm^3 mol^{-1}), of the paramagnetic species by equation (3.32), where Q is 1 for an instrument with an electromagnet and 2 for a superconducting instrument, c is the concentration of solute (M), and v_{ref} is the frequency of the pure solvent reference peak (Hz).

$$\chi_M = (477)\Delta v/(Q v_{ref} c) \tag{3.32}$$

Many solid, diamagnetic nickel(II) Schiff base complexes in coordinating solvents can become "magnetically anomalous" or partially paramagnetic. A prototypical complex is *bis*(N-methylsalicylaldiminenickel(II), Figure 3.9, which in solution at near-room temperature attains a magnetic moment in CHCl$_3$ of 1.9–2.2. The moment increases with concentration (2.7 times from 0.037–0.130 M). In solutions of strong paramagnetism, the intensity of the ~615 nm band ($^1A_{1g} \rightarrow {}^1B_{3g}$ or $^1B_{1g}$) for this complex drops due to the decreased singlet character complex.

The paramagnetic behavior of nickel Schiff bases is attributed to the diamagnetic, square planar–paramagnetic octahedral equilibrium shown in Figure 3.9. The contact shifts for nickel(II) Schiff bases are generally very large (e.g., −35 to +70 ppm) and result in simple first-order spectra with one line for each chemically distinct set of nuclei.

Whether or not a nickel(II) Schiff base complex exhibits partial paramagnetism depends on the ligand structure. For example, if the substituent group is bulkier than methyl for the complex in Figure 3.9, the equilibrium is hindered. In this experiment, you will investigate the salpd ligand and [Ni(salpd)] structures by NMR spectroscopy in both the non-coordinating CDCl$_3$ and coordinating d_6-DMSO solvents and elucidate the species observed in solution.

Procedure 3.8: Assignment of ^{1}H NMR Spectra

1. Place 50 mg salpd ligand in 1 ml CDCl$_3$ containing a reference (e.g., TMS). In another tube, place 30 mg salpd ligand in 1 ml d_6-DMSO containing a reference (e.g., DSS).
2. Place ~50 mg brown [Ni(salpd)] complex in 1 ml CDCl$_3$ with reference and in another tube, place ~10 mg (amount is limited by solubility) in 1 ml d_6-DMSO with a reference.
3. Obtain a one-dimensional ^{1}H spectrum and a two-dimensional COSY spectrum from 0 to 10 ppm for your ligand. Obtain spectra from −100 to +100 ppm for your [Ni(salpd)] complex.

Analysis of [Ni(salpd)] Magnetic Moment in Solution—Evans Method

1. Place a sealed inner tube containing pure $CDCl_3$ into your [Ni(salpd)] NMR sample tube. Record the frequency of the solvent peak of the pure solvent vs that in the sample.

2. Repeat for using pure d_6-DMSO for your sample dissolved in d_6-DMSO.

Name _____   Section _____

Results Summary for the ^{1}H NMR Analysis of N,N'-Disalicylaldehyde-1,3-propane-diiminenickel(II)

1. List the chemical shift assignments for the salpd ligand and the [Ni(salpd)] complex in Table 3.16.

Table 3.16   Chemical Shifts for salpd and [Ni(salpd)] in $CHCl_3$ and d_6-DMSO ___ °C

Compound	Structural Assignment	Chemical Shift Assignment	
		$CDCl_3$	d_6-DMSO
Salpd			
[Ni(salpd)]			

Evans Method

		$CDCl_3$	d_6-DMSO
1.	Chemical shift, v_{ref}, (pure solvent reference)	_____Hz	_____Hz
2.	Chemical shift, v (in the presence of [Ni(salpd)])	_____Hz	_____Hz
3.	Δv	_____Hz	_____Hz
4.	Magnetic susceptibility, χ_M ($cm^3 \, mol^{-1}$) see equation (3.31)	_____	_____
5.	Number of unpaired spins	_____	_____
6.	Proposed geometry for [Ni(salpd)] in solution	_____	_____

Post-lab Questions

Q3.12   Summarize and comment on any similarities and differences observed between your results in $CDCl_3$ and d_6-DMSO. Consult reference 6 on Ni Schiff base solution behavior.

REFERENCES

1. Kauffman, G. B. symposium chairman, *Werner Centennial.* Advances in chemistry Series no. 62. American Chemical Society: Washington, DC, 1967.

2. Godwin, H. A., Hoffman, B., Bowman-James, K. Report form the *Frontiers of Inorganic Chemistry Workshop*, National Science Foundation, held at Copper Mountain, CO, 8 September 2001.

3. Lifschitz, I., Bos, J. G., Dijkema, K. M. *Z. Anorg. Allg. Chem.* **1939**, *242*, 97.

4. Pauling, L. *Nature of the Chemical Bond and the Structure of Molecules and Crystals*, 2nd edn. Cornell University Press: Ithaca, NY, 1948.

5. Azevedo, F., Freire, C., De Castro, B. *Polyhedron* **2002**, *21*, 1695–1705.

6. Cotton, F. A. and Wilkinson, G. *Advanced Inorganic Chemistry*, 5th edn., Wiley: New York, 1988, pp. 743–757.

7. Huheey, J. E., Keiter, E. A., Keiter, R. L. *Inorganic Chemistry—Principles of Structure and Reactivity*, 4th edn. HarperCollins: New York, 1993, pp. 459–467.

8. Szafran, Z., Pike, R., Singh, M. M. *Microscale Inorganic Chemistry*. Wiley: Toronto, 1991.

9. Angelici, R. J. *Synthesis and Technique in Inorganic Chemistry*, 2nd edn. W.B. Saunders: Philadelphia, 1977.

10. Figgis, B. N., Lewis, J. *The Magnetochemistry of Complex Compounds*, in Lewis, J., Witkins, R. G. (eds). *Modern Coordination Chemistry*. Interscience: New York, 1960.

11. *Johnson Matthey Magnetic Susceptibility Balance Instruction Manual*. Johnson Matthey, Inc.: West Chester, PA.

12. Skoog, D. A., Holler, F. J., Nieman, T. A. *Principles of Instrumental Analysis*, 5th edn. Saunders: Philadelphia, PA 1998, Chap. 31.

13. http://www.colby.edu/chemistry/PChem/lab/DiffScanningCal.pdf

14. Nakamota, K. *Infrared Spectra of Inorganic and Coordination Compounds*, 3rd edn. New York: Wiley, 1978.

15. Skoog, D. A., Holler, F. J., Nieman, T. A. *Principles of Instrumental Analysis*, 5th edn. Saunders: Philadelphia, PA, 1998, Chap. 17.

16. Drago, R. S. *Physical Methods in Inorganic Chemistry*. Reinhold Publishing Corporation: New York, 1965.

17. Vollhardt, K. P. C. *Organic Chemistry*. Freeman: New York, 1987, Chap. 17.

18. Jeewoth, T., Bhowon, M. G., Wah, H. L. K. *Transition Metal Chem.* **1999**, *24*, 445–448.

19. Bentley, F. F., Smithson, L. D., Rozek, A. L. *Infrared Spectra and Characteristic Frequencies* ~700–$300\ cm^{-1}$. Interscience: New York, 1968.

20. Costamagna, J., Lillo, E., Matsuhiro, B., Noseda, M. D., Villagran, M. *Carbohydrate Res.* **2003**, *338*, 1535–1542.

21. Campos-Vallette, M., Figueroa, K. A., Latoore, R., Manriquez, G., Díaz, G., Costamagna, J., Otero, M. *Vib. Spectrosc.* **1992**, *4*, 77–90.

22. Garg, B. S., Kumar, D. N. *Spectrochim. Acta Part A* **2003**, *59*, 229–234.

23. Hinchliffe, A. *Computational Quantum Chemistry*. Wiley: New York, 1988.

24. Szabo, A., Ostlund, N. S. *Modern Quantum Chemistry—Introduction to Advanced Electronic Structure Theory*. Dover: Mineola, NY, 1996.

25. *Spartan '02 Windows*. Wavefunction, Inc.: Irvine, CA, 2001.

26. Job, P. *Ann. Chim.* **9**, *113*, 1928.

27. Bailar, J. C. (ed.). *Chemistry of Coordination Compounds*. Reinhold: New York, 1956.

28. Angelici, R. J. *Synthesis and Technique in Inorganic Chemistry—A Laboratory Manual*, 3rd edn. University Science Books: Sausalito, CA, 1999.

29. Martell, A. E., Hancock, R. D. *Metal Complexes in Aqueous Solution*. Plenum Press: New York, 1996.

30. Martell, A. E., Smith, R. M. *Critical Stability Constants*, Vol. 1. Plenum Press: New York, 1974.

31. Martell, A. E., Motekaitis, R. J. *Determination and Use of Stability Constants*, 2nd edn. Wiley-VCH: New York, 1992.

32. Connors, K. A. *Binding Constants—the Measurement of Molecular Complex Stability*. Wiley: New York, 1987.

33. Fleck, G. M. *Equilibria in Solution*. Holt, Reinhart and Winston: New York, 1966, pp. 129–132.

34. Webb, G. A. In *Nuclear Magnetic Resonance Spectroscopy of Paramagnetic Species*, Money, E. F. (ed). Academic Press: New York, 1975, pp. 211–259.

35. James, T. L. *Nuclear Magnetic Resonance in Biochemistry*. Academic Press: New York, 1975.

36. Thwaites, J. D., Sacconi, L. *Inorg. Chem.* **1966**, *5*, 1029–1035.

37. Holm, R. H. Spectral and magnetic studies of substituted Ni(II) salicylaldimine complexes, in *Advances in the Chemistry of the Coordination Compounds*, Kirschner, S. (ed.). Macmillan: New York, 1961.

38. Chemical Rubber Company, Dean, J. A. *Large's Handbook of Chemistry*, 14th edn. McGraw-Hill: New York, 1992.

Reactivity I: Substitution Reactions—The Reaction of Aquapentacyanoferrate(II) Ion $[Fe(CN)_5(H_2O)]^{3-}$ with Amino Acids[1,2]

(The rate at which substitution reactions take place) is, in fact, of greater importance for many of the observations which are made than is the factor of stability (Henry Taube, 1957).

PROJECT OVERVIEW[3]

A solid foothold on molecular and electronic structure and stability enabled chemists to begin thinking seriously about the factors governing coordination complex reactivity. In 1951, Taube argued strongly for the need to study inorganic substitution reactions. Bjerrum's term "robust" was still universally used to describe the behavior of a metal complex, but Taube pointed out that this mainly referred to its thermodynamic stability. Taube set straight the confusion underlying the solution behavior of metal complexes by distinguishing between the kinetic and thermodynamic properties of a complex. By introducing the term "inert" for slow substitution at a metal center in solution and "labile" to refer to rapidly reacting systems, he emphasized the point that stability does not necessarily lead to inertness. Taube further proposed correlations between reactivity and metal ion electronic structure. By the late 1950s the crystal field model, replacing Pauling's valence bond model, was used extensively to understand reactivity and mechanisms of primarily slow-reacting octahedral and square planar complexes. By the mid-1960s fast reaction data on metal systems of various geometries had been accumulated using new commercially available stopped-flow instrumentation and molecular orbital theory enhanced the purely ionic crystal field model. A basic understanding of reaction mechanisms soon laid the groundwork for metal complex applications. Organometallic and bioinorganic reaction mechanisms, as well as photochemical catalysis, were soon hot topics in coordination chemistry and remain so today.

In this set of experiments, you will investigate the substitution reactivity of two amino acids, glycine and methionine, with the iron(II) complex, $[Fe(CN)_5(H_2O)]^{3-}$, as will be demonstrated in this series of experiments. Although much pedagogical value can be

Integrated Approach to Coordination Chemistry: An Inorganic Laboratory Guide. By Rosemary A. Marusak, Kate Doan, and Scott D. Cummings
Copyright © 2007 John Wiley & Sons, Inc.

gained by examining the reactivity of only one amino acid, a comparison of glycine and methionine reactivity demonstrates the variety of Lewis bases to which the intermediate acid, iron(II) can bind as well as the resulting altered behavior. Through UV–visible and infra-red spectroscopy, semi-empirical calculations, and electrochemistry, you will investigate the binding mode of glycine and methionine to the iron(II) center. You will then study the substitution kinetics of each amino acid with $[Fe(CN)_5(H_2O)]^{3-}$. The kinetics are fast and require stopped-flow instrumentation. Today, relatively inexpensive instruments enable such rapid reactions to be investigated at the undergraduate teaching level. Another challenging and interesting feature of this system is the *lack* of color change upon modification of the metal ion environment. You will need to use spectroscopy and other achromatic methods to detect alterations in the metal ion environment.

EXPERIMENT 4.1: SYNTHESIS OF THE AMMINEPENTACYANOFERRATE(II) COMPLEX, $Na_3[Fe(CN)_5(NH_3)]\cdot 3H_2O$

Level 4

The student will need to put in extra time into this synthesis. The protocol is conceptually simple, but can be somewhat challenging.

Pre-lab 4.1: Substitution Properties of Metal Ion Complexes in Aqueous Solution[4,5]

The preparation of substituted pentacyanoferrate(II) ion complexes involves a series of ligand exchange reactions at the iron(II) metal center. Equations (4.1)–(4.3) outline the synthesis of amino acid (AA) metal complexes in aqueous solution. Starting from sodium nitroprusside ion, $[Fe(CN)_5(NO)]^{2-}$, equation (4.1), the nitrosyl ligand, NO^+, is replaced by an ammine moiety, NH_3. The aquapentacyanoferrate(II) ion, $[Fe(CN)_5(H_2O)]^{3-}$, is then generated *in situ*, equation (4.2), followed by reaction with an AA to yield the desired $[Fe(CN)_5(AA)]^{(3+n)-}$ complex, equation (4.3).

$$[Fe(CN)_5(NO)]^{2-} + NH_3 \rightleftharpoons [Fe(CN)_5(NH_3)]^{3-} + NO^+ \quad (4.1)$$

$$[Fe(CN)_5(NH_3)]^{3-} + H_2O \rightleftharpoons [Fe(CN)_5(H_2O)]^{3-} + NH_3 \quad (4.2)$$

$$[Fe(CN)_5(H_2O)]^{3-} + AA \rightleftharpoons [Fe(CN)_5(AA)]^{3-} + H_2O \quad (4.3)$$

Because many metal complex substitution reactions take place in aqueous media, a good first approximation of the rate of substitution at the metal center is the metal ion's rate of water exchange, equation (4.4). In equation (4.4) O^* is a labeled oxygen atom that allows the reaction to be followed. Figure 4.1 tabulates water exchange rate constants for many high spin metal ions. The trend seen in Figure 4.1 reflects primarily the metal ion charge to radius ratio (Z^2:r), although CFSE also plays a significant role in transition metal systems. From Figure 4.1, *high spin* octahderal aqueous iron(II) exhibits a very fast water exchange rate, with a rate constant, $k_{ex} \approx 5 \times 10^6 \text{ s}^{-1}$.

$$[M(H_2O)_x]^{n+} + yH_2O^* = [M(H_2O^*)_x]^{n+} + yH_2O \quad (4.4)$$

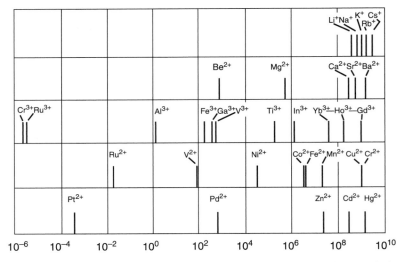

Figure 4.1   Water exchange rate constants for solvated cations. Reproduced with permission from Docommun, Y., Merbach, A. E. In *Inorganic High Pressure Chemistry*, van Eldik, R. (ed.). Elsevier: Amsterdam, 1986, p. 70.

Metal complex (ML$_n$) substitution rates depend on many factors, including the number and type of surrounding ligands (L), reaction mechanism, temperature, and ionic strength. One way to assess the lability and inertness of a metal complex is to calculate the crystal field activation energy (CFAE) for its substitution reaction. The CFAE is defined as the difference in CFSE of the original complex and that of the activated complex (transition state structure). Since substitution in many octahedral complexes proceeds by a dissociative-type mechanism, the CFAE would be the difference in stability of the octahedral and square pyramidal complexes. Table 4.1 gives CFAE values for both high- (weak field) and low-spin (strong field) configurations.

TABLE 4.1   CFAE Calculations for High-Spin (hs) and Low-Spin (ls) Otahedral Substitution Reactions (values from Basolo and Johnson[4])

Electron Configuration	CFSE, Δ Octahedral	CFSE, Δ Sq pyr	CFAE, $\delta\Delta$	Predicted Behavior
d^0	0	0	0	Labile
d^1, d^6 (hs)	0.4	0.45	−0.05	Labile
d^2, d^7 (hs)	0.8	0.91	−0.11	Labile
d^3, d^8	1.20	1.00	+0.20	Inert
d^4 (hs), d^9	0.6	0.91	−0.31	Labile
d^5 (hs), d^{10}	0		0	Labile
d^4 (ls)	1.6	1.46	+0.14	Inert
d^5 (ls)	2.0	1.91	+0.09	(Inert)
d^6 (ls)	2.4	2.0	+0.40	Inert
d^7 (ls)	1.8	1.91	−0.11	Labile

Procedure 4.1: Synthesis of Sodium Amminepentacyanoferrate(II), $Na_3[Fe(CN)_5(NH_3)] \cdot 3\,H_2O$ from Sodium Nitroprusside, $Na_2[Fe(CN)_5(NO)] \cdot 2H_2O$

Safety warning! Operations set off between the lines *must be performed in a fume hood.*

1. Dissolve 3.3 g of sodium acetate in 40 ml of concentrated aqueous ammonia (30% NH_4OH_{aq}) in a 100 ml beaker and chill in an ice bath.
2. Grind 10 g of $Na_2[Fe(CN)_5(NO)] \cdot 6H_2O$ into a fine powder using a mortar and pestle and add to the chilled NH_4OH solution.
3. Continue chilling while stirring the mixture for 1.5 h. Keep the mixture at 4 °C for up to 5 h and cold filter the yellow product.
4. Redissolve the crude product in 20 ml H_2O, filter, and rapidly pour the mixture into 20 ml of cold ethanol.
5. Filter the yellow product and wash with two 10 ml portions of absolute ethanol and suction dry. Dry in a vacuum dessicator overnight and store product under $N_{2(g)}$ and in the dark at -20 °C until needed.

Optional Recrystallization (Recommended)

6. If further recrystallization is needed (indicated by a green-yellow product), dissolve the impure product in a minimum amount of 3 M aqueous NH_4OH (approximately 1.5 ml per 1 g product) purged with an inert gas ($Ar_{(g)}$ or $N_{2(g)}$).
7. Add 10 ml cold ethanol (volume may vary) to the brown solution and place in an ice bath.[†] Filter the cold solution using a Hirsh funnel. Collect the precipitate and repeat this procedure.
8. Wash the precipitate several times with cold absolute ethanol until the product becomes powdery. Record your yield (MM of $Na_3[Fe(CN)_5(NH_3)] \cdot 3H_2O$ product is 325.9). Store under a blanket of $N_{2(g)}$ at -20°C.

Name _____   Section _____

Results Summary the Synthesis of the Amminepentacyanoferrate(II) Complex, $Na_3[Fe(CN)_5(NH_3)] \cdot 3H_2O$

1. Limiting reagent for the $[Fe(CN)_5(NH_3)]^{3-}$ synthesis: _____
2. Theoretical yield: _____ g
3. Actual yield: _____ g
4. Percentage yield: _____ g
5. What are the CFAEs for substitution involving your ammine- and aqua-pentacyanoferrate(II) complexes? Comment on expected substitution behavior.
   $[Fe(CN)_5(NH_3)]^{3-}$ _____ g
   $^-Fe(CN)_5(H_2O)]^{3-}$ _____ g
6. Does your answer in (5) agree with observations made during your synthesis?

[†]Alternatively, place the solution in the freezer. Thawing yields yellow crystals.

EXPERIMENT 4.2: UV–VISIBLE SPECTROSCOPIC ANALYSIS OF $[Fe(CN)_5(H_2O)]^{3-}$ AND ITS AMINO ACID DERIVATIVES: PENTACYANOGLYCYLFERRATE(II) ION, $[Fe(CN)_5(gly)]^{4-}$; PENTACYANOMETHIONYLFERRATE(II) ION, $[Fe(CN)_5(met)]^{3-}$

Level 3

Pre-lab 4.2: Metal Amino Acid Complexes

In this experiment you will investigate the electronic absorption spectra of your $[Fe(CN)_5(AA)]^{n-}$ complexes, which will provide the first of several pieces of information that will help you to determine the binding mode of each amino acid to your iron(II) center. The structures of the amino acids, glycinate (gly⁻) and methionine (met) are shown in Figure 4.2. Under the conditions of our experiment, Gly⁻ offers two binding sites, COO^- and NH_2, as does the zwitterionic methionine, COO^- and S (the protonated NH_3^+ has no lone pairs available).

   Figure 4.3 shows common binding modes for amino acids to metal centers. However, because only one coordination site is made available by the labile H_2O at the iron(II) center, we expect the amino acids to bind in a monodentate fashion. Recall from HSAB theory (Chapter 1) that iron(II) is classified as an intermediate acid and may therefore bind a variety of atom types. By examining the lowest energy transitions in your complexes, you will gain evidence for the amino acid binding site. Review Pre-lab 3.6.b for information on electronic transitions in metal complexes. After you have determined your binding mode, you will identify the nature of your lowest energy transitions in these complexes using semi-empirical calculations (Experiment 4.5).

Procedure 4.2.a: *In Situ* Preparations of $[Fe(CN)_5(H_2O)]^{3-}$ and AA Substituted Pentacyanoferrate(II) Ions, $[Fe(CN)_5(AA)]^{n-}$ and Spectrophotometric Analysis[1,2]

Upon dissolving $[Fe(CN)_5(NH_3)]^{3-}$ into water, rapid substitution of the NH_3 ligand by H_2O takes place [equation (4.2)]. Iron complex concentrations must be no greater than 0.1 mM. At concentrations greater than 0.3 mM the equilibrium between the ammine and aqua species is significant. Aqueous solutions of the $[Fe(CN)_5(H_2O)]^{3-}$ complex must be freshly prepared and stored in the dark. All $[Fe(CN)_5(AA)^{n-}]$ complexes form stable 1:1 complexes under conditions of excess ligand.

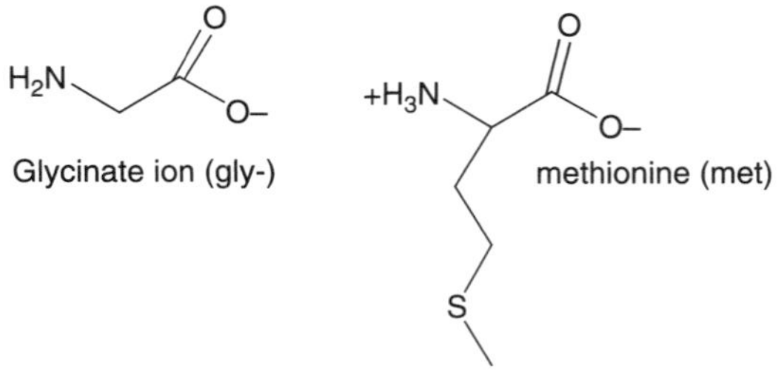

Figure 4.2 The glycinate anion (gly⁻) at pH ~11.5, and the methionine zwitterion, pH 8–9, treated as a neutral molecule.

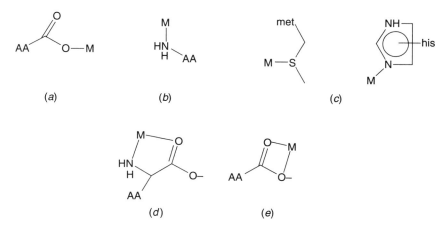

Figure 4.3 Common binding modes of amino acid metal complexes. Monodentate through (*a*) carboxy terminus, (*b*) amine terminus, and (*c*) side chains (met and his shown); bidentate through (*d*) amine and carboxy termini and (*e*) carboxy terminus.[8]

1. For each AA ligand, prepare a 10 ml solution of 5.4 mM AA. The pH of the glycine solution should be approximately 11.5, while that of methionine should be 8–9.

2. Prepare 10 ml of a 1 mM stock of $Na_3[Fe(CN)_5(NH_3)] \cdot 3H_2O$ by weighing 3.26 mg (0.01 mmol) of the complex (record exact mass) and adding this to a 10 ml volumetric flask. Fill to the mark with ddH_2O.[8]

3. Pipet 760 μl of the iron complex stock solution into a 10 ml volumetric flask. Fill to the line with either ddH_2O (for $[Fe(CN)_5(H_2O)]^{3-}$) or your AA solutions (for $[Fe(CN)_5(AA)]^{n-}$).

4. Record the final pH of each solution.

5. Prepare free AA ligand solutions as in step 3, but substitute 760 μl ddH_2O for the iron complex solution.

6. Obtain a spectrum (300–700 nm) for $[Fe(CN)_5(H_2O)]^{3-}$, each AA ligand, and each $[Fe(CN)_5(AA)]^{n-}$ complex. Record the maximum wavelength for each and determine the molar absorptivity for each complex. (To save time, only one trial for molar absorptivity determination is made. To obtain good reliable values, several trials should be made.)

Name _____   Section _____

Results Summary the Spectrophotometric Analysis of $[Fe(CN)_5(H_2O)]^{3-}$ and the
AA-pentacyanoferrate(II) Complexes, $[Fe(CN)_5(AA)]^{n-}$

Data for the 300–700 nm range:

	Trial 1	Trial 2	Trial 3
1. $[Fe(CN)_5L]^{n-}$	_____	_____	_____
2. Wavelength of maximum absorbance	_____	_____	_____
3. ε value at λ_{max} for $[Fe(CN)_5L]^n$	_____	_____	_____
4. Amino acid	_____	_____	_____
5. Wavelength of maximum absorbance	_____	_____	_____

6. List your $[Fe(CN)_5L]^{n-}$ complexes in order of greatest to least energy of absorption (use lowest energy band).

7. What type of transition do you think is responsible for the absorption(s) in the visible range? Why?

EXPERIMENT 4.3: INFRARED SPECTROSCOPIC ANALYSIS OF $[Fe(CN)_5(NH_3)]^{3-}$ AND SUBSTITUTED AA, $[Fe(CN)_5(AA)]^{n-}$ COMPLEXES

Level 4

This procedure requires isolating pure product by careful fractional precipitation.

Pre-lab 4.3: IR Spectroscopy of CN$^-$ and AA Ligands[9,10]

Pre-lab 3.4 introduced you to IR spectroscopy of metal complexes and is recommended reading. The stretching frequency for free CN^- (KCN) is 2080 cm^{-1}. In the free ion then, resonance structure **A** plays a more important role than **B**. Recall from Pre-lab 3.4 that the CN frequency will either increase or decrease depending on the predominant bonding (σ donation or π back bonding) interaction. This is dependent on the electronegativity, oxidation state, and coordination number of the metal. Metal-bound CN^- stretching frequencies are sharp and occur between 2200 and 2000 cm^{-1}. The CN^- stretch in $[Fe(CN)_6]^{3-}$ appears at 2125 cm^{-1}, whereas in the iron(II) complex, $[Fe(CN)_6]^{4-}$, this stretch occurs between 2094 and 2006 cm^{-1}. This suggests that resonance structure **C** plays a very important role in the iron(III) complex, while **D** predominates in the iron(II) complex (Fig. 4.4).

In the solid state, amino acids exist in the zwitterionic form, **E** (Fig. 4.5). The primary ammonium group, NH_3^+, exhibits a broad band 3040 cm^{-1} and two bands common to amino acids at ~ 2500 and ~ 2100 cm^{-1}. Strong bands at ~ 1665 and ~ 1590 cm^{-1} for the COO^- asymmetric and symmetric stretches, respectively, and NH_3^+ deformations at 1550 cm^{-1} are also typical of amino acids. A full assignment for glycine is found in Nakamoto.[10] The metal-bound COO^- and NH_2 groups will show characteristic shifts in these stretching frequencies.

A: :C≡N: ⟷ **B:** :C=N:

C: M—C≡N: ⟷ **D:** M=C=N:

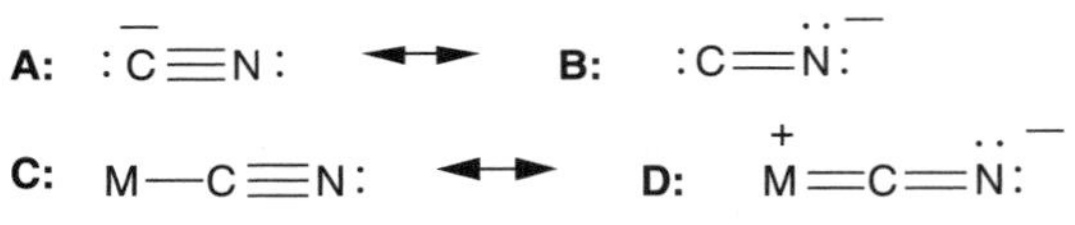

Figure 4.4 **A**, **B** resonance structures for CN$^-$ ligand; **C**, **D** resonance structures for the metal-bound CN$^-$ ligand.[8]

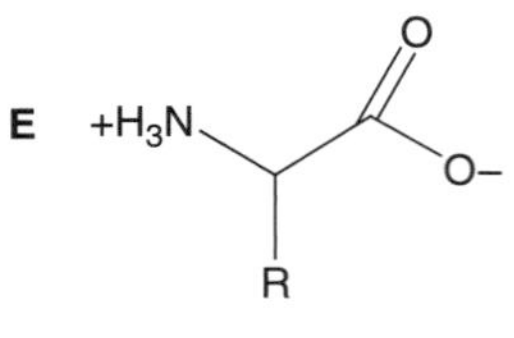

E +H$_3$N ... R ... O, O-

Figure 4.5 General structure of the zwitterionic form of an amino acid.[8]

In the present system, you will be looking for characteristic shifts of coordinated cyano $(C \equiv N)$ groups and in the amino acids, the COO^- (and NH_2) group. However, it should be cautioned that intermolecular H-bonding of the COO^- in the solid state can also affect the absorption frequency. For methionine, the S group is also a viable candidate for metal binding. M$-$S stretching bands are observed in the far-IR, $\sim 320-440$ cm^{-1} (see Experiment 3.4) and will not be investigated here.

Procedure 4.3: Preparation and Isolation of AA Substituted Pentacyano–ferrate(II) Complexes, [Fe(CN)$_5$(AA)]$^{n-}$ and Infrared Analysis[1,2]

AA = glycine$^-$ (H$_2$N—CH$_2$—COO$^-$); K$_4$[Fe(CN)$_5$(gly)]·4H$_2$O

1. Dissolve 1 g (3 mmol) of sodium amminepentacyanoferrate(II), Na$_3$[Fe(CN)$_5$ (NH$_3$)]· 3H$_2$O (MM = 325.9) and 1.24 g glycine (15.5 mmol) in 50 ml ddH$_2$O at a pH $\sim$11.5.
2. Allow the mixture to react for 15 min at room temperature.
3. Add 10 g of KI.
4. Cool the reaction mixture and precipitate the K$_4$[Fe(CN)$_5$(gly)]·4H$_2$O product with the careful addition of ethanol ($\sim$150 ml). Allow the precipitate to form and collect the first fraction by vacuum filtration. More fractions may be obtained upon further addition of ethanol. Alternatively, your product can be recrystallized from cold water containing excess ligand.

AA = methionine ($^+$H$_3$N—CH(CH$_2$CH$_2$SCH$_3$)COO$^-$); Cs$_3$[Fe(CN)$_5$(met)]·nH$_2$O

1. Dissolve 0.2 g (0.6 mmol) of sodium amminepentacyanoferrate(II), Na$_3$[Fe(CN)$_5$ (NH$_3$)]·3H$_2$O (MM = 325.9) and 0.492 g methionine (3.3 mmol) in 10 ml ddH$_2$O at a pH $\sim$8–9.
2. Allow the mixture to react for 15 min at room temperature.
3. Add 2 g of CsCl.
4. Cool the reaction mixture and precipitate the Cs$_3$[Fe(CN)$_5$(met)]·nH$_2$O product with the careful addition of ethanol ($\sim$20 ml). Allow the precipitate to form and collect the first fraction by vacuum filtration. More fractions may be obtained upon further addition of ethanol. Alternatively, your product can be recrystallized from cold water containing excess ligand.

Infrared Analysis

1. Prepare samples of your AA ligand and [Fe(CN)$_5$(AA)]$^{n-}$ complexes as well as your [Fe(CN)$_5$(NH$_3$)]$^{3-}$ starting material for IR analysis by grinding to a fine powder, a mixture of compound ($\sim$2 mg) and dry, IR-quality KBr (100 mg) and pressing into a thin, clear, pellet.
2. Obtain mid-IR spectra (400–4000 cm^{-1}) for your samples.
3. Record results in the Results Summary page table.
4. Compare your observed frequencies with those calculated in Experiment 4.5.

Name _____   Section _____

Results Summary for the Infrared Analysis of AApentacyanoferrate(II) Complexes, [Fe(CN)$_5$(AA)]$^{n-}$

TABLE 4.2 **IR assignments for [Fe(CN)$_5$(AA)]$^{n-}$ Complexes**

Compound	Functional Group	Frequency (cm^{-1})
[Fe(CN)$_5$(NH$_3$)]$^{3-}$		
Glycine (gly)		
[Fe(CN)$_5$(gly)]$^{4-}$		
Methionine (met)		
[Fe(CN)$_5$(met)]$^{3-}$		

1. List and assign all important transitions observed in the IR spectra for your [Fe(CN)$_5$(AA)]$^{n-}$ complexes in Table 4.2.
2. What, if any, frequency shifts occur?
3. Predict the binding for each amino acid to the iron(II) center.

EXPERIMENT 4.4: CYCLIC VOLTAMMETRIC ANALYSIS OF [Fe(CN)$_6$]$^{3-}$; [Fe(CN)$_5$(H$_2$O)]$^{2-}$ AND ITS SUBSTITUTED AA DERIVATIVES, [Fe(CN)$_5$(AA)]$^{n-}$

Level 4

This experiment involves advanced theory and substantial reagent preparation that requires outside-lab prep time. The goal of this experiment is to determine the standard reduction potentials (E^0, V) for a series of substituted pentacyanoferrate(II) complexes. By comparing the electrochemical behavior of each AA ligand system, information about electronic structure and solution properties will be obtained. An introduction to cyclic voltammetry is given in Appendix 2.

Pre-lab 4.4: Cyclic Voltammetric Analysis of the Fe(CN)$_6$$^{4-/3-}$ (1 M KNO$_3$) System[11]

If you are comfortable with the information in Appendix 2, continue by reviewing the plot description of the well-behaved Fe(CN)$_6$$^{4-/3-}$ system below. You will need to repeat this result before moving on to your AA substituted complexes (Fig. 4.6).

Important Parameters Anodic peak current, i_{pa}; cathodic peak current, i_{pc}; anodic peak potential, E_{pa}, cathodic peak potential, E_{pc}.

Plot Description

(a) No current: electrode too (+) for reduction to take place; $E \gg E°$ Fe^{3+} pre-dominates.

(b) Cathodic current produced due to concentration polarization from the rxn:

$$Fe(CN)_6^{3-} + e^- \leftrightarrows Fe(CN)_6^{4-}$$

Rapid rise in current, almost instantaneous reduction at

(c) As $E_{appl} = E^0$ and

$$E = E^{0'} + 0.059/1 \log[Fe^{3+}]/[Fe^{2+}]$$

(d) Maximum (all Fe^{3+} around electrode reduced [c] $\approx$ 0).

(b)–(d) Cathodic current.

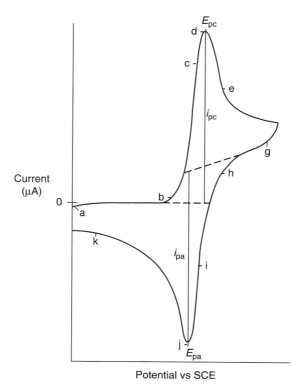

Potential vs SCE

Figure 4.6 Cyclic voltammogram of the $Fe(CN)_6^{4-/3-}$ (1 M KNO_3) couple. Descriptions of letters are given in the text. Adapted from Bott and Jackson.[11]

(e) The applied potential and rate of change become immaterial to the diffusion-controlled current (i.e., current is determined by rate of diffusion of Fe^{3+} to electrode, which decays $t^{-1/2}$ due to the concentration dependence).

(f) Applied potential is switched to (+).

(i)–(k) Anodic current.

The above description of the redox reaction for $Fe(CN)_6^{4-/3-}$ is a "textbook example" because the system nicely obeys Nernstian conditions and many cycles can be repeated without distortion of the voltammogram (we call this a *reversible system*): electron transfer is rapid and reversible at the electrode surface and complete concentration polarization is achieved under conditions of 1 M KNO_3. We know this is a reversible system because $i_{pa}/i_{pc} \approx 1$. For a rigorous check on this condition, a plot of i_p vs $v^{1/2}$ should be linear (where v = scan rate), equation (4.5).

$$i_p = (2.69 \times 10^5)n^{3/2}AD^{1/2}Cv^{1/2} \qquad (4.5)$$

Further evidence for good reversibility is that the peak separation, ΔE_p, is close to 60 mV for this one electron transfer process, equation (4.6), and is independent of v. For a two-electron transfer, this separation should be 30 mV, etc.

$$\Delta E_p = E_{pa} - E_{pc} \approx 0.059/n \qquad (4.6)$$

For such a reversible system, $E^{0'}$ (the prime indicating under experimental conditions given) is obtained using equation (4.7).

$$E^{0'} = E_{pa} + E_{pc}/2 \qquad (4.7)$$

In reality, ΔE_p is usually larger than expected due to solution resistance (between the working and reference electrodes and nonlinear diffusion). It is generally accepted that, depending on the system, ΔE_p up to even ~90 mV may constitute a reversible system. However, all means for looking into why a peak separation is large should be exhausted (e.g., polishing the working electrode, IR compensation, etc.), as larger separations may indicate irreversibility due to chemical reactions or *charge transfer polarization*. Irreversibility due to charge transfer polarization occurs when the rate of electron transfer in either direction at either electrode is not sufficiently rapid enough to yield currents of the magnitude demanded by theory. Slow electron transfer kinetics is cause for a large ΔE_p, which will increase with increasing v.

Procedure 4.4.a: Study of Ferricyanide by Cyclic Voltammetry Using the CV-50W[11]

Note: reagents and glassware of highest purity should be used throughout this procedure.

Reagent and Electrode Preparation
1. Prepare a 10 mM aqueous solution of $K_3Fe(CN)_6$ containing 1 M KNO_3 in 100 ml total volume. Label this the Stock solution.
2. Prepare four 25 ml dilutions of 2, 4, 6, and 8 mM $K_3Fe(CN)_6$ in 1 M KNO_3.
3. Use the following electrodes for your work:
    working electrode—platinum (should be lightly polished with alumina and rinsed with water before each experimental run), black;
    reference electrode—silver/silver chloride, white;
    auxiliary electrode—platinum wire, red.
4. Set your general parameters to:
    Initial E: 650 mV
    High E: 650 mV
    Low E: 0 mV
    Initial direction: negative
    Number of segments: 2
    Sensitivity: 10 μA V^{-1}
    Scan rate: 20 mV s^{-1}

Remember to **polish the Pt electrode** in between each run!

Concentration Dependence Study
5. Place your 2 mM solution in a water-jacketed (if available) sample vial. Equilibrate to 25 °C. Purge your solutions with $N_{2(g)}$ prior to each run. Although the $Fe(CN)_6{}^{4-/3-}$ reaction is insensitive to aerial oxidation, the Pt working electrode will show activity due to O_2.

6. Begin your scan by pressing **Start Run**.

7. For **Results Options** in **Analysis** menu, ensure **Peak Shape** is Tailing, **Method** is Auto, and that **Find peaks after Load data** and **Find peaks after Run** are *not* checked. Set **Half peak width** = 100 mV.

8. **Redisplay** using Results Graph under Analysis.

9. **Save Data** as (.bin) file.

10. Repeat steps 5–9 for your other solutions.

11. Calculate your ΔE_p and $E^{0'}$ vs Ag–AgCl, SCE, and NHE reference electrodes for each solution and record in Results Summary section.

12. Using your five concentrations, construct a plot of i_p vs concentration—we are interested only in the linearity of the relationship between concentration and peak current.

Effect of Scan Rate

13. Using your 10 mM solution, repeat steps 1–9 and 11 at 20, 50, 100, 150, and 200 mV s^{-1} (save files).

14. **Plot as Multi Graph** using the Graphics menu → Graphics Options and Overlay as Graph Style. The 200 mV s^{-1} run is retained in immediate memory and its axes are used.

15. **Plot** the scan rate dependence of the peak currents and peak potentials: (a) i_p vs (scan rate)$^{1/2}$; and (b) ΔE_p vs scan rate.

Check on Electrolyte Background   *Ideally* a check on the background electrochemistry of your system (i.e., a blank) should be completed *before* running your analyte so you are not wasting your time interpreting background peaks! Also, with the Pt working electrode and small analyte concentrations, dissolved O_2 can become a contaminant in a certain range.

16. Make sure your electrodes and electrochemical cell compartment are thoroughly rinsed and free from $K_3Fe(CN)_6$ and fill with 1 M KNO_3 solution. Degas your system and run a scan under the same parameters used for your analyte runs (the highest sensititivity should be used).

Procedure 4.4.b: Study of Substituted Pentacyanoferrate(II) Complexes by Cyclic Voltammetry

Stock Preparation

1. Prepare 100 ml 0.1 M $LiClO_4$.

2. Prepare 25 ml of 0.017 M $[Fe(CN)_5(NH_3)]^{3-}$ (ionic strength, $\mu = 0.1$ M) stock by weighing 0.138 g $Na_3[Fe(CN)_5(NH_3)] \cdot 3H_2O$ into a 25 ml volumetric flask and diluting to the mark with your $LiClO_4$ stock.

Reagent Preparation and CV Analysis   Prepare each reagent solution just prior to CV analysis.

3. a. Prepare 10 ml of 10^{-4} M $[Fe(CN)_5(H_2O)]^{3-}$ at $\mu = 0.1$ M by diluting an appropriate amount of your 0.017 M $[Fe(CN)_5(NH_3)]^{3-}$ stock with your prepared 0.1 M $LiClO_4$ stock. Use the electrodes defined above and set **General Parameters** according to Table 4.4. Run Scan at 10 mV s^{-1} and 25 °C.

   b. Calculate ΔE_p and $E^{0'}$ vs Ag–AgCl and NHE. Observe what happens with a change in scan rate. Record in the Results Summary section.

4. a. For each AA ligand, prepare 10 ml of 4 mM $[Fe(CN)_5(AA)]^{n-}$ solution ($\mu = 0.1$ M) containing a 10-fold excess of AA ligand. Dilute an appropriate amount of your 0.017 M $[Fe(CN)_5(NH_3)]^{3-}$ stock with your prepared 0.1 M $LiClO_4$ stock into a 10 ml volumetric that contains 0.40 mmol (resultant 40 mM) AA. For glycinate ion, the pH should be $\sim$11.5. For methionine, the pH should be 8–9. Use the electrodes defined above. Based on your knowledge of AA-substituted complexes, predict **General Parameters** for these complexes and record in Table 4.3. Run scans at 10 mV s^{-1} and 25 °C.

   b. Calculate ΔE_p and $E^{0'}$ vs Ag–AgCl and NHE. Observe what happens with a change in scan rate. Record in the Results Summary section.

5. For the glycinate system, increase the concentration of gly$^-$ to 0.2 M (50-fold excess)—note total ionic strength will be changed. Run a scan.

6. For each trial, run a blank containing only excess AA (when appropriate) and 0.1 M $LiClO_4$.

Name _____   Section _____

Results Summary for the CV Analysis of Substituted Pentacyanoferrate(II) Complexes, $[Fe(CN)_5(L)]^{n-}$

1. Fill in Table 4.4.

   For each L substituent (2–5):

2. Is your ΔE_p 60 mV? Explain.
3. Does your $E^{0'}$ vs NHE value agree with the literature?
4. What does your concentration dependence (for L $=$ CN$^-$) tell you?
5. What does your scan rate dependence (for L $=$ CN$^-$) tell you?
6. Order your observed potentials from least to most positive. Explain in terms of ligating L atoms?

TABLE 4.3 *Suggested* **General Parameters for Your Initial Scans**

General Parameters	$[Fe(CN)_5(H_2O)]^{3-}$	$[Fe(CN)_5(gly)]^{4-}$	$[Fe(CN)_5(met)]^{3-}$
Initial E	-150 mV		
High E	400 mV		
Low E	-150 mV		
Initial direction	Positive		
Number of segments	2		
Sensitivity	10 µA V^{-1}		
Scan rate	10 mV s^{-1}		

TABLE 4.4   Results Summary for CV Analysis of Substituted Pentacyano ferrate(II) Complexes, $[Fe(CN)_5(L)]^{n-}$

Substituent, L	ΔE (mV)	$E^{0'}$ (vs Ag–AgCl)	$E^{0'}$ (vs NHE)
CN^-			
H_2O			
gly$^-$ (0.04 M)			
gly$^-$ (0.2 M)			
met			

7. What occurs when the concentration of gly$^-$ is increased from 0.04 to 0.2 M? Explain.

8. Your reference electrode's potential must be checked periodically. How would you do this?

EXPERIMENT 4.5: SEMI-EMPIRICAL CALCULATIONS IN THE STUDY OF SUBSTITUTED PENTACYANOFERRATE(II) COMPLEXES

Level 3

Pre-lab 4.5: Introduction to the Semi-empirical Method[12,13]

In Pre-lab 3.5 we discussed quantum chemical calculations and specifically the Hartree–Fock approximation. Quantum chemical calculations are often referred to as first principles or *ab initio* (Latin meaning "from the beginning") calculations in that they attempt to solve the molecular electronic structure problem from the fundamental laws of quantum mechanics. Although a Hartree–Fock calculation is not very computationally intensive relative to more accurate correlation methods, no *ab initio* methods are practical for larger molecules. Instead, semi-empirical methods are typically employed to treat molecules ranging in size from tens to hundreds of atoms.

As the name implies, semi-empirical calculations simplify the computational problem presented by the Schrödinger equation. Semi-empirical approaches eliminate some of the direct "number crunching" involved in *ab initio* calculations and instead replace portions of the calculation with values that may be taken from experimental data or other calculations that are parameterized to agree with empirical data. There is a variety of semi-empirical schemes that differ in the types of parameterizations that are made. Three common semi-empirical methods that are included in Spartan02© are called AM1, PM3, and MNDO. Each has strengths and weaknesses depending on the specific molecular environment one wishes to model.

The energy reported at the end of the Hartree–Fock calculation is a "total energy", i.e., relative to separated nuclei and electrons at rest. In this section, we will see that the energy reported at the end of a semi-empirical calculation is a "heat of formation" energy rather than a total energy as reported in an *ab initio* calculation. This distinction is useful in defining the respective zero of energy for each type of calculation. In either type of calculation, the energies of an atomic orbital contributions to the molecular orbitals can lend valuable insight for interpreting UV–vis spectra and understanding the type of electronic transitions that are involved in the excitations of interest. Unfortunately, reproducing the true energetics of such a transition requires a more sophisticated quantum chemical approach to improve the excited state energetics. Vibrational energies are calculated using the second derivatives of the energy, whereas forces are calculated through the first derivatives. Although vibrational energies computed from semi-empirical and Hartree–Fock calculations tend to not agree well with experimental values, introduction of a scaling factor may improve this agreement considerably and allow assignment of modes for complicated molecules.

Procedure 4.5: Semi-empirical Calculations for AA Ligands

1. Open the Spartan program.
2. To start a new project, go to **File, New**. On the right-hand side screen are various palates from which you may construct the four iron complexes (see Fig. 4.7) and the two free amino acid ligands (see Fig. 4.8). Iron is under the **Exp**. pallet and the Gly and Met are under the **Pep**. pallet. To place the central iron on the worksheet, simply select Fe and octahedral geometry and click anywhere on the workspace. Ligands may be added by clicking any of the yellow active sites around the iron atom. After each structure is completed, save under **File**.
3. To find the heat of formation and the vibrational frequencies for one of your structures, select **Calculations** under **Setup**. Choose **Geometry Optimization** with **Semi-empirical** using **PM3**. Leave the **Symmetry** option checked. All structures should be in the **singlet** state. The pentacyano-gly and hexacyanoferrate complexes both require a total charge of −4. The aqua and met pentacyano species have a total charge of −3. The free glycine amino acid has a charge of −1, while the zwitterionic methionine is overall neutral. You will need to actually type the charges into the total charge field because the default scroll down only allows for up to dianion. Select **Freq**. In the **Compute** box. In the **Option** box, type "SCFCYCLE = 10000, GEOMETRYCYCLE = 2000," and check the **Converge**. Now click **Submit**.
4. Run similar calculations for all six structures (the four iron complexes and the two free amino acid ligands).
5. When all the calculations are finished, look at the HOMO and LUMO of the four iron complexes. To do this you must again go to **Setup**, and this time select **Surfaces**. In the **Surfaces** window click the **Add** button. Choose **HOMO** under **Surface**. Leave the **Resolution** at the default **Medium**. Click **OK**. To add the **LUMO** orbital, again click Add in the Surface window. Select **LUMO** under **Surface** and click **OK**. Under **Setup**, choose **Submit**.

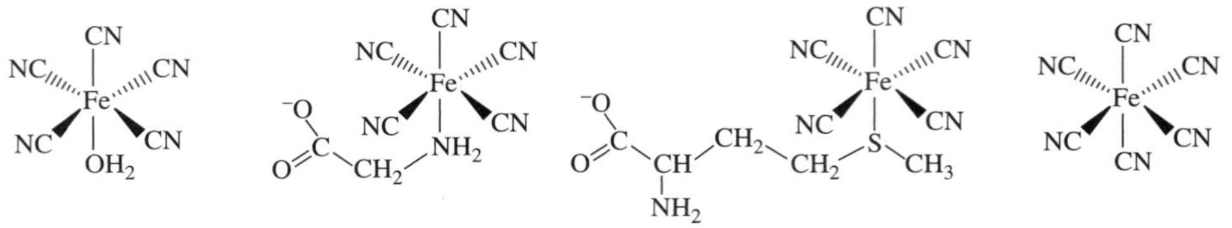

Figure 4.7 [Fe(CN)$_5$L]$^{n-}$ complexes used in the calculations: L = (a) H$_2$O; (b) gly;$^-$ (c) met; (d) CN$^-$. Note that the complexes with the gly$^-$ and met amino acids have a negative charge on the carboxylic acid group. Also note that the gly$^-$ binds through the nitrogen, whereas the met binds through the sulfur.

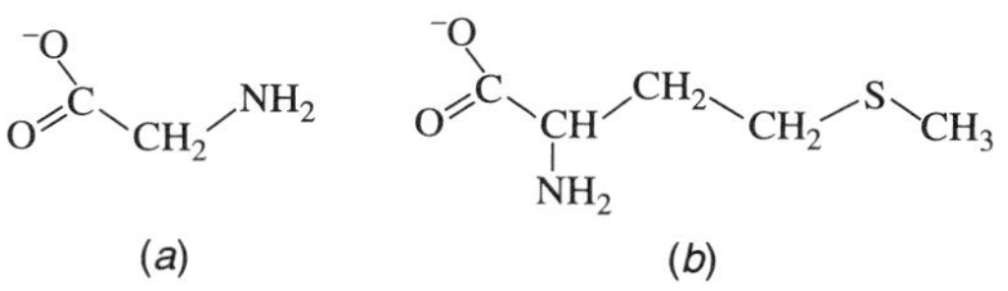

Figure 4.8 Free amino acid ligands, gly⁻ and met, which will be useful in the vibrational frequency analysis.

6. When the job is finished, go to **Display, Surface**.
7. Inspect each iron-complex's HOMO and LUMO. Look for any patterns in the type of orbital observed. Under Display, Output, you can also find more explicit information of the type of each orbital. The information is presented as shown in Table 4.5. The first column indicates the atom, in this case the one and only iron atom. The second column indicates the type of atomic orbital (e.g., s, p_x, d_{xy}, etc.). After that the columns show the molecular orbitals, such as MOs 1, 2, and 3 shown below. If an orbital has a large absolute value for a particular atomic orbital, then that atomic orbital is a significant contributor to the character of the given molecular orbital. In Table 4.5, the s atomic orbital is a major contributor to MOs 1 and 2 and the p_z to MO 3. To find which orbital is the HOMO, note the number of electrons that are also given in the output. Divide this number in half and the resultant number is the HOMO. The LUMO is then readily assigned. At the top of each molecular orbital is the energy of the orbital. How large is the HOMO/LUMO gap? Do you see any degeneracy for HOMO-2, HOMO-1, and HOMO? This degeneracy is especially clear for the hexacyanoferrate(II) ion.

TABLE 4.5 The Iron Atomic Orbital Contributions Towards the First Three Molecularorbitals for Pentacyanoaquaferrate

			MO		
			1	2	3
				Eigenvalues	
			−1.05178	−0.78897 (eV)	−0.72235
			−28.62047 A	−21.46891 A	−19.65610 A
1	Fe1	S	−0.25362	0.32667	0.00014
2	Fe1	PX	0.00251	0.00202	0.01390
3	Fe1	PY	−0.14649	−0.12718	−0.00231
4	Fe1	PZ	−0.00152	−0.00152	0.22016
5	Fe1	DZ**2	0.01209	0.01031	0.00028
6	Fe1	DXX-YY	0.03180	0.01493	0.00014
7	Fe1	DXY	0.00118	0.00048	0.00335
8	Fe1	DXZ	−0.00359	−0.00137	−0.00014
9	Fe1	DYZ	−0.00053	−0.00038	0.01430

8. Now go to **Display, Vibrations**. By clicking the box next to a listed frequency in the vibration window, you can see how the molecule vibrates at that frequency. View the CN^- and carbonyl vibrations for the $[Fe(CN)_5AA]^{n-}$ complexes and free met and gly. (Hint: met carbonyl vibrations are at approximately 560, 1550, and 1850 Hz. Gly carbonyl vibrations are at approximately 600, 1560, and 1850 Hz.) Look for any shifts in frequency for a given vibration when the AA ligand is bound as opposed to when the ligand is free.

Name _____   Section _____

Results Summary for the Semi-empirical Calculations on $[Fe(CN)_5L]^{n-}$ Complexes

---

MO Characteristics for [Fe(CN)₅L]ⁿ⁻ Complexes

1. What atomic orbitals contribute to the HOMOs and LUMOs of the $Fe(CN)_5L]^{n-}$ complexes?

L =	HOMO	LUMO
H_2O	_____	_____
gly^-	_____	_____
met	_____	_____
CN^-	_____	_____

2. Do you see any degeneracy for HOMO-2, HOMO-1, and HOMO-3 in your complexes? Discuss.

3. What is the energy of the HOMO/LUMO gap for L =

H_2O	_____eV
gly^-	_____eV
met	_____eV
CN^-	_____eV

4. How does your ordering compare with that observed experimentally (optical spectra—see Experiment 5.2)?

Vibrational Characteristics for [Fe(CN)₅L]ⁿ⁻ Complexes

5. List the calculated and observed vibrational frequencies for the metal-bound CN^- and AA ligands and their shifts (Δv) from free forms in Table 4.6.

   Comment on your results.

TABLE 4.6   Results Summary for the Vibrational Characteristics for $[Fe(CN)_5L]^{n-}$ Complexes

Compound	Functional group	Frequency (cm^{-1})		Δv(cm^{-1})
		Calculated	Observed[a]	
gly (free)				——
$Fe(CN)_5gly^{4-}$				
met (free)				——
$Fe(CN)_5met^{3-}$				

[a]From Experiment 4.3.

EXPERIMENT 4.6: SUBSTITUTION KINETICS I—DETERMINATION OF THE RATE OF FORMATION, k_F, FOR THE REACTION BETWEEN $[Fe(CN)_5(H_2O)]^{3-}$ AND AMINO ACIDS

Level 4

This experiment involves advanced theory of kinetics and mechanism.

Pre-lab 4.6.a: Substitution in Octahedral Metal Complexes[2,14]

The four types of substitution mechanisms, associative (A), dissociative (D), associative interchange (I_a), and dissociative interchange (I_d), were introduced in Chapter 1. As taught in general chemistry, the most important evidence for mechanism type is the experimental rate law (how concentrations of reactants and products change with time). Accordingly, one may expect that, by determining the kinetics of the reaction, the mechanism will be revealed; i.e., associative-type mechanisms follow second-order kinetics and dissociative-type mechanisms, first-order kinetics. In reality, however, things are not so simple. Because additional steps may take place in substitution reaction mechanisms, simple first- and second-order kinetics are often obscured. Consider the substitution of ligand X^- by ligand Y^- in the well-studied octahedral pentamminecobalt(III) complex $[Co(NH_3)_5X]^{2+}$, equation (4.5). Because this complex is coordinatively saturated, a dissociative-type mechanism with first-order kinetics is expected. In aqueous media at pH < 5, the reaction is indeed first order with respect to $[Co(NH_3)_5X]^{2+}$ and zero order with respect to Y^-. However, a thorough investigation reveals that a different mechanism, one involving solvent intervention, equations (4.9)–(4.10), is responsible for the observed first-order kinetics.

$$[Co(NH_3)_5X]^{2+} + Y^- \rightleftharpoons [Co(NH_3)_5Y]^{2+} + X^- \qquad (4.8)$$

$$[Co(NH_3)_5X]^{2+} + H_2O \rightarrow [Co(NH_3)_5H_2O]^{3+} + X^- \qquad (4.9)$$

$$[Co(NH_3)_5H_2O]^{3+} + Y^- \rightarrow [Co(NH_3)_5Y]^{2+} + H_2O \qquad (4.10)$$

Continuing with this system, consider the anation reaction of the aqueous intermediate, $[Co(NH_3)_5H_2O]^{3+}$, equation (4.10). This step is also expected to be dissociative in nature. Experimentation shows, however, a surprising dependence on the incoming Y^-; the rate $= k_f [Co(NH_3)_5H_2O^{3+}][Y^-]$. How can we account for this dependence on Y^-? There are several possible mechanisms:

1. An I_a mechanism with direct attack of Y^- on the octahedral complex, equation (4.10).
2. A D mechanism involving a preequilibrium with a five coordinate intermediate, equation (4.11). When $k_2[Y^-] < k_{-1}$, a dependence on Y^- will be seen.

$$[Co(NH_3)_5H_2O]^{3+} \underset{k_{-1}}{\overset{k_1}{\rightleftharpoons}} [Co(NH_3)_5]^{3+} + H_2O$$

$$[Co(NH_3)_5]^{3+} + Y^- \xrightarrow{k_2} [Co(NH_3)_5Y]^{2+} \qquad (4.11)$$

3. An I_d mechanism in which an ion-pair precursor complex is formed in a rapid pre-equilibrium step, equation (4.12), followed by a rate determining loss of H_2O and entry of Y^- from the outer-sphere of the complex.

$$[Co(NH_3)_5H_2O]^{3+} + Y^- \underset{k'_{-1}}{\overset{k'_1}{\rightleftharpoons}} \{[Co(NH_3)_5H_2O]^{3+}, Y^-\} \quad K_{IP} = k'_1/k'_{-1}$$

$$\{[Co(NH_3)_5H_2O]^{3+}, Y^-\} \overset{k'_2}{\longrightarrow} [Co(NH_3)_5Y]^{2+} + H_2O$$

(4.12)

Since we are dealing with a coordinatively saturated octahedral complex, we should expect (1) to be the least likely mechanism and proceed to distinguish between mechanisms (2) and (3). Because the pre-equilibria of (2) and (3) result in more complex rate laws, experimentalists simplify the mathematics and data interpretation by working under limiting conditions. When limiting conditions are used in this system, however, the two mechanisms yield the same—either first- or second-order—kinetics, making it very difficult to pin point the mechanism! As an example, let us examine the conditions that lead to second-order kinetics in both cases. For both (2) and (3), solvent H_2O (55 M) is in much greater excess than Y^-.

In (2), the rate-determining step is the loss of H_2O. Assuming a steady state for $[Co(NH_3)_5]^{3+}$ and taking the rate $= k_2[Co(NH_3)_5{}^{3+}][Y^-]$, the overall rate of formation, equation (4.13), can be derived:

$$d[Co(NH_3)_5Y^{2+}]/dt = k_1k_2 \frac{[Co(NH_3)_5H_2O^{3+}][Y^-]}{k_{-1} + k_2[Y^-]}$$

(4.13)

Under conditions of excess Y^-, $k_f = k_1k_2/(k_{-1} + k_2[Y^-])$. When $k_{-1} \gg k_2[Y^-]$, as is the case of 55 M H_2O solvent, equation (4.13) reduces to equation (4.14) and second-order kinetics are observed. For any incoming Y^-, the k_1/k_{-1} term is identical, and any difference in k_f measured reflects the k_2 step.

$$d[Co(NH_3)_5Y^{2+}]/dt = (k_1k_2/k_{-1})\,[Co(NH_3)_5H_2O^{3+}][Y^-]$$
$$k_{obs} = k_1k_2[Y^-]/k_{-1} \text{ and } k_f = k_1k_2/k_{-1}$$

(4.14)

Now consider mechanism (3). The rate determining step, k_2', is preceded by an outer-sphere association with the activated complex having a weakly bound Y^-. Using the laws of mass action and conservation of mass, the rate of formation is equation (4.15). Under conditions of excess $[Y^-]$, $k_{obs} = k_2K_{IP}[Y^-]/(1 + K_{IP}[Y^-])$ and $k_f = k_2K_{IP}/(1 + K_{IP}[Y^-])$. When $K_{IP}[Y^-]$ is small ($\ll 1$), second-order kinetics are observed. Thus we see that, under the defined conditions, both mechanisms (2) and (3) give second-order kinetics and, therefore, are not distinguishable.

$$d[Co(NH_3)_5Y^{2+}]/dt = K_{IP}k_2' \frac{[Co(NH_3)_5H_2O^{3+}][Y^-]}{1 + K_{IP}[Y^-]}$$

(4.15)

When reactions such as the above are complicated, evidence for a particular mechanism is gathered by measuring rate constants for a series of Y^- reactions under a variety of conditions (e.g., pH, ionic strength, temperature, and sterics). (More sophisticated evidence for stoichiometric mechanisms is presented in Basolo and Johnson.[4]) In experiment 4.7 you will investigate the temperature dependence of the $[Fe(CN)_5(H_2O)]^{3-}$/amino acid reaction, exemplifying a means for gathering such mechanistic evidence. For this experiment, however, you will be told the particular mechanism. The observed rate constant under pseudo-first-order conditions, k_{obs}, will be measured and the rate of formation, k_f, will be determined for the substituted pentacyanoferrate(II) complexes. Toma *et al.*[14] have shown that mechanism (3) is most appropriate for the evaluation of k_f in this amino acid system.

Pre-lab 4.6.b: The Stopped-Flow Method[15]

The stopped-flow method is used to measure rapid reactions (approximately ≥ 20 ms). Two reactant solutions of equal volume (as little as 0.1–0.2 ml) are driven into a mixing chamber and then into an observation cell, at which time the flow is abruptly stopped. When the solution comes to rest (a few milliseconds), the absorbance is monitored spectrophotometrically. A diagram of the instrumentation is shown in Figure 4.9.

Procedure 4.6.a: Solution Preparation

Wash all glassware with concentrated H_2SO_4 and plasticware with 3 M HCl. All reagents and water should be of highest purity.

Out-Flow Stock Solutions
1. Prepare 100 ml of 2.0×10^{-4} M $Na_3[Fe(CN)_5H_2O]$ by dissolving 0.0065 g $Na_3[Fe(CN)_5NH_3]\cdot 3H_2O$ (MM = 325.9 g mol^{-1}) with 100 ml ddH$_2$O in a volumetric flask (stock 1—prepare prior to use and store in dark).
2. Prepare 100 ml of 4×10^{-2} M AA solution at the appropriate pH (8–9 methionine; 11.5 glycinate); use volumetric glassware. AA = methionine (149.21 g mol^{-1}); glycinate$^-$ (75.07 g mol^{-1}) (Stock 2).
3. Prepare 50 ml of 1 M LiClO$_4$ (MM = 106.39 g mol^{-1}) by dissolving 5.32 g LiClO$_4$ with 50 ml ddH$_2$O in a volumetric flask (Stock 3).
4. Just prior to each kinetic run, prepare 10 ml AA plus LiClO$_4$ according to Table 4.7. Out-flow solutions must be prepared with twice the desired concentration since

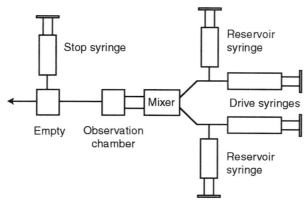

Figure 4.9   Diagram of the stopped-flow instrumentation.

TABLE 4.7   Initial or *Outflow* AA–LiClO₄ Solutions in a Total Final Volume (V_f) 10 ml

Reaction	AA (M_f)	AA (V_i) Stock (2)	LiClO₄ (M_f)	LiClO₄ (V_i) Stock (3)
1	5×10^{-3}		0.194	
2	1×10^{-2}		0.178	
3	2×10^{-2}		0.158	

equal volume reactants will be rapidly mixed together in the reaction chamber. Use $M_i V_i = M_f V_f$ to calculate appropriate amounts of each stock needed.

In-Flow Concentrations   The *final or in-flow concentrations* (i.e., after equal mixing of the two reactant solutions) for the reactants are given in Table 4.8. Calculate the ionic strength, μ(M) of the reaction solution using equation (4.16) where c and z are the concentration and charge of the ith ion in solution. (See Chapter 7, experiment 7.4, for a more rigorous consideration and calculation of ionic strength effects.)

$$\text{Ionic strength, } \mu(\text{M}) = 1/2 \sum_i c_i z_i^2 \tag{4.16}$$

Procedure 4.6.b: Determination of the Formation Constant, k_f, for the Reaction Involving [Fe(CN)₅(H₂O)]³⁻ and AA Ligands, $\mu = 0.1$ M (LiClO₄), 25 °C

1. Load one of the stopped-flow syringes with Stock 1 and the other with the Reaction 1 solution from Table 4.7.
2. Observe the absorbance decrease at 444 nm (the absorbance maximum for the [Fe(CN)₅(H₂O)]³⁻ complex) over at least three reaction half-lives. Carry out three trial runs.
3. Repeat for Reaction Solutions 2 and 3.
4. For each run, plot $\ln |A_\infty - A_t|$ vs t, where A_t is the measured absorbance at time, t. Obtain the pseudo-first-order rate constant, k_{OBS} (s⁻¹) from the slope of the plot.
5. Report the average of your three trials for each reaction 1, 2, and 3.
6. To obtain the formation rate constant, k_f (M⁻¹ s⁻¹), plot k_{OBS} vs [AA].

Name _____   Section _____

Results Summary for the Determination of the Rate of Formation, k_f, of [Fe(CN)₅(AA)]ⁿ⁻ Complexes

TABLE 4.8   Values for the Final (In-Flow) Reactant Concentrations

Reaction	AA(M)	Fe(CN)₅H₂O³⁻ (M)	LiClO(M)₄	μ(M)
1	1×10^{-3}	$\leq 1 \times 10^{-4}$		0.097
2	5×10^{-3}	$\leq 1 \times 10^{-4}$		0.089
3	1×10^{-2}	1×10^{-4}		0.079

TABLE 4.9   Results Summary for Ligand

			Reaction						
	1			2			3		
				Trial					
	1	2	3	1	2	3	1	2	3
pH_f									
k_{obs} (s^{-1})									
Average k_{obs} (s^{-1})									
k_f $(M^{-1}s^{-1})$									

1. Complete Table 4.9 for each ligand tested.
2. Comment on your k_f values.
3. Under the experimental conditions, the formation rate constant, $k_f = k_2 K_{IP}/(1+K_{IP}[Y^-])$. Assuming $K_{IP} \ll 1$, comment on the value of k_2 for your AA ligands.
4. Other negatively charged amino acids tend to react with a rate close to that of gly^-. Met^-, however, has a $k_f = 169$ M^{-1} s^{-1}. Comment on the high rate of reaction for methionine.

EXPERIMENT 4.7: SUBSTITUTION KINETICS II—DETERMINATION OF THE ACTIVATION PARAMETERS FOR THE REACTION BETWEEN $[Fe(CN)_5(H_2O)]^{3-}$ AND AMINO ACIDS

Level 4

This experiment involves advanced theory of kinetics and mechanism.

Pre-lab 4.7: Indirect Evidence for Mechanism Type: Activation Parameters[14,16]

In Experiment 4.6, you determined the rates of substitution of the H_2O ligand in $[Fe(CN)_5(H_2O)]^{3-}$ by two different amino acid ligands. Because of the mechanistic complexity (pre-equilibrium steps), to calculate the k_f values for these reactions you were given an appropriate mechanism from which the rate law was derived. How is a particular mechanism decided upon? The answer is, additional evidence must be gathered. For example, you may try to isolate or kinetically observe an intermediate giving evidence for a D or A mechanism. Most often, however, indirect evidence is gathered by carrying out reaction kinetics under a variety of conditions. Consider the general substitution reaction in equation (4.17). Table 4.10 gives expected mechanism characteristics that can be revealed by changing variables in the reaction and seeing how the rate is affected. Table 4.10 is by no means exhaustive. Many other variables including overall metal complex charge affect rates in a predictable way. Langford and Gray[14] discusses experimental manipulations in detail.

$$MX_n + Y^- = MX_{n-1} + Y + X^- \tag{4.17}$$

Activation parameters, which describe the energy difference between the reactants and transition state, Figure 4.10, can provide indirect evidence for mechanism type. These parameters, defined by equations (4.18)–(4.20), are derived from transition state theory. $\Delta V^{\ddagger}$

TABLE 4.10 *Expected* **Characteristics Giving Indirect Evidence for Substitution Reaction Mechanisms**

Mechanism	Dependence on Y	Dependence on X	Limiting Rate Behavior	Activation Parameters
D	None	Steric bulk increases rate	Equal to solvent exchange exactly	$\Delta H^{\ddagger}$ (+), relatively large $\Delta S^{\ddagger}$ (+) large $\Delta V^{\ddagger} \sim 9 \text{ cm}^3 \text{ mol}^{-1}$
A	Strong (base strength and nucleophilicity)			$\Delta H^{\ddagger}$ relatively small $\Delta S^{\ddagger}$ (−) $\Delta V^{\ddagger}$ $\sim (-)9 \text{ cm}^3 \text{ mol}^{-1}$
I_d	Relatively independent of nucleophilicity depending on charge	Steric bulk	Close but not equal to solvent exchange	Similar to D $0 < \Delta V^{\ddagger}$ $< (+)7 \text{ cm}^3 \text{ mol}^{-1}$
I_a	Nucleophilicity charge			Similar to A $-7 < \Delta V^{\ddagger}$ $< 0 \text{ cm}^3 \text{ mol}^{-1}$

values are obtained from pressure (p) dependence studies. $\Delta H^{\ddagger}$ and $\Delta S^{\ddagger}$ values are related to the Arrhenius expression, equation (4.21), by equations (4.22) and (4.23). Therefore, by studying the temperature dependence of the reaction rate, activation parameters can be obtained. Although Table 4.10 provides a good guideline for expected activation parameter values during bond-breaking (D-type) and bond-making (A-type) processes, solvation effects in the transition state can often make interpretation of $\Delta H^{\ddagger}$ and $\Delta S^{\ddagger}$ difficult. This is especially true when changes in charged species are involved.

$$k_f = RT/Nh \exp(-\Delta G^{\ddagger}/RT) \tag{4.18}$$

$$k_f = RT/Nh \exp(-\Delta H^{\ddagger}/RT) \exp(-\Delta S^{\ddagger}/R) \tag{4.19}$$

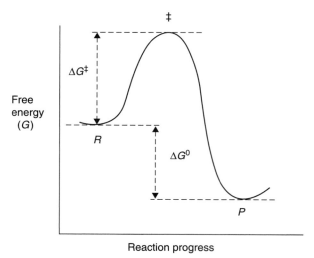

Figure 4.10 Reaction coordinate diagram defining free energy activation change, $\Delta G^{\ddagger}$.

$$\Delta V^{\ddagger} = -RT \ln (k_2/k_1)/(p_2 - p_1) \tag{4.20}$$

$$k_f = Ae^{-E_a/RT} \tag{4.21}$$

$$\Delta H^{\ddagger} = E_a - RT \, (\text{kJ mol}^{-1}) \tag{4.22}$$

$$\Delta S^{\ddagger} = 19.15(\log A - 13.23) \, (\text{J mol}^{-1}) \tag{4.23}$$

where E_a is the activation energy and A is the pre-exponential factor.

Procedure 4.7.a: Solution Preparation

Wash all glassware with concentrated H_2SO_4 and plasticware with 3 M HCl. Follow all directions given in Procedure 4.6.a for Reaction 1.

Procedure 4.7.b: Determination of $\Delta H^{\ddagger}$ and $\Delta S^{\ddagger}$ for the Reaction Involving $[Fe(CN)_5(H_2O)]^{3-}$ and AA Ligands, $\mu = 0.1$ M (LiClO$_4$)[1,2]

1. Load one of the stopped-flow syringes with Stock 1 and the other with the Reaction 1 solution.
2. Observe the absorbance decrease at 444 nm (the absorbance maximum for the $[Fe(CN)_5(H_2O)]^{3-}$ complex) over at least three reaction half lives at 30 °C. Carry out three trial runs.
3. Repeat step 2 at 20 and 10 °C.
4. For each run, plot $\ln|A_\infty - A_t|$ vs t, where A_t is the measured absorbance at time, t. Obtain the pseudo-first-order rate constant, k_{OBS} (s^{-1}) from the slope of the plot.
5. Report the average of your three trials for each temperature run.
6. Obtain the formation rate constant, k_f (M^{-1} s^{-1}), using equation (4.24).

$$k_{OBS} = k_f[AA] \tag{4.24}$$

7. For each ligand, evaluate the activation parameters, $\Delta H^{\ddagger}$ and $\Delta S^{\ddagger}$.
   a. Plot $\ln k_f$ vs $1/T$ according to the rearranged Arrhenius equation (4.25)

$$\ln(k_f) = -E_a/R(1/T) + \ln A \tag{4.25}$$

   b. Record and use values of E_a and A to calculate $\Delta H^{\ddagger}$ (kJ mol^{-1}) and $\Delta S^{\ddagger}$ (J mol^{-1}) according to equations (4.22) and (4.23), respectively.

Name _____   Section _____

Results Summary for the Determination of $\Delta H^{\ddagger}$ and $\Delta S^{\ddagger}$ in the Formation of
$[Fe(CN)_5(AA)]^{n-}$ Complexes

1. Complete Table 4.11 for each ligand tested.
2. Comment on your values obtained.

TABLE 4.11   Results for Ligand

	Reaction Temperature								
	10 °C			20 °C			30 °C		
				Trial					
Reaction Temperature	1	2	3	1	2	3	1	2	3
pH_f									
k_{obs} (s^{-1})									
Average k_{obs} (s^{-1})									
k_f (M^{-1}s^{-1})									
Activation parameters	E_a			A			$\Delta H^{\ddagger}$		$\Delta S^{\ddagger}$

3. Use CFAE calculations to compare dissociative-type and associative-type substitution mechanisms for the $[Fe(CN)_5(H_2O)]^{3-}$ complex.

Post-lab Questions

Q4.7   Consider all data obtained for your substitution kinetics throughout this chapter. Comment on the lability/inertness of your low spin iron(II) complex.

REFERENCES

1. Toma, H. E., Martins, J. M., Giesbrecht, E. *J. Chem. Soc. Dalton Trans.* **1978**, 1610–1617.
2. Toma, H. E., Batista, A. A., Gray, H. B. *J. Am. Chem. Soc.* **1982**, *104*, 7509–7515.
3. Basolo, F., Pearson, R. G., *Mechanisms of Inorganic Reactions—a Study of Metal Complexes in Solution*, 2nd edn. Wiley: New York, 1967.
4. Basolo, F., Johnson, R. *Coordination Chemistry*. Benjamin: New York, 1964.
5. Huheey, J. E., Keiter, E. A., Keiter, R. L. *Inorganic Chemistry—Principles of Structure and Reactivity*, 4th edn. HarperCollins: New York, 1993, p. 549.
6. Brauer, G. *Handbook of Preparative Inorganic Chemistry*, 2nd edn. Academic Press: New York, 1963.
7. Gray, H. B. *J. Am. Chem. Soc.* **1982**, *26*, 782.
8. Cowan, J. A. *Inorganic Biochemistry—an Introduction*. 2nd edn. Wiley: New York, 1997.
9. Williams, D. H., Fleming, I. *Spectroscopic Methods in Organic Chemistry*, 3rd edn. McGraw-Hill: New York, 1980.
10. Nakamoto, K. *Infrared Spectra of Inorganic and Coordination Compounds*, 2nd edn. Wiley Interscience: New York, 1970.
11. Bott, A. W., Jackson, B. P. BAS CV-1B Operation/Instruction Manual and CV-27 Operation/Instruction Manual. Bioanalytical Systems: West Lafayette, IN; *http://www.bioanalytical.com/cursep/cs151f4.html*.
12. Hinchliffe, A. *Computational Quantum Chemistry*. Wiley: New York, 1988.
13. Szabo, A., Ostlund, N. S. *Modern Quantum Chemistry—Introduction to Advanced Electronic Structure Theory*. Dover: Mineola, NY, 1996.
14. Langford, C. H., Gray, H. B. *Ligand Substitution Processes*. Benjamin: Reading, MA, 1966.
15. Sykes, A. G. *Kinetics of Inorganic Reactions*, Pergamon Press: New York, 1966.
16. Lappin, A. G. *Redox Mechanisms in Inorganic Chemistry*. Ellis Horwood: New York, 1994.

Electron Transfer Reactions— Structure, Properties and Reactivity of Tris(bidentate chelate) cobalt(II/III) Complexes

Life would not be possible in a Marcusian World (Henry Taube, early 1980s).

PROJECT OVERVIEW[1-3]

Oxidation–reduction (redox) reactions comprise the second major class of coordination chemistry reactions and are classified as proceeding through either an inner-sphere or outer-sphere mechanism (see Chapter 1). Since the 1950s, a wealth of data have been accumulated, allowing new details about these events—whose reaction times span orders of magnitude—to be uncovered. Electron transfer theory developed by Marcus in the 1950s describes and predicts everyday general interest events, including corrosion, chemiluminescence ("cold light") photochemical fuel production, and natural biological electron transfer events. Perhaps one of the theory's greatest achievements is its contribution to the latter, providing an understanding of photosynthesis. Electron transfer mechanisms may appear deceivingly simple: as Taube states in his Nobel address, 'Admittedly, all the possible electron transfer processes are governed by the same principles, at least when these are stated in a general enough way. But, as these principles manifest themselves in the different subclasses, the descriptive chemistry can be quite different, and these differences are the fabric of chemistry.' To fully understand electron transfer mechanisms one must be familiar with electronic structure as it governs rates of substitution, stability of reaction intermediates, barriers to bond length changes to achieve the activated complex, and the rates of conversion of precursor complexes to successor complexes. The study of electron transfer has come a long way since Taube's early years, about which he notes, "I had failed to interest any of my graduate students in the work, because, of course, no one foresaw what it might lead to."

Both inner- and outer-sphere electron transfer mechanisms will be investigated in this series of experiments. The chapter begins the synthesis of four cobalt(III) coordination complexes followed by analysis and reactivity studies. Electronic structure will be investigated using visible spectroscopy and the redox chemistry of two of the complexes will be examined

Integrated Approach to Coordination Chemistry: An Inorganic Laboratory Guide. By Rosemary A. Marusak, Kate Doan, and Scott D. Cummings
Copyright © 2007 John Wiley & Sons, Inc.

using cyclic voltammetry. Using these data, the potentials of the others will be predicted. The information gathered will be important for understanding the solution electron transfer reactivity. The electron transfer reaction between $[Co(en)_3]^{2+}$ and $[Co(ox)_2(en)]^-$ will then be investigated. After the stoichiometry of the redox reaction is determined, a reaction mechanism will be proposed and a rate law derived. Kinetic measurements using conventional visible spectrophotometry will be made and based on the given rate law, the outer-sphere rate constant will be determined. The culminating experiment compares the experimental outer-sphere electron transfer rate constant to that calculated by Marcus theory.

EXPERIMENT 5.1: SYNTHESIS OF TRIS(BIDENTATE CHELATE)COBALT(III) COMPLEXES

Level 4

This excepts the synthesis of $Ca[Co(ox)_2(en)]_2 \cdot 4H_2O$, which because of its challenging nature and time requirement, is level 5.

Pre-lab 5.1: $Ca[Co(ox)_2(en)]_2 \cdot 4H_2O$: pH and Co-precipitate Considerations[4]

The syntheses of $[Co(en)_3]Cl_3 \cdot H_2O$, $[Co(en)_2(ox)]Cl \cdot H_2O$, and $K_3[Co(ox)_3] \cdot 3.5H_2O$ are straightforward, however the synthesis of $Ca[Co(ox)_2(en)]_2 \cdot 4H_2O$ is a bit trickier! Close attention to co-precipitate impurities (calcium oxalate) discussed in Chapter 1 is highly recommended and, in light of this, it is imperative that you adhere to timing, temperature, and pH restrictions as written in the procedure. Yields are generally low. Following its preparation the calcium salt product is converted to $Na_2[Co(ox)_2(en)]_2 \cdot H_2O$, which is needed for electron transfer reactivity studies. You don not need much product, but it is mandatory that it is pure. The purity of your product will be checked by UV–visible spectroscopy.

   The co-precipitate problems in the $Ca[Co(ox)_2(en)]_2 \cdot 4H_2O$ synthesis occur because of the similar solubilities of the product and calcium oxalate under the given conditions. Calcium oxalate is an example of a sparingly soluble salt of a weak acid and provides a unique opportunity to examine solubility and pH. Sparingly soluble salts of weak acids can be dissolved in dilute solutions of strong acids. The protonation of oxalate(2−), equations (5.1) and (5.2), lowers the concentration of oxalate anions in solution; solubility can therefore be controlled by pH.

$$ox^{2-} + H^+ \rightleftharpoons Hox^- \qquad K_1 = 6.5 \times 10^{-2} \tag{5.1}$$

$$Hox^- + H^+ \rightleftharpoons H_2ox \qquad K_2 = 6.1 \times 10^{-5} \tag{5.2}$$

$$[Ca^{2+}][ox^{2-}] = K_{sp} = 2.0 \times 10^{-9} \tag{5.3}$$

Assuming activity coefficients are unity, and using the solubility expression for Ca(ox), equation (5.3), the species distribution diagram for oxalate ions can be generated, Figure 5.1. You can see that, with excess oxalate in solution, pH will play a critical role in balancing the co-precipitation of $H_2ox_{(s)}$, Ca(ox), and the cobalt complex product.

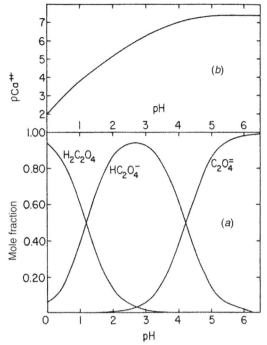

Figure 5.1   (a) Mole fraction of oxalate species as a function of pH. (b) -log[Ca^{2+}] in equilibrium with solid calcium oxalate and 0.05 M aqueous oxalate. [Reproduced with permission from Walton, H. F. *Principles and Methods of Chemical Analysis*, 2nd edn. Prentice-Hall: Englewood Cliffs, NJ, 1964.]

Procedure 5.1.a: Tris(ethylenediamine)cobalt(III) Chloride, [Co(en)$_3$]Cl$_3$ · H$_2$O[5]

Safety warning! Operations set off between the lines must be performed in a fume hood.

1. Partly neutralize 30.5 g of 30% ethylenediamine with 8.5 ml of 6 M HCl. Pour the resulting mixture into a solution of 12 g of CoCl$_2$ · 6H$_2$O in 37.5 ml of H$_2$O.

2. Oxidize the cobalt(II) by bubbling a vigorous stream of air through the solution for 3 h. A water aspirator vacuum set-up is shown in Figure 2.3.

3. Transfer your mixture to a beaker and evaporate the solution by suspending the beaker in a boiling water bath. Cease the evaporation process when a crust of product begins to form over the surface (~7.5–10 ml). Add 7.5 ml of HCl$_{conc}$ and 15 ml of ethanol to the mixture.

4. After cooling in an ice bath, filter the crystals of [Co(en)$_3$]Cl$_3$ · H$_2$O and wash with alcohol until the washings are colorless. Finally, wash the product with ether or dry in the oven.

5. Recrystallize your product from H$_2$O−methanol mixture. Dissolve 5.5 g of product in approximately 50 ml warm H$_2$O. Add 100 ml cold MeOH. Cool the solution and filter orange crystals. Record your yield.

Procedure 5.1.b: Bis(ethylenediamine)(oxalato)cobalt(III) Chloride Monohydrate, [Co(en)$_2$(ox)]Cl · H$_2$O[6]

Safety warning! Operations set off between the lines must be performed in a fume hood.

---

1. Prepare a solution of 20 g of cobalt(II) acetate tetrahydrate in 100 ml H$_2$O at 60 °C and add this to a mixture of 15 g of oxalic acid dihydrate and 15 ml of 99% ethylenediamine in 100 ml H$_2$O at 70 °C. Rapidly heat the reaction mixture to 80 °C with continuous stirring and add 10 g lead(IV)oxide.

2. Gently boil the resulting mixture for 30 min. After 10 min boiling, add an additional 2 g PbO$_2$ and then another 2 g PbO$_2$ after 20 min.

3. Cool to RT and treat with 10 ml of 10 M H$_2$SO$_4$. Filter the solution. Add 25 ml of 10 M HCl to the filtrate and evaporate to 100 ml in a water bath using an air stream.

4. Cool the solution on ice and filter to obtain red crystals, which are washed with 80% methanol, methanol, and diethyl ether, and dried on the funnel by suction.

5. Recrystallize by dissolving 4.00 g of the complex in 47.5 ml hot H$_2$O and 15.8 ml concentrated HCl. Cool the solution on ice and filter the pink powdery crystals. Wash the product as before. Record your yield.

Procedure 5.1.c: D,L-Sodium bis(oxalato)ethylenediamine Cobaltate(III) Monohydrate Na$_2$[Co(ox)$_2$(en)] H$_2$O[7]

Safety warning! Operations set off between the lines must be performed in a fume hood. Two alternative approaches are given in the instructor notes.

Calcium Bis(oxalato)(ethylenediamine)cobalt(III) 4 hydrate, Ca[Co(ox)$_2$(en)]$_2$ · 4H$_2$O Timing is critical to the success of this synthesis. Read all instructions carefully and prepare all reagents and equipment *prior* to carrying out the synthesis

---

1. Prepare ~10 g ethylenediamine dihydrochloride (H$_2$en · 2HCl) by *carefully* adding two equivalents of HCl to ethylenediamine (en, free base) and filtering the white crystalline product.

2. To 100 ml ddH$_2$O in a 400 ml beaker, add 25 g potassium oxalate (K$_2$ox). Heat with stirring until dissolved.

3. Add 10 g cobalt acetate (Co(CH$_3$COO)$_2$) slowly over 1–2 min. Record the pH of the solution: it should be between 5 and 6.5. The solution will be reddish.

4. Once the cobalt acetate has dissolved, add the 5.76 g ethylenediamine dihydrochloride slowly. Bring the solution to a full boil; the solution should turn more purple.

5. Add 25 g lead(IV)oxide (PbO$_2$) over 5–10 min—the liquid will bubble and turn a darker color. Place a watch glass over the beaker, bring the solution back to a boil and boil for 30 min.

6. Meanwhile, dissolve 16.76 g calcium chloride (CaCl$_2$, hydrated) in about 10 ml dH$_2$O in a small beaker.[†] Pour this solution into a graduated cylinder and adjust the volume to 30 ml.

[†]The original procedure recommends 25 g calcium chloride in a small amount of cold water. We have found that this is in excess and causes more calcium oxalate co-precipitate problems.

7. After the cobalt solution has boiled for 30 min, filter it through Whatman filter paper on a Buchner funnel to remove the lead(IV)oxide from the solution. The resulting filtrate should be a vivid purple color. Rinse the filter cake several times with small portions of water until the filtrate passing through the filter is no longer purple.

8. Dilute the filtrate to a total of 200 ml and pour into a clean 400 ml beaker.

9. Add the 2.21 g oxalic acid (H_2Ox) over 1–2 min;[‡] place a watch glass over the top of the beaker. Bring to a boil and let the solution boil for 15 min.

10. Remove the solution from the heat and let cool on the countertop (no ice, no water bath) *all the way to room temperature*. This will probably take 30–40 min.[§] The pH is ~4.5.

If you perform the next step without waiting for the solution to cool fully, your purple product will form an inseparable precipitate with calcium oxalate. Waiting too long to add the calcium chloride solution, however, will result in the formation of a red powdery precipitate. This has λ_{max} of 375 and 525 nm and is highly insoluble in H_2O. Attempts to revive the solution by addition of more calcium chloride and by the application of heat result in decomposition to a similar red precipitate.

11. Meanwhile, take three pieces of 11 cm filter paper, weigh them, and write their masses on one side with a nonwater-soluble pen (for convenience of measuring yield; the product is often difficult to remove from the filter paper).

12. Immediately after the cobalt solution has cooled, add the calcium chloride solution (16.75 g in 30 ml dH_2O prepared previously) quickly and uniformly with stirring. The solution will turn a milky purple color. Turn off the stirring. Let sit for 4 min without stirring.

Longer crystallization times cause much calcium oxalate to separate.

13. Using the weighed pieces of filter paper and the large Buchner funnel, vacuum filter the solution. It is advisable to pour the solution into the funnel uniformly over about 1 min to avoid clogging. A dark purple product should stay on the filter paper; a milky solution, still purple, will pass through.

14. Once all the solution has been filtered, immediately remove the filter paper to a large watch glass. Pour the filtrate back into the beaker, place a new piece of filter paper on the funnel, and filter this solution through again as above. Repeat a third time with the third piece of filter paper. This third filter cake may be thick and milky. Calcium oxalate precipitate causes this color and causes the filter to clog. If this occurs, the third filter cake should be discarded.

Note that, after 10–15 min, it is no longer possible to separate calcium oxalate from the desired cobalt product. Therefore, the first two filtrations must be accomplished in less than 10 min.

[‡]The original procedure calls for 9 g oxalic acid to be added.
[§]The original procedure recommends adding sufficient HCl (10 M) to make the solution approximately 2 M. We found that this step furthered contamination problems with oxalate.

15. Wash the dark purplish-red crystalline powder with 1 M HCl and ethyl alcohol.

16. The product may be further freed from calcium oxalate by suspending in normal HCl (25 ml at 40 °C) and filtering while warm. Wash with warm 1 M HCl, H_2O, ethyl alcohol and acetone and air dry. Record your yield.

If this attempt to free the product from calcium oxalate is not successful, conversion to the sodium salt followed by recrystallization is recommended.

D,L-*Sodium Bis(oxalato)ethylenediamine Cobaltate(III) Monohydrate Na$_2$[Co(ox)$_2$(en)] · H$_2$O*

1. Suspend 5 g of the calcium salt in hot water (100 ml, 65 °C) containing sodium oxalate (1.24 g) and heat to boiling with rapid stirring. Continue boiling and stirring until decomposition of the calcium salt is complete (15–20 min) and cool and filter the mixture.

2. Evaporate the filtrate using a water bath and after cooling, filter and add ethanol gradually until crystallization is complete.

3. This product may be recrystallized from a minimum amount of water by adding ethanol. Record your yield.

Procedure 6.1.d: Potassium Tri(oxalato)cobaltate(III), K$_3$[Co(ox)$_3$] · 3.5H$_2$O[8]

Safety warning! Operations set off between the lines must be performed in a fume hood.

---

1. Dissolve 11.9 g cobalt carbonate ($CoCO_3$) in a solution of 12.6 g of oxalic acid (H_2ox) and 36.85 g of potassium oxalate (K_2ox) in 250 ml of hot dH_2O.

2. When the solution has cooled to 40 °C, while vigorously stirring, slowly add 11.95 g of PbO_2 followed by 12.5 ml of glacial acetic acid a drop at a time.

3. Continue stirring for 1 h, during which time the color changes from red to deep green.

4. After the unused PbO_2 is filtered out, precipitate the tri(oxalato)cobaltate(III) product by adding 250 ml of ethyl alcohol. *The product is sensitive to heat and light.*

5. The product may be recrystallized by dissolving 5.55 g of the complex in 35 ml cold H_2O and adding 35 ml cold ethyl alcohol (95%). Place the solution on ice and filter the green crystals. Record your yield.

Name _____   Section _____

Results Summary for the Synthesis of Coordination Compound

1. Write a balanced chemical equation describing the reaction to form this product, based on your procedure.

2. Calculate the moles of the reagents used in your procedure (concentrated HCl is 12 M):

3. Determine the limiting reagent: _____

4. a. What was your product yield? _____ g
   b. Calculate the theoretical yield _____ g
   c. Calculate your percent yield _____ g

Post-lab Questions

Q5.1 Why is $[Co(ox)]_3^{3-}$ sensitive to light and heat and the other complexes not?

Q5.2 Note the colors of all four cobalt complexes prepared in the lab. Provide an explanation for this.

Q5.3 All four syntheses begin with a cobalt(II) salt, which after addition of appropriate ligands, is oxidized to the cobalt(III) product. Why is this done this way?

Q5.4 Only $[Co(en)_3]^{3+}$ is prepared by aerial oxidation (O_2) while the other procedures use $Pb(IV)O_2$. Offer an explanation for this.

Q5.5 If the pH is too low during the synthesis or work-up of your metal complexes, what complications can arise?

EXPERIMENT 5.2: VISIBLE SPECTROSCOPY OF TRIS(BIDENTATE CHELATE)COBALT(III) COMPLEXES

Level 4

Pre-lab 5.2: Electronic Spectra of Metal Complexes—Polyelectronic Systems

This discussion builds upon Pre-lab 3.6.b.[9] To understand spectral transitions in metal complexes, we must think in terms of polyelectronic systems since we are often concerned about metal ions with more than one d electron. Instead of the familiar t_{2g} and e_g nomenclature (labels for the atomic orbitals in CFT), *term symbols* are used to describe the different ways in which electrons can be arranged in the d orbitals. One arrangement, the ground state, will be most favorable and lowest in energy. Other electronic arrangements lead to higher energy excited states (less favorable arrangements of the electrons). Basic definitions and rules for determining states and term symbols for polyelectronic systems are given in Appendix 3. For now, simply be familiar with the nomenclature. The terms that arise are like those given for atomic orbitals, but are now capitalized: S, P, D, F, G, ... , similar to atomic orbitals, s, p, d, f, g, ... Like the atomic orbitals, these terms are split in an octahedral field with different energies depending on coulombic repulsions of the electrons:

$$S \longrightarrow A_{1g}$$
$$P \longrightarrow T_{1g}$$
$$D \longrightarrow E_g + T_{2g}$$
$$F \longrightarrow A_{2g} + T_{1g} + T_{2g}$$
$$G \longrightarrow A_{1g} + E_g + T_{1g} + T_{2g}$$

The spin multiplicities (noted as superscripts—not shown above) are conserved. For example, the free ion ground state for a d^6 metal ion such as cobalt(III) is ^5D (5 is the spin multiplicity). In an octahedral field, the ^5D state is split into $^5E_g + {}^5T_{2g}$. The energies of these new states vary with the ligand field strength, Dq ($10\, Dq = \Delta_o$). The higher energy excited states for the d^6 configuration, ^3H, ^3F, ^3G, ^1I, and ^1F, are also split in an octahedral field, their energies varying with Dq.

These concepts are best visualized using what are known as *Tanabe–Sugano diagrams*, which plot the energy of the states, E/B (where B is a repulsion term) as a function of field strength, Dq/B. Figure 5.1 gives the diagram for a d^6 metal ion. In a weak ligand field, the ground state is $^5T_{2g}$; however, upon increasing field strength (larger Dq), the $^1A_{1g}$ state, derived from the excited state free ion ^1I term, rapidly drops in energy. At $Dq/B = 2$, a discontinuity arises in the plot as the spin state changes from high to low and the $^1A_{1g}$ becomes the ground state. It is the transitions from this ground state to higher energy states *of the same multiplicity* that we are concerned with in our low-spin (high field) cobalt(III) complexes. After you acquire your UV–visible spectrum of your complex, you will use the Tanabe–Sugano diagram in Figure 5.2 to assign the visible d–d transitions of your complex, and calculate Dq and B.[†]

Procedure 5.2.1: Visible Spectroscopic Analysis of Tris(bidentate chelate)cobalt(III) Complexes

1. Quantitatively prepare *three* solutions of $\sim 5 \times 10^{-3}$ M (*record exact concentration*) cobalt(III) complex solution (dH$_2$O) using an analytical balance and volumetric flasks.

2. Obtain the spectrum of each solution from 300 to 700 nm.

3. Calculate the molar absorptivity for each absorption at its maximum wavelength for your complex using Beer's law (see Results Summary sheet).

4. Calculate $10\, Dq$ (Δ_o) for your complex (see Results Summary sheet).

5. Include error analysis.

Name _____   Section _____

Results Summary for the UV–Vis Analysis of Coordination Compound

1. Molar mass of metal complex _____
2. Theoretical mass needed (mg) in 10 ml total volume _____

	Trial 1	Trial 2	Trial 3
3. Actual mass used (mg)	_____	_____	_____
4. Concentration (M)	_____	_____	_____
5. Molar absorptivity (ε, M^{-1} cm^{-1})			
Transition (ν_1) λ_{max} = _____	_____	_____	_____

Average _____

[†] The electron arrangement for $^1A_1$ is $t_2{}^6$ with an energy equal to $-24\, Dq + 5B + 8C - 120\, B2/10\, Dq$ ($C = 4B$). Excited state configurations for low-spin d^6 are: $t_2{}^5e$ ($^3T_1$, $^3T_2$, $^1T_1$, $^1T_2$); $t_2{}^4e^2$ (5T2); $t_2{}^3e^3$ (5E).

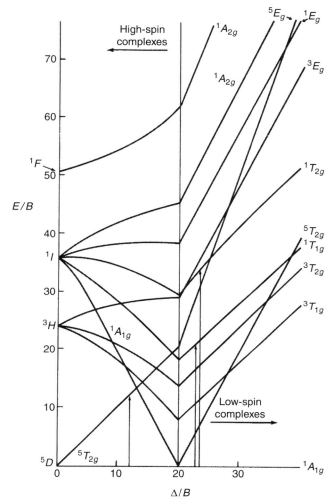

Figure 5.2 Tanabe–Sugano diagram for the d^6 electron configuration.

Transition (ν_2) λ_{max} = _____ _____ _____ _____

Average _____

[Note: ν_1 (=1/λ) is your lowest energy transition,]

6. Value of 10 Dq (Δ_o), cm^{-1} _____

Method: use the Tanabe–Sugano diagram in Figure 5.1.

a. List the transition(s) you observe for your low spin cobalt(III) complex. [*Note*: ν_1 (=1/λ) is your lowest energy transition.] Provide *both* the transition assignment and the energy at maximum absorbance in cm^{-1}.

ν_1 = _____ ν_2 = _____ ν_3 = _____

TABLE 5.1   Class Composite Data

Compound	Color Transmitted	Color Absorbed	λ (nm)	ε ($M^{-1}cm^{-1}$)	λ (nm)	ε ($M^{-1}cm^{-1}$)	Δ_o (cm^{-1})
$[Co(en)_3]Cl_3 \cdot H_2O$							
$[Co(en)_2(ox)]Cl \cdot H_2O$							
$Na[Co(ox)_2(en)] \cdot H_2O$							
$K_3[Co(ox)_3] \cdot 3.5\ H_2O$							

Now determine Δ_o for your complex. If a particular absorption corresponds directly to 10 Dq, the energy of the absorption maximum is Δ_o. This, however, is not the case for low-spin cobalt(III). A fairly accurate value for Δ_o is obtained by determining the ratio of the transitions (v_2/v_1) and matching this ratio to the ratio of energy separations on the Tanabe–Sugano diagram. The actual energies for the found transitions are: $E(v_1) = 10\ Dq - C$; $E(v_2) = 10\ Dq + 16B - C$. Using $C = 4B\ Dq$ and B can be calculated; however, this is dangerous due to the possibility of inaccurate assumptions.

b. Determine the value for Dq/B at this ratio:

$Dq/B =$ _____ $\Delta_o/B =$ _____

c. Record the value of E/B (actually B' since B is the electronic repulsion term for the free ion) for the lowest energy transition, v_1:

$E/B' =$ _____

d. Solve for B' using $v_1/B' = E/B'$

$B' =$ _____

e. B for the free Co^{3+} ion is 1100 cm^{-1}. How does your value compare to this? Explain.

f. Calculate Δ_o using your Δ_o/B value above:

$\Delta_o =$ _____ $10\Delta_o =$ _____

7. Fill in Table 5.1

Literature values:

Post-lab Questions

Q5.6   Consider the structures of the cobalt(III) complexes prepared and the relative order of magnitude of the molar absorptivities, ε, for the absorptions in the visible region. Discuss.

Q5.7   Consider the structures of the cobalt(III) complexes prepared and the relative order of magnitude of $10Dq(\Delta_o)$. Discuss.

Q5.8   Compare your ε values with the literature values for your complex. The ratio of ε for $\lambda_2 : \lambda_1$ is indicative of the purity of your complex. Why? How do your values compare to the literature?

Q 5.9   If your ε ratio is different from the literature, what could be the cause?

Q 5.10   On what *specific* instrument did you use to obtain your electronic spectra?

EXPERIMENT 5.3: INVESTIGATION OF THE ELECTROCHEMISTY OF TRIS(BIDENTATE CHELATE)COBALT(III) COMPLEXES USING CYCLIC VOLTAMMETRY[10]

Level 4

Pre-lab 5.3.a: Thermodynamic Stability Considerations

In this experiment, the electrochemistry of both $[Co(en)_3]^{3+/2+}$ and $[Co(ox)_3]^{3+/2+}$ will be investigated using cyclic voltammetry, and the standard reduction potential (E^0, V) for the $[Co(en)_3]^{3+/2+}$ couple will be measured. For metal complex stability reasons discussed below, it is not possible to use this technique to obtain reduction potentials for the mixed ligand cobalt systems; an exercise at the end of this experiment helps to estimate these. The E^0 values obtained will be important for experiment 5.6, in which outer-sphere electron transfer rate constants between $[Co(en)_3]^{2+}$ and $[Co(en)_2(ox)]^+$ will be mathematically modeled using Marcus theory.

The equilibrium expression and accompanying Nernst equation for the $Co(en)_3^{3+/2+}$ couple in this experiment are given in equations (5.4) and (5.5):

$$[Co(en)_3]^{3+} + e^- \rightleftharpoons [Co(en)_3]^{2+} E^0 = ? \tag{5.4}$$

$$E = E^0 - RT/nF \ \ln[[Co(en)_3]^{3+}]/[[Co(en)_3]^{2+}] \tag{5.5}$$

Remember that, although cobalt(III) complexes are very kinetically stable and inert to substitution, the reduced cobalt(II), d^7, complexes are less stable and extremely labile in aqueous solution. Because of this, we must use stability constant data to determine the excess concentration of ligand that will be necessary to maintain the binding of all three bidentate ligands (as opposed to two bidentate ligands and two H_2O ligands for example) to the metal center. We consider the $Co(en)_3^{2+}$ system below.

The formation constants for $[Co(en)_3]^{2+}$ ($\mu = 0.1$ M) are given in equations (5.6)–(5.8) and the acid dissociation constants for en are given in equations (5.9) and (5.10).

$$Co^{2+} + en \rightleftharpoons Co(en)^{2+} \ \ \log K_1 = 5.6 \tag{5.6}$$

$$Co(en)^{2+} + en \rightleftharpoons Co(en)_2^{2+} \ \ \log K_2 = 4.9 \tag{5.7}$$

$$Co(en)_2^{2+} + en \rightleftharpoons Co(en)_3^{2+} \ \ \log K_3 = 3.3 \tag{5.8}$$

$$H_2en \rightleftharpoons H^+ + Hen \ \ pK_{a1} = 7.08 \tag{5.9}$$

$$Hen \rightleftharpoons H^+ + en \ \ pK_{a2} = 9.89 \tag{5.10}$$

If we assume that a ratio of *tris:bis* complex of 100:1 is good enough for our experimental purposes, we can calculate the excess en needed to maintain this ratio. Since:

$$[Co(en)_2]^{2+} + en \rightleftharpoons [Co(en)_3]^{2+} K_3 = 2000 \ \text{M}^{-1} \tag{5.10}$$

then for

$$[Co(en)_3]^{2+}/[Co(en)_2]^{2+} \ \text{to be } 100/1$$

solving for [en] (unbound or free in solution): [en] = 0.05 M.

This is the *excess* [en] that must be in solution over and above the en bound to the metal center. To be on the safe side, and also to leave room for buffering capacity, 0.1 M en will be used in the $[Co(en)_3]^{3+/2+}$ system. Taking advantage of the first protonation of en, the solution will be buffered by the addition of a small amount of HNO_3 (0.01 M final concentration). In Procedure 5.3.b you will solve for the excess oxalate^{2-} needed for thermodynamic stability in the $Co(ox)_3{}^{3-/4-}$ system.

Pre-lab 5.3.b: Buffer and pH Considerations

The equilibria in equations (5.6)–(5.10) show the importance of pH in the $[Co(en)_3]^{3+/2+}$ system. The pH can be calculated using the Henderson–Hasselbach equation (5.11), where A^- refers to en and HA to Hen$^+$. The actual pH of the experiment must be measured following each run.

$$pH = pK_a + \log[A^-]/[HA] \tag{5.11}$$

Pre-lab 5.3.c: Ionic Strength Considerations

Pre-lab 4.4 discusses the importance of having an electrolyte in your electrochemical set-up. In addition, as with all thermodynamic measurements, the ionic strength must be constant in electrochemical analysis. The concentration of electrolyte (KNO_3) needed to maintain a constant ionic strength ($\mu = 0.1$ M) is calculated using the equation (5.12), where c is the concentration of the ith ion with charge z. Note that a more rigorous treatment of ion activity is used in Chapter 7 (experiment 7.4).

$$\mu = 1/2 \sum_i c_i z^2 i \tag{5.12}$$

Pre-lab 5.3.d: Oxygen Sensitivity in Cobalt(II) Systems

When the cobalt(III) complex has been reduced to cobalt(II), the metal center may become sensitive to oxidation by atmospheric O_2 and will compete with the oxidation of the complex at the electrode surface. For this reason, all samples must be rigorously purged with $Ar_{(g)}$ or $N_{2(g)}$ and solutions kept under a blanket of inert gas during the experiment.

Pre-lab Procedure

It is advisable to begin this experiment by reading Pre-lab 4.4 and carrying out Procedure 4.4.a to ensure that your system is working properly.

Procedure 5.3.a: Electrochemistry of $[Co(en)_3]^{3+/2+}$

All glassware and plasticware must be acid-washed with concentrated H_2SO_4 and 3 M HCl, respectively, to remove trace metal ions. Use purest H_2O available for solutions.

Safety warning! Operations set off between the lines must be performed in a fume hood.

Prelab Work
1. Calculate the amount of KNO_3 needed to achieve an ionic strength (μ) = 0.1 M for the $[Co(en)_3]^{3+/2+}$ system, given: 5×10^{-3} M $[Co(en)_3]^{3+}$, 0.015 M Cl$^-$, 0.01 M Hen$^+$, 0.01 M NO_3-.

TABLE 5.2 Preparation of [Co(en)$_3$]$^{3+}$ Solution and Background Electrolyte (25 ml Total Volume)

Complex	Reaction Concentration	MW	Stock Concentration	d. (g ml^{-1})	Amount Needed
[Co(en)$_3$]Cl$_3$ · H$_2$O	5 E–3 M	363.3	—	—	g
Ethylenediamine^a	0.1 M	60.02	—	0.899	g
HNO$_3$, 1 M^a (prepared from fuming ~14.3 M)	0.01 M	63.01	1 M	1.574 (14.3 M stock)	ml
KNO$_3$	0.06 M	101.11	—	—	g

^{a}Work with reagents in hood!

2. Calculate the amounts of reagents needed to prepare the following reaction solution in 25 ml total volume: 5×10^{-3} M [Co(en)$_3$]$^{3+}$; 0.01 M HNO$_3$; 0.06 M KNO$_3$; 0.1 M en. Record in Table 5.2.

Solution Preparation

1. Prepare your [Co(en)$_3$]$^{3+}$ solution using N$_{2(g)}$ (or Ar$_{(g)}$) purged nanopure or ddH$_2$O, according to Table 5.2.

Cyclic Voltammetry

Working electrode: platinum (should be lightly polished with alumina and rinsed with water before *each* experimental run).

Reference electrode: silver–silver chloride (Ag–AgCl).

Auxilary electrode: platinum wire.

All measurements will be carried out under an inert atmosphere at 25 °C. Be sure to purge your solution with your inert gas for ~5 min prior to scanning. Polish your Pt working electrode, and perform an IR compensation prior to each run.

1. Set your instrument to the following general parameters:

Initial E: -200 mV
High E: -200 mV
Low E: -650 mV
Initial direction: negative
Number of segments: 2
Sensitivity: 10 μA V^{-1}
Scan rate: 2 mV s^{-1}
Temperature: 25 °C

2. Perform a Run Scan, Plot, and Save Data.
3. Repeat the above for scan rates 5, 10, and 100 mV s^{-1} (save files)—plot as a multi graph.
4. Repeat your best run on high sensitivity for your background electrolyte solution.

Optional Experiments

Obtain cyclic voltammograms using other suitable working electrodes.

Obtain cyclic voltammograms at varying temperatures.

Procedure 5.3.b: Electrochemistry of $[Co(ox)_3]^{3-/4-}$

All glassware and plasticware must be acid-washed with concentrated H_2SO_4 and 3 M HCl, respectively, to remove trace metal ions. Use purest H_2O available for solutions.

Pre-lab Work

1. Given the following equilibria ($\mu = 1$ M), calculate the excess oxalate ion (ox^{2-}) needed to thermodynamically maintain a ratio of $Co(ox)_3{}^{4-}$: aq. $Co^{2+} = 1000:1$

$$Co^{2+} + 3\,ox^{2-} \rightleftharpoons Co(ox)_3{}^{4-} \quad \log \beta_3 \approx 8$$

$$H_2ox \rightleftharpoons H^+ + Hox^- \quad pK_{a1} = 1.04$$

$$Hox^- \rightleftharpoons H^+ + ox^{2-} \quad pK_{a2} = 3.55\,(3.82\,@\,\mu = 0.1\,\text{M})$$

2. Calculate the amount of KCl needed to achieve an ionic strength (μ) = 1.0 M for the $[Co(ox)_3]^{3-/4-}$ system given: 5×10^{-3} M $Co(ox)_3{}^{4-}$, 0.1 M $K_2C_2O_4$, 0.1 M KHC_2O_4. What would be the concentrations of $K_2C_2O_4$ and KHC_2O_4 if the pH were adjusted (with KOH) to 7.0?

3. Calculate the amounts of reagents needed to prepare the following reaction solution in 25 ml total volume: 5×10^{-3} M $Co(ox)_3{}^{4-}$, 0.1 M $K_2C_2O_4$, 0.1 M KHC_2O_4, $\mu = 1$ M (KCl). Record your results in Table 5.3.

Solution Preparation

1. Prepare your $[Co(ox)_3]^{3-}$ solution using $N_{2(g)}$ (or $Ar_{(g)}$) purged nanopure or ddH_2O according to Table 5.3.

Cyclic Voltammetry See Procedure 5.3.a for electrode specifications.

1. Set your instrument to the following general parameters:

Initial E: 750 mV

High E: 750 mV

Low E: 100 mV

Initial direction: negative

Number of segments: 2

TABLE 5.3 Preparation of $[Co(ox)_3]^{3-}$ Solution (25 ml Total Volume)

Complex	Reaction Concentration	MW	Amount Needed
$K_3[Co(ox)_3] \cdot 3.5H_2O$	5×10^{-3} M		g
$K_2C_2O_4$ (ox^{2-})	0.1 M		g
KHC_2O_4 (Hox^-)	0.1 M		g
KCl	M^a		g

^{a}Record concentration from Pre-lab work question 3.

Sensitivity: 100 $\mu A\ V^{-1}$
Scan rate: 20 mV s^{-1}
Temperature: 25 °C

2. Perform a Run Scan, Plot, and Save Data.
3. Repeat the above for scan rates 5, 10, and 100 mV s^{-1} (save files)—plot as a multi graph.
4. Repeat your best run on high sensitivity for your background electrolyte solution.

Name _____   Section _____

Results Summary for the Electrochemistry of the $[Co(en)_3]^{3+/2+}$ and $[Co(ox)_3]^{2-/4-}$ Systems

1. Use your data from run 1 to calculate your ΔE_p for the $[Co(en)_3]^{3+/2+}$ couple. Is it 60 mV? Explain.
2. Use your data from run 1 to calculate $E^{0'}$ vs Ag–AgCl and NHE reference electrodes for the $[Co(en)_3]^{3+/2+}$ couple. Do your values agree with textbook references? If not, what could be the problem?
3. For the $[Co(en)_3]^{3+/2+}$ couple plot the scan rate dependence of the peak currents and peak potentials:

   (a) i_p vs (scan rate)$^{1/2}$ (b) ΔE_p vs scan rate

   Describe your plots. Explain why they look the way they do. If they do not agree with theory, offer possible explanations for this.
4. Compare your E^0 value for $[Co(en)_3]^{3+/2+}$ with that reported in the literature $(-0.21$ V vs NHE.[10] Discuss sources of error.
5. Would you consider the $[Co(en)_3]^{3+/2+}$ electrochemistry to be reversible, quasi-reversible, or irreversible. Explain. Give reasons for the behavior observed.
6. Would you consider the $[Co(ox)_3]^{3-/4-}$ electrochemistry to be reversible, quasi-reversible, or irreversible. Explain. Give reasons for the behavior observed.
7. The literature value for the $[Co(ox)_3]^{3-/4-}$ couple is $+0.57$ V vs NHE (*J. Chem. Soc. A* **1967**, 298; *Dalton Trans.* **1985**, 1665). It is clear that the ligands play a crucial role in governing the redox potential of the cobalt(III) complexes. Using CFT, suggest a plausible explanation for the differences in the $[Co(en)_3]^{3+/2+}$ and $[Co(ox)_3]^{3-/4-}$ values.
8. The E^0 for $[Co(en)_2(ox)]^+$ and $[Co(ox)_2(en)]^-$ cannot be measured by cyclic voltammetry. Why not? Use your values for $[Co(ox)_3]^{3-/4-}$ and $[Co(en)_3]^{3+/2+}$ to predict redox potentials for the mixed ligand complexes.

EXPERIMENT 5.4: ELECTRON TRANSFER REACTION BETWEEN $[CO(en)_3]^{2+}$ AND $[Co(ox)_2(en)]^-$—STOICHIOMETRY DETERMINATION USING ION-EXCHANGE CHROMATOGRAPHY[11]

Level 5

The inert atmosphere technique and the time required outside the laboratory for column chromatography increase the difficulty of this experiment.

Pre-lab 5.4.a: Air-sensitive Technique for the $[Co(en)_3]^{2+}$–$[Co(ox)_2(en)]^-$ System

In this exercise, the stoichiometry of the electron transfer reaction between $[Co(en)_3]^{2+}$ and $[Co(ox)_2(en)]^-$ will be investigated. Product isolation will be by ion-exchange chromatography and identification and quantification by visible spectrophotometry. The $[Co(en)_3]^{2+}$ reactant is highly sensitive to atmospheric oxidation and must be handled at all times under inert conditions. The following is an outline for effectively handling this system as well as Experiment 5.5. Use Figure 5.2 for guidance.

Safety warning! Operations set off between the lines must be performed in a fume hood.

Stock Solution Preparation   All glassware and plasticware must be acid-washed with concentrated H_2SO_4 and 3 M HCl, respectively to remove trace metals. All reagents and H_2O should be of highest quality. Do not use metal spatulas to weigh out reagents. Use luer lock fittings and tubings for all purposes. *Outflow* stocks are prepared at twice the concentration of the reaction (or *inflow*) concentration. Use the concentrations in Table 5.4 for this experiment. It is advisable to practice steps 1–5 using only ddH$_2$O to get the feel for inert atmosphere solution transfer and maintenance.

1. Prepare *outflow* $[Co(ox)_2(en)]^-$ stock in a volumetric flask equipped with a rubber septum and one 18–22 gage needle. Label Stock 1. Push a teflon syringe needle (18 gage, 24 in. long with Kel-F luer hub) through the septum and degas for ~5–10 min. (The tip of the needle should be cut at a slant for easy entry. Further, it is advisable to first puncture the septum with a larger 13 gage needle and to use gloves for a sure grip.) After purging the sample, raise the tubing above the solution, allowing a slow stream of inert gas to blanket the solution [Fig. 5.2(a)].

2. Prepare a solution of cobalt(II) salt and electrolyte needed for maintaining ionic strength *four* times the concentration needed for the reaction in an Erlenmeyer flask containing a stir bar and equipped with a rubber septum and needles as above. Label Stock 2A. Purge the solution with inert gas for ~5–10 min as in step 1. After this time, raise the tubing above the solution, allowing a slow stream of inert gas to blanket the solution [Fig. 5.2(a)].

---

3. Prepare a solution of ethylenediamine and acid for buffering (if appropriate) that is four times the concentration needed for the reaction in an Erlenmeyer flask equipped with a rubber septum and needles as above. Label Stock 2B.

Purge the solution with inert gas for ~5–10 min. as in step 1. After this time, raise the tubing above the solution allowing a slow stream of inert gas to blanket the solution [Fig. 5.3(a)].

4. Using a 10 ml gas-tight Hamilton© syringe, transfer an equal volume of Stock 2B into Stock 2A. (You should purge your syringe with inert gas three times prior to drawing up liquid.) Stir the solution while maintaining an inert atmosphere by blanketing the solution with inert gas throughout this process. This is now the *outflow* $[Co(en)_3]^{2+}$ Stock 3 [Fig. 5.3(b)]. [The $[Co(en)_3]^{2+}$ solution is a very pale yellow

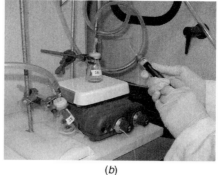

(a)   (b)

Figure 5.3   (a) Set-up for inert gas purging of electron transfer solutions. (b) Transfer of contents from Stock 2B into Stock 2A. Note that inert gas inline is placed onto the steel needles while the Teflon needles are used for transfer. Also, an extra needle is placed in each septum to allow for gas outflow. Once the transfer has been made, Flask 2A becomes Stock 3.

color. If you obtain a deep yellow color (that continues to get darker with time), your mixture is contaminated with O_2.]

Stoichiometry Experiment

5. Using inert technique (gas tight syringe), transfer an equal volume of *outflow* $[Co(ox)_2(en)]^-$ Stock 1 into Stock 3, while maintaining a blanket of inert gas over the reaction mixture. This is time zero for the *inflow* reaction.

Procedure 5.4: Determination of the Electron Transfer Reaction Stoichiometry—$[Co(en)_3)]^{2+}$ and $[Co(ox)_2(en)]^-$, $\mu = 0.1$ M (KNO_3), 25 °C

Note: the stoichiometry reaction needs to be carried out twice for the separation and isolation of both products A and B.

Pre-lab Preparation

1. Prepare 100 ml of 1 M $Co(NO_3)_2 \cdot 6H_2O$ (MM = 291.03 g mol^{-1}). Dilute to 0.2 M and determine the actual concentration by visible spectroscopy ($\varepsilon_{512} = 4.84$ M^{-1} cm^{-1}).
2. Prepare 100 ml of 1 M KNO_3 (MM = 101.11 g mol^{-1}).
3. Prepare 50 ml of 2 M ethylendiamine (60.10 g mol^{-1}, $d = 0.899$ g ml^{-1}, 14.96 M).
4. Prepare 100 ml of 0.01 M HNO_3 (14 M concentrated).
5. Prepare 50 ml 6 M HCl (chill on ice).
6. Complete calculations for Table 5.4.

In-flow Reactant Concentrations

$[Co(ox)_2(en)]^-$ 1×10^{-3} M

$[Co(en)_3]^{2+}$ 0.01 M

$[en]_{xs}$ 0.22 M

KNO_3 0.1 M

Out-flow Reactant Preparation (to be Done during the Lab Period)

1. Follow instructions 1–4 for inert technique given in Pre-lab 5.4 using the concentrations of reactants in Table 5.4. Stocks 1 and 3 are the reactant *out-flow* solutions.

2. After completing step 5 of the inert technique instructions, let the reaction stir under a blanket of inert gas for 10 min.

3. Cool the reaction mixture in ice and add 4 ml ice cold 6 M HCl such that the reaction mixture has a pH < 1.

Product A—Preparation of the SP-C-25 Sephadex Column

1. Add a small amount of SP-C-25 Sephadex resin to ddH_2O to create a slurry of resin.

2. Push a small amount of glass wool into the base of a pasteur pipet and fill the pipet with ddH_2O. As the ddH_2O flows out, pipet in the slurry and allow it to pack evenly, making a 1×10 cm column.

3. Connect the column to a 500 ml separatory funnel using a piece of tygon tubing (see Fig. 5.4).

4. Add $\sim$50 ml ddH_2O to the separatory funnel and allow this to flow through the column (pinch and release the tygon tubing to fill the connecting stem). When the ddH_2O has finished flushing the column, turn the column off until you are ready to load your diluted reaction mixture (do not allow the column to run dry). The column may be stored at 4 °C for longer storage periods.

Product A—Separation and Isolation

5. Dilute the reaction mixture to 450 ml ddH_2O in the 500 ml separatory funnel connected to the prepared 1×10 cm Pasteur pipet column containing SP-C-25 Sephadex cation exchange resin. Allow the mixture to drip through the column ($\sim$8 h). Some column contraction will occur. For stopping periods, the column may be shut off, the bottom sealed with parafilm and stored at 4 °C. Some band diffusion may occur.

6. When the mixture has finished loading, a tight dark yellow band remains at the top of the resin (red color will pass through). Rinse the column with 75 ml ddH_2O, and elute the yellow band with 0.5 M HCl.

7. Evaporate to dryness using a rotovap.

TABLE 5.4   Calculations for Out-Flow Reactant Concentrations

Stock no.	Reagent	Concentration (M)	Total Volume (ml)	Amount Needed	Actual Amount
1	$Na[Co(ox)_2(en)] \cdot H_2O$	2×10^{-3}	10	g	g
2a	$Co(NO_3)_2 \cdot 6H_2O$	0.04	2.5	ml of original stock	ml
	KNO_3	0.4	—	ml of original stock	ml
2b	Ethylenediamine (en)	1 M	2.5	ml of original stock	ml
3	$[Co(en)_3]^{2+}$–en–KNO_3	0.02–0.5–0.2 M	5	—	—

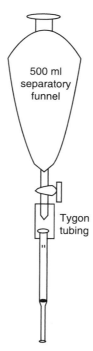

Figure 5.4 Set-up of the 500 ml separatory funnel with tygon tubing connecting the SP-C-25 Sephadex Pasteur pipet column.

8. Redissolve the yellow product in a small amount ddH_2O and quantitatively transfer to a 10 ml volumetric flask. Fill to the mark with ddH_2O.

9. Obtain a UV–visible spectrum scanning from 300 to 600 nm and determine the product identity and concentration. Calculate the percentage product formed.

Product B—Preparation of the CM-C-25 Sephadex Resin Column

1. Add CM-C-25 Sephadex resin to ddH_2O to create a slurry of resin.

2. Add ddH_2O to a 2 × 30 cm (or longer) column. As the ddH_2O flows out, pour in the slurry and allow it to pack evenly making a 2 × 20 cm column. Do not allow the column to run dry. Turn the column off until you are ready to load your diluted reaction mixture. The column may be shut off, and stored at 4 °C.

Product B—Separation and Isolation

3. Dilute the reaction mixture to 1200 ml ddH_2O in a 2000 ml separatory funnel. Allow the mixture to pass through the column (~6–8 h). Some column contraction will occur. For stopping periods, the column may be shut off, and stored at 4 °C. Some band diffusion will occur.

4. When the mixture has finished loading, wash the column with ~150 ml ddH_2O. You should notice three bands beginning to pass through the column (see Fig. 5.5). A very faint pink band 1 travels the furthest upon loading and washing, and a

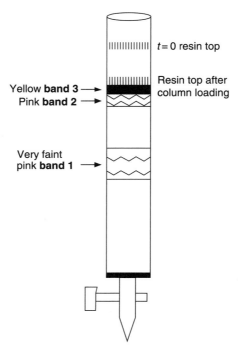

Yellow **band 3** ➝
Pink **band 2** ➝

Very faint
pink **band 1** ➝

$t = 0$ resin top

Resin top after
column loading

Figure 5.5   The CM-C-25 Sephadex resin before and after loading reaction mixture.

second pink band 2 lies below a dark yellow band 3. You are interested in collecting
band 1.

5. Separate the bands further with 500 ml 0.01 M HCl and elute and collect band 1 with
   0.05 M HCl. (*Note*: band 1 is diffuse and pale; careful observation will ensure
   quantitative collection!)

6. Evaporate your eluate to dryness using a rotovap.

7. Redissolve the pink product in a small amount ddH$_2$O and quantitatively transfer to
   a 5 ml volumetric flask. Fill to the mark with ddH$_2$O.

8. Obtain a UV–visible spectrum scanning from 300 to 600 nm and determine the
   product identity and concentration. Calculate the percentage product formed.

Name_____   Section_____

Results Summary for the Stoichiometry Determination of the Electron
Transfer Reaction between $[Co(en)_3]^{2+}$ and $[Co(ox)_2(en)]^-$

1. Identify product A by comparing the UV–visible spectrum to those obtained in
   experiment 5.1   _____

2. Calculate the percentage product A formed in the reaction based on your limiting
   reagent.   _____

3. Identify product B by comparing the UV–visible spectrum to those obtained in
   experiment 5.1   _____

4. Calculate the percentage product B formed in the reaction based on your limiting reagent. _____

5. Write a balanced reaction for the electron transfer reaction between $[Co(en)_3]^{2+}$ and $[Co(ox)_2(en)]^-$.

6. Describe the electron transfer processes that led to products A and B and comment on the percentages of product obtained in the reaction.

7. Describe the loading/washing and elution of your metal complex from the column. If other species were present, try to identify these.

Post-lab Questions

Q5.11 Describe cation and anion exchange chromatography. What ion exchange chromatography are you doing in this lab?

Q5.12 Look up and draw the functional groups of the SP-C-25 Sephadex and CM-C-25 Sephadex resins.

Q5.13 Predict the product(s) of the reaction between $[Cr(H_2O)_6]^{2+}$ and $[Co(NH_3)_5O_2CCH_3]^{2+}$.

EXPERIMENT 5.5: CONVENTIONAL KINETIC MEASUREMENT OF THE REACTION BETWEEN $[Co(en)_3]^{2+}$ AND $[Co(ox)_2(en)]^-$— DETERMINATION OF k_{so} AND k_{os}[11]

Level 5

The inert atmosphere technique requires dedicated patience.

Pre-lab 5.5: Derivation of the Rate Law for the Electron Transfer Reaction Between $[Co(en)_3]^{2+}$ and $[Co(ox)_2en]^-$

In general, many kinetics data are accumulated prior to proposing a reaction mechanism. In our case, we will simply use the stoichiometry information obtained in Experiment 5.4 along with intuition based on past work in the field. The following is an interactive pre-lab exercise for proposing the rate law for the electron transfer between $[Co(en)_3]^{2+}$ and $[Co(ox)_2(en)]^-$. The kinetics will then be investigated using conventional visible spectroscopy. Experimental data, in combination with the rate law, will be used to determine the outer-sphere electron rate constant.

For a simple one-step, bimolecular electron transfer reaction involving $Co^{II}L$ and $Co^{III}L$ complexes the rate expression is equation (5.13) where k_{so} is the second-order rate constant.

$$\text{rate} = k_{so}[Co^{III}L][Co^{II}L] \tag{5.13}$$

Experimentally, to simplify data analysis we carry out such reactions under pseudo-first-order conditions. Under pseudo-first-order conditions with excess $[Co^{II}L]$, the rate law becomes:

$$\text{rate} = k_{obs}[Co^{III}L] \text{ where } k_{obs} = k_{so}[Co^{II}L] \tag{5.14}$$

Most metal complex reactions are more complex however, and the k_{so} does not reflect a simple bimolecular collision step. Our reaction between $[Co(en)_3]^{2+}$ and $[Co(ox)_2en]^-$ is an example of such a complex reaction. Recall from the stoichiometric measurements in Experiment 5.4 that this reaction yields two products: $[Co(en)_3]^{3+}$ (80%) and $[Co(en)_2ox]^+$ (20%), indicating that the reaction proceeds by at least two different pathways. Further, the reaction involves a rapid pre-equilibrium step during which the two reactants are brought together. This complexity is embedded in the second-order rate constant. As an exercise, follow steps 1–7 below to derive the second-order rate constant.

1. Write the overall reaction corresponding to the stoichiometry data.
2. Express k_{obs} in terms of both the cobalt(II) inner- and outer-sphere reactants.
3. Propose a three-step reaction mechanism (involving K_3 for $[Co(en)_3]^{2+}$ and the outer- and inner-sphere rate constants, k_{os} and k_{is}, respectively) that is consistent with the stoichiometry information.

The rate expression must be in terms of *total* cobalt(II) and cobalt(III) concentrations and must also contain both the inner-sphere (k_{is}) and outer-sphere (k_{os}) rate constants.

4. Define the *total* cobalt(II) concentration, Co_T^{II}.
5. Using the equilibrium constant, K_3, solve for $[Co(en)_2]^{2+}$ and $[Co(en)_3]^{2+}$ in terms of Co_T^{II}.
6. Substitute your expressions for $[Co(en)_2]^{2+}$ and $[Co(en)_3]^{2+}$ into the rate expression for k_{obs}.
7. To simplify, obtain a common denominator by substituting into the first term: $1 = K_3[en]_f/K_3[en]_f$.

The second-order rate constant k_{so} should now be:

$$k_{so} = (k_{os}K_3[en]_f + k_{is})/(1 + K_3[en]_f) \tag{5.15}$$

and the reaction rate

$$\text{rate} = \{(k_{os}K_3[en]_f + k_{is})/(1 + K_3[en]_f)\}\, [Co_T^{II}][Co(ox)_2en]^- \tag{5.16}$$

Because our en concentration will be in large excess, we can assume $[en]_f = [en]_T$. You will be asked to calculate $[en]_f$ in Table 5.5 to show that this assumption holds.

How do we obtain k_{is} and k_{os} from such a complicated rate expression? As indicated in equations (5.13) and (5.14), we can determine k_{so} experimentally by carrying out the reaction between $[Co(ox)_2(en)]^-$ and $[Co(en)_3]^{2+}$ under pseudo-first-order conditions with varying excess $[Co(en)_3]^{2+}$. A plot of k_{obs} vs $[Co(en)_3]^{2+}$ will give the second-order rate constant (k_{so}) for the reaction. Using equation (5.15), and given the value for the inner-sphere rate constant, k_{is}, which has been measured independently, the outer-sphere rate constant, k_{os}, will be determined. In Experiment 5.6, Marcus theory will be used to model this reaction and the calculated vs the observed outer-sphere rate constants will be compared.

TABLE 5.5   Values for the Final (In-Flow) Reactant Concentrations. Fill in [KNO₃] and [en]ₑ Values Before Starting Lab and pHₑ, and k_{obs} values after analysis

$[Co(ox)_2(en)]^-$: 5×10^{-4} M
Ionic strength $= 0.25$ M (KNO_3)

	Co^{2+} (M)	$[en]_T$ (M)^a	HNO₃ $[en]_f$ (M)	KNO₃ (M)	pH$_f$ (M)	*I* Calculated	*I* Actual (act)	k_{obs} (s^{-1})
Rxn 1	1×10^{-2}	0.18	0.002					
Rxn 2	2.5×10^{-2}	0.225	0.002					
Rxn 3	5×10^{-2}	0.3	0.002					

^{a}Amount of en needed to form the Tris complex plus 0.15 M excess.

Procedure 5.5: Inert Atmosphere Kinetics Using Conventional Visible Spectroscopy Methods

In this procedure you will determine the second-order (k_{so}) and outer-sphere (k_{os}) electron transfer rate constants for the reaction between $[Co(en)_3]^{2+}$ and $[Co(ox)_2(en)]^-$, $\mu = 0.25$ M (KNO_3), 25 °C

Pre-lab Calculations

1. Given $K_3 = 2000$ for the reaction: $[Co(en)_2]^{2+} + en = [Co(en)_3]^{2+}$ calculate the amount of $[en]_{xs}$ needed to maintain a 100:1 ratio of $[Co(en)_3]^{2+}:[Co(en)_2]^{2+}$ for this experiment.

2. Calculate the $[en]_f$ that will be in solution during the reactions and record values in Table 5.5. Do these meet the requirements above for formation of the Tris complex?

3. Given that en will be used to buffer your system, use the Henderson–Hasselbach equation calculate the expected final pH for the reaction and place your value in Table 5.5.

4. Calculate the molar amount of KNO_3 needed to maintain an ionic strength $= 0.25$ M for each reaction and record this in Table 5.5.

Pre-lab Solution Preparation

1. Prepare 100 ml of 1 M $Co(NO_3)_2 \cdot 6H_2O$ (MM $= 291.03$ g mol^{-1}). Dilute to 0.2 M and determine the actual concentration by visible spectroscopy ($\varepsilon_{512} = 4.84$ M^{-1} cm^{-1})

2. Prepare 100 ml of 1 M KNO_3 (MM $= 101.11$ g mol^{-1}).

3. Prepare 50 ml of 2 M ethylendiamine (60.10 g mol^{-1}, $d = 0.899$ g ml^{-1}, 14.96 M).

4. Prepare 100 ml of 0.1 M HNO_3 (14 M concentrated).

In-lab Preparation   The *final or in-flow concentrations* (i.e., after equal mixing of the two reactant solutions) for the oxidant, reductant, and HNO_3 (for maintaining buffer capacity) are given in Table 5.5. Out-flow solutions must be prepared with twice the desired concentration since equal volumes of metal complex stock solutions will be rapidly mixed together in the reaction chamber.

TABLE 5.6   Calculations for Out-Flow Reactant Concentrations

Stock no.	Reagent	Concentration (M)	Total Volume (ml)	Amount Needed	Actual Amount
1	Na[Co(ox)$_2$(en)] · H$_2$O (2x in-flow concentration)	1×10^{-3}	10	g	g
2a	Co(NO$_3$)$_2$ · 6H$_2$O (4× in-flow concentration)	Rxn 1—0.04	5	ml^a	ml
		Rxn 2—0.1		ml^a	ml
		Rxn 3—0.2		ml^a	ml
	KNO$_3$ (4× in-flow concentration)	Rxn 1—		ml^a	ml
		Rxn 2—		ml^a	ml
		Rxn 3—		ml^a	ml
2b	Ethylenediamine (en) (4× in-flow concentration)	Rxn 1—0.72	5	ml^a	ml
		Rxn 2—0.90		ml^a	ml
		Rxn 3—1.2		ml^a	ml
	HNO$_3$ (4× in-flow concentration)	Rxn 1—0.008		ml^a	ml
		Rxn 2—0.008		ml^a	ml
		Rxn 3—0.008		ml^a	ml
3	[Co(en)$_3$]$^{2+}$/en/KNO$_3$		10	—	—
		Rxn 1:	0.01/0.118/0.002		
		Rxn 2:	0.025/0.225/0.002		
		Rxn 3:	0.05/0.3/0.002		

^{a}ml of original stock.

Out-flow Reactant Preparation   To be done by students during the lab period: Before beginning, review steps 1–4 of Pre-lab 5.4 (inert technique).

1–4. Using inert, metal-free technique and following steps 1–4 of Pre-lab 5.4, prepare solutions for Rxn 1 according to Table 5.6. Note that Stocks 1 and 3 are your reactant *out-flow* solutions.

5. *Electron transfer kinetics*: for kinetics experiments, equal amounts of Stock 1 and Stock 3 will need to be delivered into a cuvette equipped with a rubber septum and *two* needles (22 gage) for blanketing the reaction with an inert gas. The stocks can be delivered with two Hamilton gas-tight syringes, or alternatively, a rapid mixing system such as that shown in Figure 5.4 can be constructed.

6. To a 3 ml 1 cm cuvette that is continually purged with inert gas (N$_2$ or Ar), add 1.25 ml of each Stock 1 and Stock 3 using Hamilton gas-tight syringes. Start the kinetics run. Maintain a blanket of inert gas over the cuvette throughout the run. *Note*: this is best accomplished by keeping two needles (for gas in-flow and out-flow) in the septum capping the cuvette. Due to the small reaction volume, the stream of inert gas must be *carefully* regulated. If the gas stream is too vigorous, significant evaporation will occur and en will condense on the outside of the cell. If the gas stream is too tepid, oxygen will leak into the system. Both will result in an unwanted increase in absorption during the kinetics run. Patience! This technique requires some practice but it is worth the effort!

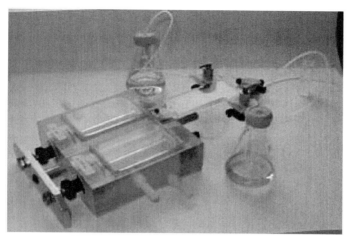

Figure 5.6 This home-made rapid mixer houses two 10 ml gas-tight Hamilton syringes that are encased in a block of plexiglass equipped for a temperature-controlled water bath. Three-way taps allow for solution uptake into the syringes and then equal volume delivery into a small mixing chamber before delivery into the cuvette.

7. Monitor the reaction at 541 nm where the oxidant has an $\varepsilon = 112 \, \mathrm{M}^{-1} \, \mathrm{cm}^{-1}$. The approximate time for the Rxn 1 run is 5000 s.

8. Repeat for reactions 2 and 3. Approximate run times are 1600 s (Rxn 2) and 1200 s (rxn 3).

9. Plot $-\ln[\mathrm{Abs}_{541}(t) - \mathrm{Abs}_{541}(\inf)]$. For each reaction, determine the first-order rate constant, k_{obs}, by plotting $\ln[\mathrm{Abs}_{541}(t) - \mathrm{Abs}_{541}(\inf)]$ vs t. The slope is equal to k_{obs}.

10. Determine k_{so} by plotting k_{obs} vs $[\mathrm{Co(en)_3}]^{2+}$.

11. Calculate k_{os}, given k_{is} (see calculation 3 in the Results Summary section).

Name _____   Section _____

Results Summary for the Electron Transfer Reaction between $[\mathrm{Co(en)_3}]^{2+}$ and $[\mathrm{Co(ox)_2(en)}]^{-}$

1. vs time (s) for each reaction. The slope equals k_{obs}.
   k_{obs} Reaction 1: _____   Reaction 2: _____   Reaction 3: _____

2. Plot k_{obs} vs [Co(II)]. The slope equals k_{so}. k_{so} = _____

3. Use equation (5.15) to calculate the outer-sphere electron transfer rate constant, k_{os} (where $k_{\mathrm{is}} = 9 \, \mathrm{M}^{-1} \, \mathrm{s}^{-1}$)
   k_{os} = _____

Post-lab Questions

Q5.14 Compare your k_{os} value to that found in the literature and comment.

Q5.15 Compare the magnitudes of the inner-sphere and outer-sphere rate constants. Comment.

Q5.16 Propose a structure for: (a) the outer-sphere precursor complex, and (b) the inner-sphere precursor complex.

Q5.17 Propose a method for elucidating the structure of the outer-sphere precursor complex.

Q5.18 It is possible to determine k_{is} experimentally. Rearrange equation (5.15) to obtain an expression in $y = mx + b$ form that enables you to determine k_{is}. Explain how you would execute this experiment.

EXPERIMENT 5.6: MARCUS THEORY: THEORETICAL CALCULATION OF THE OUTER-SPHERE RATE CONSTANT, k_{os}, FOR THE REACTION BETWEEN [Co(ox)$_2$(EN)]$^-$ AND [Co(en)$_3$]$^{2+11}$

Level 5—Advanced Theory

Pre-lab 5.6: Brief Introduction to Marcus Theory[12]

In this experiment, the outer-sphere electron transfer rate constant for the reaction between $[Co(en)_3]^{2+}$ and $[Co(ox)_2(en)]^-$ will be calculated using Marcus theory. The calculated value will then be compared with the rate constant determined in Experiment 5.5.

Because outer-sphere reactions are mechanistically simple (no bond making and breaking occurs), they are amenable to theoretical treatment. Marcus theory, which allows the prediction of outer-sphere electron transfer rates between two species, was developed in the 1960s by Rudolph Marcus, who won the Nobel prize for his work in 1992. His simple equation is given below:

$$k_{AB} = (k_{AA}k_{BB}K_{AB}f_{AB})^{1/2} \tag{5.17}$$

where K_{AB} is the rate constant for the electron transfer reaction (called the cross reaction); k_{AA} and k_{BB} are the self-exchange rate constants for reactants A and B as defined below:

$$A^{red} + {}^*A^{ox} \rightarrow A^{ox} + {}^*A^{red} \quad k_{AA} \tag{5.18}$$

$$B^{red} + {}^*B^{ox} \rightarrow B^{ox} + {}^*B^{red} \quad k_{BB} \tag{5.19}$$

K_{AB} is the equilibrium constant for the overall reaction, and $f_{AB} = (\log K_{AB})^2/4 \log(k_{AA}k_{BB}/Z^2)$. Z^2, a collision term, and for uncharged particles, is 10^{11} M^{-1} s^{-1}. Modifications must be made for charged species.

You will notice that the Marcus equation brings together kinetics and thermodynamics. For the system you will be studying, the reaction takes place in what is called the "normal" region where an increase in K_{AB} leads to an increase in k_{AB}. Some reactions, however, take place in the "Marcus inverted region" where an increase in thermodynamic potential of the reaction leads to a decrease in rate.

Marcus theory has been so successful in predicting cross reaction rates in outer-sphere electron transfer reactions that it is often used to give evidence that an outer-sphere reaction is taking place. If the observed rate constant is within an order of magnitude of the calculated rate constant, then it is likely that an outer-sphere mechanism is occurring. In this way, you will use Marcus theory to give evidence for the reaction mechanism assigned to your rate constant, k_{os}, determined in Experiment 5.5.

Procedure 5.6: Equations for the Marcus Theory Calculation

As noted in the previous section, for $[Co(ox)_2(en)]^-$ (species A) and $[Co(en)_3]^{2+}$ (species B) $\mu = 0.1$ M, $25\,°C$, the Marcus expression is:

$$k_{AB} = (k_{AA}\, k_{BB}\, K_{AB}\, f_{AB})^{1/2}$$

$$\log f = (\log K_{AB})^2 / 4 \log(k_{AA}\, k_{BB}/Z^2) \tag{5.20}$$

The collision frequency, Z, for uncharged molecules is taken as 1×10^{11}. However, to account for the two species being charged, we must include a work term, $\omega(r)$, for bringing the two reactants together. To do this, we go back a step in the Marcus derivation, solving for $\Delta G_{AB}^{\ddagger}$ and then converting to k_{AB} using transition state theory: $k = RT/Nh\, e^{(-\Delta G^{\ddagger}/RT)}$.

The expressions you will need for the calculation of $\Delta G_{AB}^{\ddagger}$ are:

$$\Delta G_{AB}^* = 1/2(\Delta G_{AA}^* + \Delta G_{BB}^*)$$

$$+ 1/2(\Delta G_{AB}^0 + \omega_{AB} + \omega_{BA})(1 + \alpha) \tag{5.21}$$

Since

$$\Delta G_{AB}^* = \Delta G_{AB}^{\ddagger} - \omega_{AB}; \; \Delta G_{AA}^* = \Delta G_{AA}^{\ddagger} - \omega_{AA}; \; \Delta G_{BB}^* = \Delta G_{BB}^{\ddagger} - \omega_{BB}$$

then by substitution,

$$\Delta G_{AB}^{\ddagger} - \omega_{AB} = 1/2[(\Delta G_{AA}^{\ddagger} - \omega_{AA}) + (\Delta G_{BB}^{\ddagger} - \omega_{BB})]$$

$$+ 1/2(\Delta G_{AB}^0 + \omega_{AB} + \omega_{BA})(1 + \alpha) \tag{5.22}$$

where

$$\alpha = (\Delta G_{AB}^0 + \omega_{AB} + \omega_{BA})/4[(\Delta G_{AA}^{\ddagger} - \omega_{AA})$$

$$+ (\Delta G_{BB}^{\ddagger} - \omega_{BB})] \tag{5.23}$$

138 ELECTRON TRANSFER REACTIONS OF TRIS(BIDENTATE CHELATE)COBALT(II/III)

TABLE 5.7 Data and Results for the Marcus Theory Calculation

Species A: [Co(ox)₂(en)]⁻ Species B: [Co(en)₃]²⁺	$[Co(ox)_2(en)]^{-/2-}$	$[Co(en)_3]^{3+/2+}$	$[Co(ox)_2(en)]^-/[Co(en)_3]^{2+}$
E^0 (V)			
ΔG^0 (kcal)	—	—	
a (Å)	5.0^a	4.0^a	
k_{AA} (M^{-1} s^{-1})	1×10^{-9} ^{b}		
$\Delta G_{AA}^{\ddagger}$			
k_{BB} (M^{-1} s^{-1})		7.7×10^{-5}	
$\Delta G_{BB}^{\ddagger}$			
ω_{AA}			
ω_{BB}			
ω_{AB}			
ω_{BA}			
$\Delta G_{AB}^{\ddagger}$			
k_{AB} (M^{-1} s^{-1})			

^{a}Approximate values.
^{b}Calculated from Marcus theory.

All of the terms in the above equation are readily solved using the values given in Table 5.7. From transition state theory (above),

$$\Delta G_{ii}^{\ddagger} = 0.5921(25.33 - \ln k_{ii}) \tag{5.24}$$

or

$$k_{ii} = 1 \times 10^{11}\, e^{-\Delta G_{ii}^{\ddagger}/0.5921} \tag{5.25}$$

The work expression, $\omega(r)$, is given as:

$$\omega(r) = z_A z_B e^2 / D_s r(1 + \beta r) \tag{5.26}$$

where z_i is the charge on the ith species, e is the electronic charge, D_s is the static dielectric constant for the medium, r is the distance between the metal centers (often taken as the sum of the ion radii, a (*approximations* given in Table 5.7), and $\beta = (8\pi Ne^2/D_s kT)^{1/2}$. Substituting in the appropriate values for the constants, equation (5.26) becomes:

$$\omega(r) = 4.24\, z_A z_B / r(1 + 0.104r) \tag{5.27}$$

In Table 5.7, ω_{AB} refers to the thermodynamically favorable reaction (i.e., $[Co(ox)_2(en)]^- - [Co(en)_3]^{2+}$) and ω_{BA} refers to the reverse, $[Co(en)_3]^{3+} - [Co(ox)_2(en)]^{2-}$ couple.

Finally, you need to remember that:

$$\Delta G_{AB}{}^0 = -nF\Delta E^0 \text{ in kJ} \tag{5.28}$$

Name _____   Section _____

Results Summary for the Marcus Theory Calculation of the Electron Transfer
Rate Constant for the reaction between $[Co(en)_3)]^{2+}$ and $[Co(ox)_2(en)]^-$

1. Fill in Table 5.7 using equations given in Procedure 5.6. Show your work.
2. Compare and comment on your calculated value for k_{os} and the value obtained experimentally in Experiment 5.5.
3. Compare and comment on the self-exchange rates for the $[Co(ox)_2(en)]^{-/2-}$ and $[Co(en)_3]^{3+/2+}$ couples.
4. Note the values for your work terms (ω_{AB} and ω_{BA}). What is the significance of these?

REFERENCES

1. http://www.geocities.com/bioelectrochemistry/marcus.htm
2. http://www.nobel.se/chemistry/educational/poster/1992/marcus.html
3. Taube, H. *Electron Transfer between Metal Complexes—Retrospective*. The Nobel Foundation; http://nobelprice.org/ (last accessed November 2006).
4. Walton, H. F. *Principles and Methods of Chemical Analysis*. Prentice-Hall: New York, 1952, pp. 13–18.
5. Work, J. B. *Inorg. Synth.* **1946**, 2, 221–222.
6. Jordan, W. T., Froebe, L. R. *Inorg. Synth.* **1978**, 18, 96–99.
7. Dwyer, F. P., Reid, I. K., Garvan, F. L. *J. Am. Chem. Soc.* **1961**, 83, 1285–1287.
8. Bailar, J. C., Jones, E. M. *Inorg. Synth.* **1939**, 1, 37.
9. Huheey, J. E., Keiter, E. A., Keiter, R. L. *Inorganic Chemistry—Principles of Structure and Reactivity*, 4th edn. Harper-Collins: New York, 1993.
10. Marusak, R. A., Osvath, P., Kemper, M., Lappin, A. G. *Inorg. Chem.* **1989**, 28, 1542.
11. Warren, R. M. L., Tatehata, A., Lappin, A. G. *Inorg. Chem.* **1993**, 32, 1191–1196.
12. Lappin, A. G. *Redox Mechanisms in Inorganic Chemistry*. Ellis Horwood: New York, 1994.

Metals in Medicine: Synthesis and Biological Reactivity of the Platinum Anticancer Drug, *cis*-Platin and its Isomer, *trans*-Platin

I have no words of wisdom . . . I simply do my damndest (Barnett Rosenberg, 2005)

PROJECT OVERVIEW[1–5]

cis-Diamminedichloroplatinum(II), *cis*-Pt(NH$_3$)$_2$Cl$_2$, was first synthesized in 1845 by Peyrone and its structure, along with that of its *trans* isomer, was elucidated by Werner in 1893. In the mid-1960s Rosenberg, a biophysicist at Michigan State University, serendipitously discovered the remarkable anticancer activity of the *cis* analog. Today, *cis*-platin is one of our most widely used chemotherapeutic agents, successfully treating ovarian, bladder, head and neck, testicular, and cervical cancers.

Since its discovery, much information has been obtained on *cis*-platin's biological reactivity. Its primary target is DNA. Cytotoxicity was originally thought to be caused by the blocking of DNA replication and cell division, but recent evidence suggests that apoptosis resulting from *cis*-platin-damaged DNA is the mechanism by which the drug causes cell death. Cellular *cis*-platin resistance as well as toxic side effects have resulted in an intense search for more effective platinum analogs. The most successful second-generation drug has been carboplatin [Pt(NH$_3$)$_2$(cbdca) cbdca = cyclobutane1,1-dicarboxylate], which reduces toxic side effects. Other promising analogs include orally active platinum(IV) complexes, multinuclear, and water-soluble platinum compounds, and complexes with biologically active carrier ligands. Further, although *trans*-platin is biologically ineffective, new *trans*-platinum analogs are finding use in the field.

In this set of experiments you will synthesize both *cis*- and *trans*-platin complexes and study their substitution kinetics in DMSO. The primary focus of the chapter is on *cis*-platin biological reactivity; you will investigate the binding of *cis*-platin to DNA adducts by NMR and molecular mechanics methods and assess drug cytotoxicity using a chinese hamster ovary (CHO) cell line.

Integrated Approach to Coordination Chemistry: An Inorganic Laboratory Guide. By Rosemary A. Marusak, Kate Doan, and Scott D. Cummings

EXPERIMENT 6.1: SYNTHESIS OF *CIS*- AND *TRANS*-DIAMMINE-DICHLOROPLATINUM(II) COMPLEXES, [Pt(NH₃)₂Cl₂]

Level 4

Pre-lab 6.1: The *trans* Effect[6]

In this exercise, you will synthesize both *cis* and *trans* forms of diamminedichloro-platinum(II) [Pt(NH₃)Cl₂], **I** and **II**, respectively (Fig. 6.1). These preparations represent a textbook example of the *trans effect* in inorganic chemistry. The trans effect refers to the labilizing effect one ligand (L₁) has on another ligand, (L₂) that is situated 180° from L₁. The *trans*-influencing series is shown below:

$$CN^-, CO, NO, C_2H_4 > PR_3, H^- > CH_3^-, C_6H_5^-, SC(NH_2)_2, SR_2 > SO_3H^- > NO_2^-,$$
$$I^-, SCN^- > Br^- > Cl^- > py > RNH_2, NH_3 > OH^- > H_2O$$

If you are given two starting materials, $PtCl_4{}^{2-}$ and $Pt(NH_3)_4{}^{2+}$, you can predict which will give you the *cis* and which the *trans* isomer of [Pt(NH₃)Cl₂] upon substitution with the appropriate ligand. Two factors contribute to the *trans* effect. The first is a thermodynamic effect, Figure 6.2(*a*). Strong σ donation from *trans*-influencing ligand L₁ into the metal d_{x2-y2} orbital lengthens bond L₂. The effect is a lowering of the activation energy E_a because of a ground state destabilization. The second is a kinetic effect, Figure 6.2(*b*). Ligands that are good π acceptors are able to stabilize a 5 coordinate transition state. The *trans* effect leads to substitution with retention of stereochemistry.

Procedural Note $[PtCl_4]^{2-}$ is available commercially. The synthesis may be started at Procedures 6.1.b and c.[7,8]

Procedure 6.1.a: Synthesis of Potassium Tetrachloroplatinate(II)

$$2K_2[PtCl_6] + N_2H_4 \cdot 2HCl \rightarrow 2K_2[PtCl_4] + N_2 + 6HCl$$

1. Add 0.972 g (0.002 mol) K₂[PtCl₆] to a 25 ml round-bottom flask equipped with a stir bar, heating. Set up in hood.
2. Add to this flask, 10 ml ddH₂O.
3. With stirring, add 0.990 g of N₂H₄ · 2 HCl (hydrazine dihydrochloride) *slowly* in ~0.01 g increments.

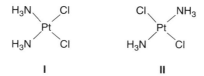

Figure 6.1   Structures of *cis*-(**I**) and *trans*-(**II**) platin complexes.

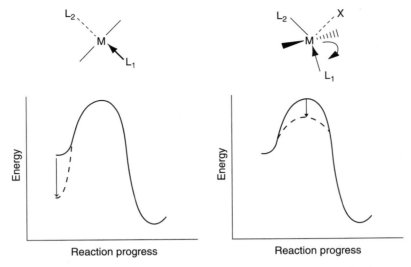

Figure 6.2   (*a*) Thermodynamic and (*b*) kinetic factors of the *trans* effect. L₁ Ø is the *trans* influencing ligand. L₂ is the leaving group and X is the incoming group.

4. After the addition, bring the temperature to 50–65 °C over 5 mins. Maintain this temperature for approximately 2 h until only a small amount of yellow starting material remains. The solution should be deep red.

5. Raise the temperature to 80–90 °C to complete the reaction.

6. Cool to room temperature and then in an ice bath. Filter the solution to remove unreacted K$_2$[PtCl$_6$]. Wash the removed K$_2$[PtCl$_6$] with 1 ml portions of ice-cold water until the filtrate is colorless.

7. The reaction may be purged with N$_{2(g)}$ and stored at 4 °C or rotovapped and stored at 4 °C.

Proceed to either Procedure 6.1.b (*trans*-platin) or 6.1.c (*cis*-platin)—the reaction solution may be divided into half and each half used in 6.1.b and c.

Procedure 6.1.b: Synthesis of *trans*-dichlorodiammineplatinum(II)

$$K_2[PtCl_4] + 4NH_3 \rightarrow [Pt(NH_3)_4]Cl_2 + 2KCl$$

1. Pour the resulting solution (containing ~0.930 g K$_2$[PtCl$_4$]) from 6.1.a into a 100 ml beaker. Heat to a boil and add 2 ml concentrated aqueous NH$_3$.

2. Carefully evaporate the solution using a hot water bath to approximately 4 ml.

$$[Pt(NH_3)_4]Cl_2 + 2HCl \xrightarrow{\Delta} trans\text{-}[Pt(NH_3)_2Cl_2] + 2NH_4Cl$$

3. Add 80 ml 6 M HCl.

4. Evaporate the solution as in step 2 to a final volume of 6–8 ml.

5. Cool the mixture in an ice bath for 15 min and vacuum filter the product. Wash with several 2 ml portions of ice water and air dry.
6. The product may be recrystallized from a *minimum* amount of boiling 0.1 M HCl and filter hot. The product crystallizes upon cooling.
7. Record yield.

Procedure 6.1.c: Synthesis of Diamminediiodoplatinum(II)

$$K_2[PtCl_4] + 4KI \rightarrow K_2[PtI_4] + 4KCl$$

1. Evaporate the solution from procedure 6.1.a to dryness using a rotoevaporator ($\sim$0.930 g $K_2[PtCl_4]$).
2. Add 1.5 ml ddH$_2$O to the 25 ml round-bottom flask containing the solid $K_2[PtCl_4]$.
3. Transfer the contents to a 100 ml beaker equipped with a stir bar and heat the solution to 40 °C.
4. Dissolve 2.232 g KI in 3.5 ml warm ddH$_2$O. Add this solution to the 100 ml beaker containing the platinum complex. The mixture should turn from red to dark brown.
5. Heat the solution to 70 °C (do not overheat) with stirring. After reaching 70 °C, cool to room temperature.

$$K_2[PtI_4] + 2NH_3 \rightarrow cis\text{-}[Pt(NH_3)_2I_2] + 2KI$$

6. Vacuum filter the solution and add 3.7 ml of $\sim$2 M aqueous NH$_3$ solution. Let the reaction stand at room temperature for 20 min.
7. Vacuum filter the yellow product and wash with a small amount ($\sim$5 ml) of ice-cold ethanol and then ether ($\sim$10 ml). Air dry.
8. Record yield.

Procedure 6.1.d: Synthesis of *cis*-Diamminedichloroplatinum(II)

$$cis\text{-}[Pt(NH_3)_2I_2] + Ag_2SO_{4(aq)} \rightarrow cis\text{-}[Pt(NH_3)_2(H_2O)_2] + 2AgI$$

1. Add 0.1575 g Ag$_2$SO$_4$ to 25 ml ddH$_2$O in a 100 ml beaker.
2. Add slowly, 0.25 g *cis*-[Pt(NH$_3$)$_2$I$_2$] to the solution.
3. Heat the mixture to 70–80 °C for 10–12 min and filter off the AgI.

$$cis\text{-}[Pt(NH_3)_2(H_2O)_2] + 2KCl \rightarrow cis\text{-}[Pt(NH_3)_2Cl_2] + 2K^+$$

4. Reduce the volume of solution with a rotovap to $\sim$5 ml.
5. Add a large excess of KCl ($\sim$0.825 g) and heat to 70–80 °C for 8–10 min.
6. Cool the mixture to 0 °C using an ice bath and vacuum filter the yellow product crystals.
7. Wash the product with 1.25 ml of cold ethanol and then ether ($\sim$2.5 ml). Air dry.
8. Record yield.

Name _____   Section _____

Results Summary for the Synthesis of *cis*- or *trans*-Diamminedichloroplatinum(II)

1. Platinum complex isomer synthesized: _____

2. Show how the *trans* effect worked in your synthesis.

3. Limiting reagent for the $K_2[PtCl_4]$ synthesis: _____

4. Theoretical yield: _____
   Actual yield: _____
   Percentage yield: _____

5. Limiting reagent for the $[Pt(NH_3)_2Cl_2]$ synthesis: _____

6. Theoretical yield: _____
   Actual yield: _____
   Percentage yield: _____

7. Which part of your synthesis was most expensive (in terms of percentage yield)?

EXPERIMENT 6.2: SUBSTITUTION KINETICS OF *TRANS*-PLATIN: DETERMINATION OF THE RATE OF FORMATION, k_{DMSO}, AND ACTIVATION PARAMETERS FOR THE REACTION BETWEEN *CIS*- AND *TRANS*-[Pt(NH₃)₂Cl₂] AND DMSO SOLVENT

Level 4

Pre-lab 6.2: Substitution in Square Planar Metal Complexes[9,10]

Square planar complexes are coordinatively unsaturated, and generally undergo A-type substitution involving a 5 coordinate intermediate/transition state. For mainly steric reasons, a trigonal bipyramidal structure is expected to be favored over the square pyramidal structure, although the two structures differ little energetically. The mechanism for an $[ML_3Y]$ complex is shown in Figure 6.3. Square planar complexes generally show second-order kinetics regardless of the incoming ligand, X^-. However, plots of k_{obs} vs $[X^-]$ reveal something interesting. The expected plots of the data for a reaction involving $[ML_3Y]$ and varying $[X^-]$ are shown in Figure 6.4. The y-intercept indicates that another pathway, independent of $[X^-]$, exists. The rate law describing these data is given in equation (6.1), where k_a is the rate constant for the $[X^-]$ independent pathway and k_b for the pathway dependent on $[X^-]$. The $[X^-]$-independent pathway is of interest to us in this experiment. Atypical for A-type mechanisms, the kinetics for this pathway are

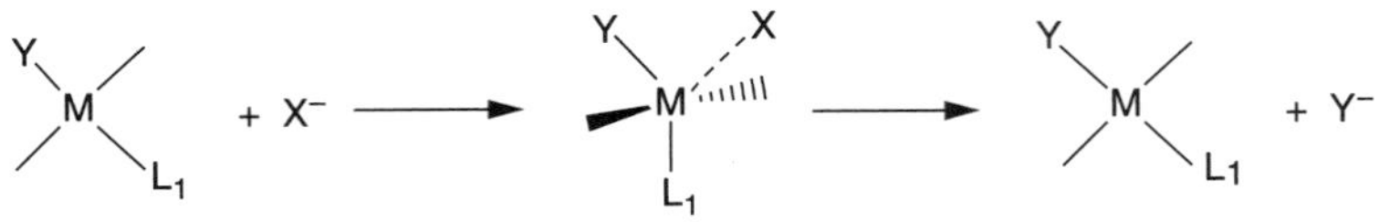

Figure 6.3 General mechanism of substitution in square planar complexes highlighting the 5 coordinate intermediate.

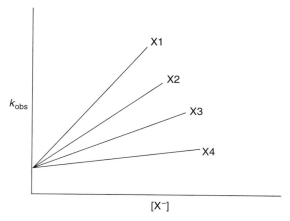

Figure 6.4 Plot showing k_{obs} as a function of $[X^-]$ in a substitution reaction involving square planar complexes.

first order,

$$rate = k_a[ML_3Y] + k_b[ML_3Y][X^-]$$ (6.1)

so

$$k_{obs} = k_a + k_b[X^-]$$

There is no rate dependence on the overall charge of the metal complex, and reactions show a dependence on the nature of the leaving group (behavior is seen with I_a mechanisms). However, like A-type mechanisms, bulkier L ligands slow down the rate of substitution. Lastly, the rate shows a dependence on the nature of the solvent. The above evidence indicates that the k_a mechanism is solvent intervention (see Pre-lab 4.1.a).

Solvent intervention plays a major role in the biochemistry of platinum-based anticancer agents. The binding of *cis*-platin to DNA, for example, involves a sequential substitution of the Cl^- ligands by solvent H_2O followed by binding to the N7 of a guanine base, N7G. The rate determining step is Cl^- substitution by solvent H_2O. For *cis*-platin, $k_{H2O} = 2.5 \times 10^{-5}$ s^{-1} ($t_{1/2} = 7.7$ h) and for *trans*-platin, $k_{H2O} = 9.8 \times 10^{-5}$ s^{-1} ($t_{1/2} = 1.96$ h).

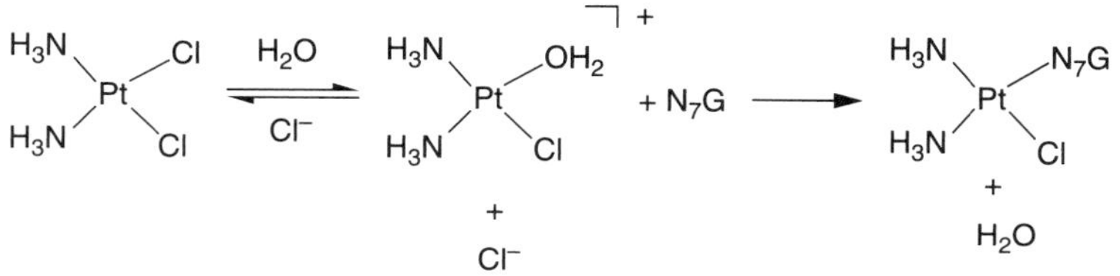

Figure 6.5 Substitution of the *cis*-platin Cl^- by H_2O followed by binding to the N7 of a guanine base. The second substitution at an adjacent guanine site proceeds similarly. Adapted from Kozelka *et al.*[10]

In this experiment, we will study the kinetics of the first Cl^- substitution in the *cis*- and *trans*-platin complexes by DMSO solvent. Temperature dependence studies will enable the calculation of activation parameters (see Experiment 4.7). DMSO has been chosen for two reasons. First, the substitution kinetics are faster than in aqueous media and are more amenable for study in the teaching laboratory. Second, many drugs like *cis*- and *trans*-platin that have low aqueous solubility are dissolved in DMSO for cytotoxicity assays. Lippard's group first pointed out that, for the platinum(II) complexes, substitution is relatively rapid in DMSO and requires caution in interpreting cytotoxicity results. We too will use DMSO as a carrier solvent in Experiment 6.5. The results obtained in this kinetics experiment will alert you to the limitations of your cytotoxicity assay.

Procedure 6.2: DMSO Substitution Kinetics of *cis*- and *trans*-Platin: Temperature Dependence Study[11]

Wash all glassware with concentrated H_2SO_4 and plasticware with 3 M HCl.

1. Your reaction will be monitored at 302 nm for *trans*-platin and 276 nm for *cis*-platin.

2. Reactions should be carried out at the temperatures and times indicated in Table 6.1.

3. To initiate each reaction, completely dissolve 7.5–9 mg of platinum complex in 3 ml DMSO. Pipet the solution into a cuvette and begin the reaction run.

4. For each run, plot $\ln|A_\infty - A_t|$ vs t, where A_t is the measured absorbance at time, t. Obtain the pseudo-first-order rate constant, k_{obs} (s^{-1}) from the slope of the plot. The substitution rate constant, k_{DMSO} $(s^{-1}) = k_{obs}$ (s^{-1}). Ideally, several trials of each should be run and k_{obs} average reported. However, due to the time constraints of the reactions, only one run per temperature study is recommended.

5. For each complex, evaluate the activation parameters, $\Delta H^\ddagger$ and $\Delta S^\ddagger$.

   a. Plot $\ln(k_{DMSO}/T)$ vs $1/T$ according to the Eyring equation from transition state theory, equation (6.2). $\Delta H^\ddagger$ is obtained from the slope and $\Delta S^\ddagger$ from the intercept.

   $$\ln(k_{DMSO}/T) = \ln(k_B/h) + \Delta S^\ddagger/R - (\Delta H^\ddagger/R)(1/T) \qquad (6.2)$$

   b. Record your values for $\Delta H^\ddagger$ (kJ mol^{-1}) and $\Delta S^\ddagger$ (J mol^{-1}) and calculate from these, $\Delta G_{298}^\ddagger$.

TABLE 6.1 Parameters for Studying Platinum Complex Kinetics by UV–vis Spectroscopy

Complex	Temperature, °C	Reaction Run Time (s)
Trans-platin	30	7,200
	37	3,600
	45	2,250
Cis-platin	37	31,930
	45	13,390
	50	9,270

TABLE 6.2   Results for *trans*-Platin

	Reaction Temperature		
	30 °C	37 °C	45°C
k_{obs} (s^{-1}) $= k_{DMSO}$ Activation parameters	$\Delta H^{\ddagger}$	$\Delta S^{\ddagger}$	$\Delta G^{\ddagger}$

TABLE 6.3   Results for *cis*-Platin

	Reaction Temperature		
	37 °C	45 °C	50 °C
k_{obs} (s^{-1}) $= k_{DMSO}$ Activation parameters	$\Delta H^{\ddagger}$	$\Delta S^{\ddagger}$	$\Delta G^{\ddagger}$

Name _____   Section _____

Results Summary for the DMSO Solvent Substitution Kinetics of *cis*- and *trans*-Platin Complexes

1. Complete Tables 6.2 and 6.3 for each complex.

Post-lab Questions

Q6.1   Compare your k_{DMSO} values for *cis*-platin with those in H_2O solvent. Which solvent reacts faster and why?

Q6.2   Why is substitution in the *trans* complex much faster than in the *cis* complex?

EXPERIMENT 6.3: NMR INVESTIGATION OF *cis*-PLATIN BINDING TO 2′-DEOXYRIBONUCLEOSIDE-5′-MONOPHOSPHATES, dGMP AND dAMP

Level 4

Time is required outside of the lab period and the advanced theory also makes this experiment a little more challenging.

Pre-lab 6.3: DNA—the Target Molecule of *cis*-Platin[12,13]

cis-Platin binds to many biomolecules, but the target responsible for its cytotoxic action is DNA. *cis*-Platin binds to DNA mainly through the N7 atoms of purines and Ho *et al.* have grouped the binding types into six categories: 1,2-intrastrand d(GpG), two adjacent guanines; 1,2-intrastrand d(ApG), adjacent adenine and guanine bases; 1,x-intrastrand purine binding where $x > 2$; monodentate purine binding; interstrand crosslinking; and DNA–protein crosslinking. Representative adducts are shown in Figure 6.6. These *cis*-platin–DNA adducts lead to arrest at various cell cycle stages (G1, S, and G2 phases) with the intent to repair the DNA–metal complex lesion. The absence of adequate repair ultimately signals apoptosis, or regulated cell suicide.

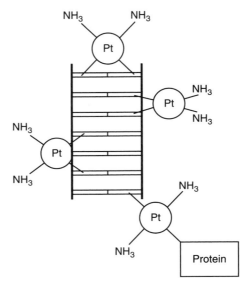

Figure 6.6 Main DNA-*cis*-platin adducts (a) interstrand cross-link; (b) 1,2-intrastrand cross-link; (c) 1,3-intrastrand cross-link; (d) protein–DNA cross-link. Adapted from Ho *et al.*[12]

Multinuclear NMR spectroscopy (including ^{1}H, ^{13}C, ^{31}P, ^{15}N, and ^{195}Pt) has been valuable in assigning binding modes and mechanisms to *cis*-platin–DNA (nucleotide) adduct formation. The platinum aqua complex $[Pt(NH_3)_2(H_2O)_2]$ rapidly binds one purine and over time forms bifunctional adducts. When Pt^{2+} binds to the N7 of G, the anomeric effect of guanine is strengthened and electron delocalization directs a sugar–phosphate conformational change that contributes to the bending of the DNA backbone. Platination shifts the N/S equilibrium of the ribose ring and influences the *syn/anti* base conformation (**I–III**). In duplex oligonucleotides, the N and S conformation has been monitored by measuring the ribose proton coupling constants, $J_{H1'-H2'}$ and $J_{H1'-H2''}$ (note that, in **II–III**, H2′ is not distinguished from H2″). The sum of these, $\Sigma^3 J$, which is measured as the separation between the outermost peaks of the 1H′ pseudotriplet, is 7.5 Hz for the N conformation and 13.6 Hz for the S conformation. Information on *anti* vs *syn* conformations is obtained from nuclear Overhauser effect (nOe) data. An nOe is observed between H8 and H1′ protons when the base is in the *anti* conformation.

In this experiment, we will investigate *cis*-platin binding to 2′-deoxyribonucleoside,5′-monophosphates by ^{1}H NMR spectroscopy. The structures of deoxyguanosine (dGMP, **II**) and deoxyadenosine (dAMP, **III**) monophosphate are shown in Figure 6.8. Note that G is arbitrarily shown in the *anti* configuration and A is in the *syn* configuration. Upon Pt(II) binding, you will look for its affect on the nonexchangeable protons (H8 of G and H2 and H8 of A) as well as on the sugar moiety protons. Appendix 1 provides an introduction to the pulsed one- and two-dimensional NMR technique.

Procedure 6.3: Solution Preparation and NMR Spectroscopy[14]

Preparation of cis-[Pt(ND₃)₂(D₂O)₂] Prepare no more than one day prior to experiment.

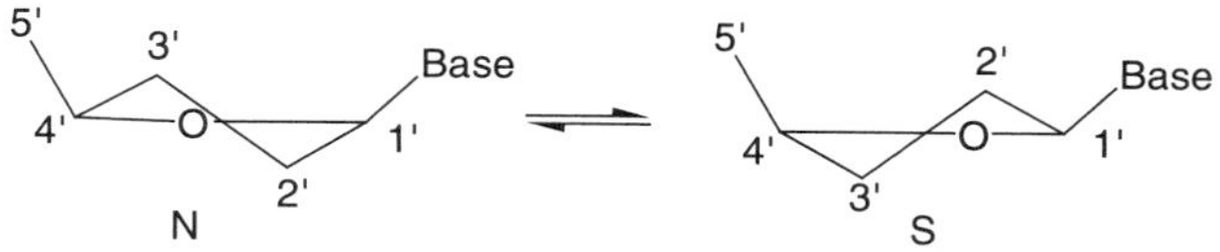

Figure 6.7   The N–S equilibrium exhibited by the ribose ring.

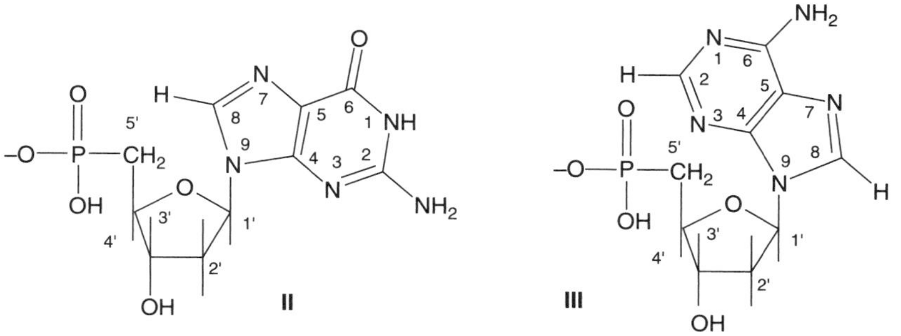

Figure 6.8   The *anti-* and *syn*-base conformations shown for dGMP(**II**) and dAMP(**III**), respectively.

1. Prepare a 20 mM solution of *cis*-[Pt(ND$_3$)$_2$(D$_2$O)$_2$] by dissolving 0.027 g *cis*-[Pt(NH$_3$)$_2$Cl$_2$] (MM = 300.00 g mol^{-1}) and 0.0312 g (2 equiv.) AgNO$_3$ (MM = 169.9 g mol^{-1}) in 4.5 ml D$_2$O in a 25 ml round-bottom flask equipped with a stir bar.
2. Put under a blanket of N$_2$ and stir vigorously overnight.
3. Vacuum filter the solution to remove AgCl$_{(s)}$ and store in a closed vial at 4 °C until needed. Alternatively, the *cis*-[Pt(ND$_3$)$_2$(D$_2$O)$_2$] solution can be rotovapped to dryness and the solid redissolved in 4.5 ml D$_2$O to ensure maximum H–D exchange.

Preparation of cis-[Pt(ND₃)₂(dNMP)₂]

1. Weigh out 0.011 g dAMP sodium salt (MM = 354.2 g mol^{-1}) and 0.011 g dGMP sodium salt (MM = 370.2 g mol^{-1}) and place each into a 1.5 ml conical vial. Add 0.75 ml of your *cis*-[Pt(ND$_3$)$_2$(D$_2$O)$_2$] solution to each vial. The dNMP ligands dissolve at ∼pH* 7. If needed, the pH* can be brought to 7 with 4–8% NaOD.
2. Wrap the vials in foil and vortex the mixtures at 25 °C for at least 20 h.

Preparation of dNMP Solutions

1. Weigh out 0.011 g dGMP sodium salt and dissolve in 0.75 ml D$_2$O.
2. Weigh out 0.011 g dAMP sodium salt and place in 0.75 ml D$_2$O.
3. Bring pH* to ∼7 with NaOD if needed.

Assignment of ¹H NMR Spectra

1. Pipet your solutions into individual NMR tubes containing a small amount of reference (e.g., DSS).

TABLE 6.4   **^{1}H NMR Assignments for dNMPs and *cis*-[Pt(ND$_3$)$_2$(dNMP)$_2$] and Frequency Shift (Δv) from Free to Pt-Bound Forms**

Compound	Structural Assignment	Chemical Shift Assignment	Frequency Shift, Δv, from Free to Pt-Bound dNMP
DGMP			—
cis-[Pt(ND$_3$)$_2$(dGMP)$_2$]			
Damp			—
cis-[Pt(ND$_3$)$_2$(dAMP)$_2$]			

2. Obtain one-dimensional ^{1}H spectrum and a two-dimensional COSY spectrum from 0 to 10 ppm. Assign and analyze your peaks and the shifts brought about by platinum complex binding.

An optional experiment is to repeat this with the *trans*-[Pt(NH$_3$)$_2$Cl$_2$] complex.

Name _____   Section _____

Result Summary for the ^{1}H NMR analysis of dGMP and dAMP and their *cis*-platin Adducts

1. List the chemical shift assignments for the free deoxynucleoside monophosphates (dNMP) and their Pt(II) bound adducts in Table 6.4.
2. Comment on your observed shifts.
3. Measure the $\sum^3 J$ coupling constants at the anomeric 1'H of both dGMP and dAMP free and bound to Pt(II). Comment.

EXPERIMENT 6.4: MOLECULAR MODELING OF THE *Cis*-PLATIN-OLIGONUCLEOTIDE, CCGGATC

Level 4

Pre-lab 6.4.a: Introduction to Molecular Mechanics Calculations[15−19]

In Chapters 3 and 4 we learned about *ab initio* and semi-empirical calculations. In this chapter, we make use of empirical calculations, which are particularly valuable when the molecular system of interest is too large to treat practically with other approaches and quantum effects are expected to play a very minor role. Molecular mechanics (MM) calculations, also known as empirical force-field calculations, are commonly employed in such cases.

It is necessary to specify atomic connectivity at the beginning of a molecular mechanics calculation. The goal of the calculation is to find the structure and corresponding energy for various conformations of the molecule, generally with the intended purpose of estimating a lowest energy conformational structure, i.e., an optimized geometry subject to the preset atomic connectivity. Forces between atoms in the input structures are defined by potential energy functions, equation (6.3). The resulting energy reflects the sterics of the system.

$$E_{\text{tot}} = \sum_{\text{molecule}} (E_{\text{bond}} + E_{\text{angles}} + E_{\text{torsion}} + E_{\text{VDW}} + E_{\text{electrostatics}} + \cdots) \qquad (6.3)$$

The *force field* for a calculation is the chosen potential energy functions and their parameterization. A mechanics calculation, then, relies on empirical "ideal" or expected bond lengths and angles with strain arising from deviations in these. As expected, a strained structure is less energetically favorable. Dipolar and van der Waals interactions between nonbonded atoms can lead to attractive and/or repulsive forces that contribute to the energy of a given structure and inform conformational optimization to an energetic minimum. Information on the fundamental principles of MM and force field parameterization in inorganic and bioinorganic systems can be found in Comba and Remenyi.[17]

Good results from MM calculations, then, rely on a suitable force. Force fields for *cis*-platin binding to DNA have been aggressively developed over the past 15 years. Parameter refinements include the deviation of the Pt–N7 bond from the plane of the coordinated purine, planarity of the platinum moiety, the van der Waal radius, and charges of the Pt atom. The combination of structural information from techniques such as two-dimensional NMR spectroscopy and X-ray crystallography with MM calculations are now providing valuable insight into the role of *cis*-platin.

In this exercise, we will use a very basic approach to studying the binding of *cis*-platin to the oligonucleotide, CCGGATC. The lab is written for the molecular modeling software 02 Spartan Essential PC (Wavefunction Inc.), and the force field used for calculating equilibrium geometries and strain energies of the DNA fragment, the Pt complex, and their interaction will be the Merck Molecular Force Field (MMFF94, Merck Pharmaceuticals). Other force fields, such as AMBER and MM2, may yield better results. If these are available to you, a comparison with the MMFF study would be fruitful. Further, although for this lab no significant modifications to the force field will be made, the parameters can be modified by the user, and new parameters may be added. After completing your calculations and reviewing your structures, you are encouraged to review the article by Hambley and Jones[18] and make suggestions for parameter opitmization.

Procedure 6.4: Molecular Mechanics Calculations on the *Cis*-platin–Oligonucleotide Interactions[20]

1. Double click on the Spartan shortcut icon on the desktop. This should open the Spartan program and you should have a blank, workspace window in front of you.

2. Select **New** under **File** on the top toolbar. Your building tools should appear on the right hand of the screen.

3. Click on the **Nuc.** tab of the building tools. A new panel displaying nucleotides should come to the front of the building tools.

4. In this lab we wish to build a DNA double helix. Make sure **DNA** and **α** are selected.

5. Click the **Sequence** square. A check should appear in this box.

6. You can simply create your DNA strand by clicking on the **A, G, T,** and **C** boxes. Create a CCGGATC strand. As you click on the nucleotides, Spartan will automatically display the sequence of your strand in the white field underneath **Sequence**.

7. Once you have finished encoding your strand, click anywhere on the workspace window. You should now have a DNA strand on the workspace window, complete with complementary Watson–Crick base pairs.

8. The DNA strand is probably too large to view effectively. To reduce the molecule to a more manageable size, (a) hold down the shift key, (b) hold down the right button on the mouse, and (c) drag the mouse down (i.e., towards you).

9. Experiment with moving the molecule around. The left mouse button allows you to rotate the molecule, allowing you to view it from different angles. The right mouse button allows you to pick up the entire molecule and move it around the screen. Practice moving and rotating your DNA strand.

10. You can view the molecule using many different models by going to **Model** under the top toolbar. Your choices are **Wire, Ball and Wire, Tube, Ball and Spoke**, and **Space Filling**. Experiment with these options, viewing your molecule in several different modes. When finished, select the **Wire** model, as this model works best for the lab.

11. To get the lowest energy geometry for your DNA molecule, you need to run a geometry optimization calculation. Select **Calculations** . . . under **Setup** in the top tool bar.

12. When the calculation menu appears, you want to specify **Equilibrium Geometry** with **Molecular Mechanics** using **MMFF**. Have the computer start from initial geometry, neutral total charge and singlet multiplicity. Click on **Submit** (not **OK**). If you have not already saved the file, the program will have you do so now.

13. Depending on the speed of your computer, this calculation may take a while. To watch the progress of the calculation, select **Monitor** under **Options** on the far right side of the tool bar. If you double click on the job you submitted, you can see the calculation output. Notice how with each cycle of the calculations the energy of the structure decreases.

14. When the calculation is finally done you can view your new, improved structure on the workspace window. This may look less like the familiar double helix structure than your initial DNA molecule, but it is actually the more stable structure in a no-solvent environment. Under **Display** on the toolbar, select **Output**. Record the final energy of the last cycle.

15. Next attach platinum to your geometry-optimized structure. Studies have shown that the platinum prefers to bind to the two adjacent purines, in particular, guanines in the CCGGATC sequence. Identify the neighboring guanines in your strand (see Fig. 6.9).

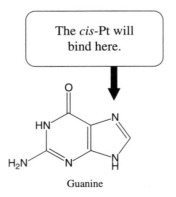

Figure 6.9   The N7 position on guanine.

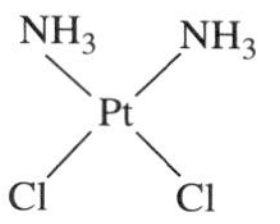

Figure 6.10 Square-planar *cis*-platin.

16. Now that you have identified the guanines, you will need to identify position N7, which is shown in Figure 6.9. Rotate the molecule so that the two N7 (one on each guanine) are easily accessible.

17. To add platinum, you must exchange both of the sp N7 nitrogens for sp^2 nitrogens. The first step in the exchange is deleting the already existing nitrogens. To do this go to the top tool bar and select the **Delete** icon that resembles a red dotted line with four diagonals. Once selected, you may click on the N7 nitrogens. They should disappear and in their place should be yellow or white bonds designating active sites to which you may add. Also note that there is an **Undo** command under **Edit** in the upper tool bar.

18. Still using the expert mode building tools, select **N** and the **trigonal planar** geometry. Now click on one of the active, yellow *double* bonds. This should insert nitrogen and create a double bond. You can bond another nitrogen active site to the other carbon active site by clicking on the bond icon on the top toolbar. This icon looks like a stick with two balls on the end. Next, click on one of the active yellow or white ends. A little ball should appear at the tip of the bond. Next click on the other active yellow or white end. This should connect the two ends together.

19. Repeat the nitrogen exchange for the other guanine.

20. These two nitrogen exchanges should leave you with one remaining active site off the end of each N7 nitrogen. You are ready to add the *cis*-platin (Fig. 6.10) to these sites, bridging the two guanines. Still under expert mode, click on the **Pt** (in the last row of elements). Select **square-planar geometry**. Now click on a yellow or white active site of one of the guanine nitrogens. This should insert the platinum atom. Now select the **bonding** icon in the upper toolbar. Select one of the platinum active sites adjacent to the already bound Pt site. Bond this to the other N7 active site, giving *cis* conformation.

21. You should still have two active sites left on the platinum. In the expert toolbox, click on **Ligands**. Select **Ammonia** as your ligand. Add ammonia groups to the two remaining active sites on platinum.

22. Your DNA–*cis*-platin molecule is complete! Run geometry optimization on this molecule using the same procedure used in steps 12–14. This calculation should not take as long as the first because you are starting with a molecule already mostly optimized. Record the final optimized energy and save your new molecule.

23. Now finished with the DNA–*cis*-platin molecule, you may hide the molecule by selecting **Close** under **File** in the top toolbar.

24. Reopen the DNA without *cis*-platin using the **Open** option under **File**.

25. Repeat all the directions for adding platinum as in steps 17–21; however, this time bridge the second guanine and the neighboring *adenine*. Continue to use N7 as the binding location for guanine and in adenine.

26. Run an energy optimization, again recording the final energy and saving the molecule. You may now close this molecule as before.

27. During the final comparison of energies, it will be beneficial to have the optimized energy of platinum with its ammonia ligands but without any DNA interactions. To create such a molecule select **New** under **File**.

28. Under the expert toolbox, under the **Exp.** tab, select **Pt** and click anywhere on the workspace window. This will create the platinum atom.

29. Add ammonia ligands to two adjacent active cites of platinum in the same manner as before.

30. Delete the extra two active cites by selecting the **Delete** icon on the above toolbar and clicking on each yellow or white active site.

31. Run a final energy optimization on this individual platinum with its two ammonia ligands. Record and save the optimized energy and structure.

32. Finally, we will quantitatively look at the structure of the final optimized geometry. Go to **Geometry**, **Measure Distance**. On the side of the DNA strand with the two adjacent guanines, click on the very first phosphate and then click on the very last phosphate. The distance between the two phosphates is given in angstroms in the lower right hand corner of the screen. Repeat this procedure for all three DNA strands (DNA, DNA with *cis*-platin bound to two guanines, and DNA with *cis*-platin bound to one guanine and one adenine).

Name _____   Section _____

Results   Summary   for   the   Molecular   Mechanics   Calculations   on   the *cis*-Platin-oligonucleotide, CCGGATC

After completing the above procedure, you should have four different optimized energies and structures: a DNA strand, a DNA strand with platinum bound to two guanines, a DNA strand with platinum bound to guanine and adenine, and an unbound platinum.

Energies of Molecules

Final energy of initial DNA molecule   _____

Energy of Pt(II) compound   _____

Energy of DNA + energy of Pt(II) compound   _____

Energy of DNA with bound (GG) *cis*-platin   _____

1. Which is lower in energy, the energy of the DNA strand with platinum bound to the two guanines or sum of the energies of the individual DNA strand and the individual platinum molecule? (*Note*: for a molecular mechanics calculation, meaningful energetic comparisons exist only for situations with identical numbers of atoms *and* bond types.)

2. Are there any major structural changes between the three DNA strands? Which strand is the longest, and which strand is the shortest?

3. Which is lower in energy, the energy of the DNA strand with platinum bound to two guanines or the energy of the DNA strand with the platinum bound to guanine and adenine?

4. Describe the conformations of the bases (*syn/anti*) as well as the ribose sugar ring conformations (N/S) as discussed in Pre-lab 6.3. Comment on your results.

5. Given your resulting structures and literature information, what parameters need to be optimized in the calculation?

EXPERIMENT 6.5: IC$_{50}$ DETERMINATION OF *CIS*-PLATIN USING A CHO CELL LINE CYTOTOXICITY ASSAY

Level 5

Although this experiment is appropriate for undergraduates, students will need some training in cell biology techniques.

Pre-lab 6.5: *cis*-Platin Cytotoxicity Assay using a CHO Cell Model[21]

Cancer occurs when normally regulated cells grow without responding to cellular controls. In other words, *transformed* cells continue to divide even when a normal cell would stop (because of contact inhibition or limited access to nutrients for example). Chemotherapy uses cytotoxic drugs to destroy cancer cells by disrupting the cell cycle and cell growth. To assess the cell killing potential of a compound, a cytotoxicity assay is carried out. A cytotoxicity assay measures the ability of a compound to inhibit cell growth.

Cell cultures are *in vitro* models used for carrying out cytotoxicity assays. They are generally made up of: *fibroblasts*, which have properties of mesodermal stem cells; *cultured epithelial cells*, which represent those from ecto and endodermal cell layers; or *blood, spleen, and bone marrow cells*, which are used to study leukemia. Cultured cells that have an indefinite life span are called a *cell line*. In this experiment, we will be using the CHO cell line to study the cytotoxicity of *cis*-platin. CHO cells are immortal and are easy to maintain in culture. Apparatus and instructions for maintaining a CHO cell culture are given in Appendix 4.

Cis-platin's cytotoxicity will be assessed by determining its IC$_{50}$ value, or the concentration that causes 50% growth inhibition. First, individual wells containing a known amount of CHO cells are incubated with varying concentrations of the *cis*-platin drug for a specified period of time. Following drug removal, an MTT assay is performed. MTT is a yellow tetrazolium dye, **I**, that is reduced by cytosolic reductases of healthy cells to form a blue formazan crystal. The blue crystal dissolves readily in DMSO and

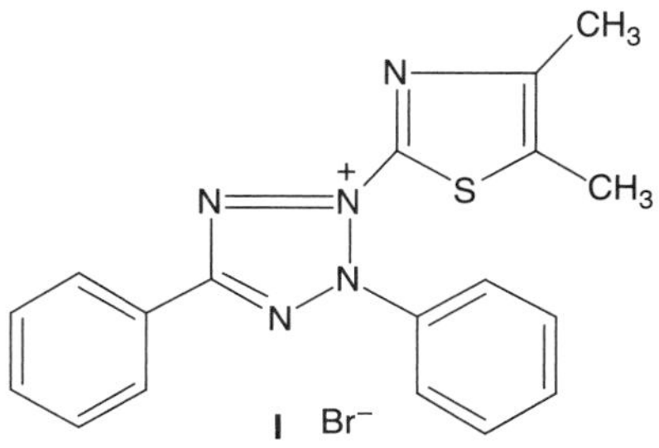

Figure 6.11   Structure of MTT.

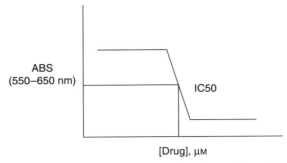

Figure 6.12   Schematic of an MTT plot showing the reading of the IC_{50} value.

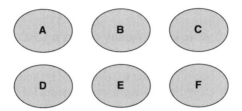

Figure 6.13   Six-well titer plate.

absorbs at 550 nm. A schematic of an MTT plot (A_{550} vs [drug]) is shown in Figure 6.12. It can be seen that, at low concentrations of drug, the cells are healthy and absorbance at 550 nm is high. As the drug concentration increases, cell death begins to occur. The IC_{50} can be read directly from the plot as shown, or this sigmoidal plot can be fit mathematically.

Procedure 6.5.a: *cis*-Platin Cytotoxicity Assay Protocol[22]

Seeding your Wells (24 h Incubation Required)

1. After passing your CHO cells, count your cells to determine the cell density in your stock cell suspension (see Appendix 4).

2. Using two, six-well titer plates, Figure 6.13, seed 11 wells at 30,000 cells per well (α-MEM, pH 7.4). The total volume of seed solution per well should be 3.4 ml and therefore you will need a total of 40 ml for 11 wells with a concentration of 8825 cells ml^{-1} [30,000 (cells/well)/3.4 (ml/well)].
   Use equation (6.4) ($M_i V_i = M_f V_f$) to determine the volume, x ml, of your stock to be diluted into α-MEM. Prepare this diluted solution in a 50 ml sterile conical vial.

$$\text{(cell density of stock, cells/ml)}\,(x\,\text{ml}) = (8825\ \text{cells/ml})\,(40\,\text{ml total volume})$$

$$(6.4)$$

3. Label 11 wells A–K (shown for A–F).

4. Pipet 3.4 ml of your prepared seed solution into each well and incubate at 37 °C, 5% CO_2 for 24 h. The cells will attach to the well bottom and double during this time.

TABLE 6.5 Information for a Single Drug (*cis*-Platin) Cytotoxicity Assay

Date	Cell Line CHO		Drug Name (*cis*-Platin)			Drug MW (300 g mol⁻¹)		Solvent for Dissolving *DMSO*	
			Stock Preparation						
Drug Stock (vol)	Wt Drug (mg)	[drug] Stock (mM)	Prep from Stock	V_{stock} (µl)	$V_{solvent}$ (µl)	$V_{Drug\ in\ Well}$ (µl)	Media Added to Well	[drug]$_{well}$ (µM)
A (100 µl)	3.0	100	—	—	—	17.5	82.5	500
B (25 µl)		40	A	10	15	17.5	82.5	200
C (50 µl)		20	A	10	40	17.5	82.5	100
D (100 µl)		10	A	10	90	17.5	82.5	50
E (50 µl)		2	D	10	40	17.5	82.5	10
F (100 µl)		1	D	10	90	17.5	82.5	5
G (50 µl)		0.2	F	10	40	17.5	82.5	1
H (100 µl)		0.02	F	20	80	17.5	82.5	0.1
I (100 µl)		0.01	H	50	50	17.5	82.5	0.05
J (media control)		0	—	—	—	—	100	0
K (DMSO control)		0	—	—	—	17.5 DMSO only	82.5	0

Drugging Procedure (4 h plus 72 h Incubation)

5. Your final drug concentrations will be (A) 500 μM, (B) 200 μM, (C) 100 μM, (D) 50 μM, (E) 10 μM, (F) 5 μM, (G) 1 μM, (H) 0.1 μM, (I) 0.05 μM, (J) 0 μM (α-MEM media control), (K) 0 μM, (DMSO control). *Cis*-platin will be dissolved in DMSO for drug delivery. To avoid cell death by DMSO, its concentration cannot be greater than 0.5% of the total reaction volume. Since your total volume is 3.5 ml, your drug volume will need to be ≤17.5 μl.

6. Prepare a concentrated Stock A (3.0 mg/100 μl) of *cis*-platin from which Stocks B–I will be prepared.[†] Table 6.5 details the preparation of your Stocks for this experiment. To efficiently prepare your solutions in a timely fashion to minimize DMSO exchange (see Experiment 6.2), label 11 150 μl conical vials A–K. Add the appropriate volume of DMSO solvent indicated in column 6 to each vial. Prepare Stock A. *Quickly* make serial dilutions as indicated in Table 6.5, vortexing for rapid mixing (recall the rapid substitution kinetics of *cis*platin in DMSO—Experiment 6.2!). Note that the drug STOCK labels are identical to the well labels. After the addition of drug and medium the total volume in each well is 3.5 ml.

7. Deliver the specified amount (17.5 μl) of Stock to each well, followed by the appropriate amount of medium. Place the lid on your wells and gently swirl to mix the solution.

8. Incubate your plates for 1–2 h (timing must be consistent for each experiment) at 37 °C, 5% CO_2.

9. Remove the solution from each well with a pipet being careful not to disturb the cells on the bottom of the well (tipping the well plate helps).

10. Wash the cells by *gently* pipeting in 2.5 ml α-MEM medium into each well and incubating the plate for 20 min.

TABLE 6.6   MTT Absorbance Readings

Well no. (Stock)	[*cis*-platin], μM	A_{550nm}	A_{650nm}	$A_{550nm} - A_{650nm}$
A	500			
B	200			
C	100			
D	50			
E	10			
F	5			
G	1			
H	0.1			
I	0.05			
J	0			
K	0			

[†]Note that because *cis*-platin is prepared in DMSO, no filter sterilization is needed. However, if drugs are ever prepared in ddH$_2$O or medium, you must filter sterilize these after preparation using a nonpyrogenic 0.2 μm Sterile-R filter.

11. Repeat steps 8–9 two more times.
12. After the third wash, leave the α-MEM medium in the wells and incubate for 72 h.

MTT Assay MTT solutions are prepared by dissolving 250 mg MTT in 100 ml of a 50:50 mixture of PBS:filter sterilized ddH$_2$O. Partition the solution into 10 ml portions in 15 ml conical vials. Wrap in foil and store at 4 °C until needed.

13. Add 300 µl MTT solution to each well and incubate as before for 4 h.

14. Remove the MTT solution from each well and replace with 1.5 ml DMSO. Gently swirl the plate and allow the blue formazan crystals to dissolve ($\sim$0.5 h).
15. Read and record the absorbance of each solution at 550 and 650 nm.
16. Plot $A_{550-650nm}$ vs [cis-platin] and determine the IC$_{50}$ value.

Name _____   Section _____

Results Summary for the Cytotoxicity Assay for *Cis*-platin using a CHO cell line

1. Record your MTT absorbances in Table 6.6.
2. IC$_{50}$ for *cis*-platin   _____ (µM)
3. How does this value compare to that of other chemotherapeutic drugs?

REFERENCES

1. Wong, E., Giandomenico, C. M. *Chem Rev.* **1999**, *99*, 2451–2466.
2. Lippert, B. *Coord. Chem. Rev.* **1999**, *182*, 263–295.
3. To, K. K. W. *Med. Res. Rev.* **2003**, *23*, 633–655.
4. Jakupec, M. A., Galanski, M., Keppler, B. K. *Rev. Physiol. Biochem. Pharmac.* **2003**, *146*, 1–53.
5. Perez, J. M., Fuertes, M. A., Alonso, C., Navarro-Ranninger, C. *Crit. Rev. Oncol./Hematol.* **2000**, *35*, 109–120.
6. Huheey, J. E., Keiter, E. A., Keiter, R. L. *Inorganic Chemistry—Principles of Structure and Reactivity*, 4th edn. Harper-Collins: New York, 1993.
7. Kauffaman, G. B., Cowan, D. O. *Inorganic Syntheses*, Vol. VII, Kleinberg, J. (ed.). McGraw-Hill: New York, 1963.
8. Szafran, Z., Pike, R. M., Singh, M. M. *Microscale Inorganic Chemistry: a Comprehensive Laboratory Experience* Wiley: New York, 1991.
9. Langford, C. H., Gray, H. B. *Ligand Substitution Processes*. Benjamin, Reading, MA, 1966.
10. Kozelka, J., Legendre, F., Reeder, F., Chottard, J.-C. *Coord. Chem. Rev.* **1999**, *190–192*, 61–82.
11. Sundquist, W. I., Ahmed, K. J., Hollis, L. S., Lippard, S. J. *Inorg. Chem.* **1987**, *26*, 1524–1528.
12. Ho, Y.-P., Au-Yeung, S. C. F., To, K. K. W. *Med. Res. Rev.* **2003**, *23*, 633–655.
13. Bertini, I., Gray, H. B., Lippard, S. J., Valentine, J. S. *Bioinorganic Chemistry*. University Science Books: Mill Valley, CA, 1994.
14. Polak, M., Plavec, J., Trifonova, A., Foldesi, A., Chattopadhyaya, J. *J. Chem. Soc. Perkin Trans.* **1999**, *1*, 2835–2843.

15. Hinchliffe, A. *Computational Quantum Chemistry*. Wiley: New York, 1988.

16. Szabo, A., Ostlund, N. S. *Modern Quantum Chemistry—Introduction to Advanced Electronic Structure Theory*. Dover: Mineola, NY, 1996.

17. Comba, P., Remenyi, R. *Coord. Chem. Rev.* **2003**, *238–239*, 9–20.

18. Kozelka, J., Chottart, J.-C. *Biophys. Chem.* **1990**, *35*, 165–178.

19. Hambley, T. W., Jones, A. R. *Coord. Chem. Rev.* **2001**, *212*, 35–59.

20. *Spartan '02*. Windows Wavefunction: Irvine, CA, 2001.

21. Lodish, H., Berk, A., Zipursky, S. L., Matsudaira, P., Baltimore, D., Darnell, J. *Molecular Cell Biology*. Freeman: New York, 2000, Chap. 24.

22. Hasinoff, B. B., Yalowich, J. C., Ling, Y., Buss, J. L. *Anti-Cancer Drugs*, **1996**, *7*, 558–567.

Metals in the Environment—Cd^{2+} Sequestration by Phytochelatins and Bioremediation

Chemistry is all around us, and it really does make a difference

(TG Spiro and WM Stigliani, 2003).

PROJECT OVERVIEW[1]

In the 1950s chemists used to wash their hands in benzene to "take the organics off." In the 1980s chemists still washed toxic water-soluble chemicals down the sink . . . as long as "plenty of water" was used for dilution. Not one decade passes without discovery of the dangers of practices we once thought safe. As Spiro and Stigliani write, "the biosphere has evolved in close association with all the elements of the periodic table . . . , [but] there is an optimum dose." Copper, iron, chromium, and zinc are all essential metal ions for us, but too much can be lethal. Heavy metals are found naturally in the environment and certain bacteria and plants have developed defense mechanisms against these toxic substances. Many plants, including rice, corn, and tobacco, have a mechanism for copper and cadmium uptake and transport, and safe storage. In fact, Cd^{2+} accumulated in the vacuoles of tobacco leaves—where it is safely sequestered from the vital areas of the tobacco plant's cells—is released when tobacco leaves are burned and is responsible for the increased amounts of cadmium found in the blood of tobacco smokers. Although relatively harmless to a tobacco plant, high levels of cadmium can result in bone, heart, lung, liver, and kidney disease, and immune system suppression in humans.

With increased technology in agriculture and other industry, we are experiencing even greater heavy metal pollution in the environment. To help clean up these toxic substances, chemists are taking advantage of the natural detoxification mechanisms of both

---

[†]Some Experiments (e.g., Experiment 7.1) in this chapter adopt a more investigative approach. Initial theory and instruction are given and students are asked to design and carry out their own investigative experiments. Others (e.g., Experiment 7.2) are at an advanced level. The independent nature of the protocols and advanced techniques make these experiments more challenging and time-consuming, and are pedagogically sound and rewarding for undergraduates. Working with partners or in teams is recommended.

Integrated Approach to Coordination Chemistry: An Inorganic Laboratory Guide. By Rosemary A. Marusak, Kate Doan, and Scott D. Cummings
Copyright © 2007 John Wiley & Sons, Inc.

bioorganisms and plants in processes called bioremediation and phytoremediation, respectively. In this exercise, we will look at one small player in the remediation effort: phytochelatins (PC_n). Phytochelatins are small cysteine-rich peptides that are produced by plants and fungi in response to heavy metal stress. In effect, phytochelatins are metal-chelating peptides, and thus form coordination compounds. We will begin the chapter by using fluorescence spectroscopy to detect trace metals in aqueous solution. We will then synthesize and investigate the chemistry of the smallest member of the PC_n group, (γ-Glu-Cys)$_2$-Gly or PC_2. Following solid-phase peptide synthesis and chromatographic purification, the Cd^{2+} binding ability of the PC_2 peptide will be explored using optical spectroscopy. Potentiometric titration involving cysteine and glycine amino acids will provide a simplified model for understanding Cd^{2+}–phytochelatin binding. An advanced Cd–PC_2 titration is offered as an optional experiment. Lastly, PC_n biosynthesis in yeast cells will be carried out and Cd^{2+} analyzed by atomic absorption spectroscopy.

EXPERIMENT 7.1: WHAT'S IN THE WATER? FLUORESCENCE DETECTION OF TRACE METAL IONS IN AQUEOUS SOLUTION[2–7]

Level 4

This experiment takes an investigative approach. Students need to become familiar with the theory and technique of fluorescence spectroscopy, solution preparation and experimental design.

Pre-lab 7.1.a: 8-Hydroxyquinoline and 8-Hydroxyquinoline-5-sulfonic acid

How much lead is in your municipal drinking water? Is platinum present in the urine of a chemotherapy patient? Can plants purify water through cadmium uptake? Can the amount and location of zinc ions be detected in neurons? The detection of trace amounts of metal ions is a central analytical method in environmental and biological chemistry. Often, a reagent that can coordinate to a metal ion of interest is added to a sample solution and the resulting metal complex is detected using analytical instrumentation. Electrochemical and fluorescence detection are two of the most sensitive methods.

One analytical reagent that has received a large amount of attention is 8-hydroxy-quinoline (HQ), also known as "oxine," which is a heterocyclic chelating ligand (Fig. 7.1). Several metal complexes of HQ display intense fluorescence in solution, allowing for easy detection of trace metal ions and also providing the basis for other technologies based on fluorescence. The complex Al(HQ)$_3$, Tris(8-hydroxyquinoline) aluminum(III), is part of a new family of materials for organic light-emitting diode (OLED) technology for flat panel display screens. In addition, numerous derivatives of HQ have also been synthesized and investigated for their metal ion binding and fluorescence properties. For example, Barbara Imperiali's research team at the Massachusetts Institute of Technology has developed a sensor for the detection of Zn^{2+} in cells. The sensor, consisting of a peptide with an HQ ligand attached, binds Zn^{2+}, which leads to a large enhancement in fluorescence intensity. The sensor can detect zinc ions down to nanomolar solution concentrations.

For some applications, the *qualitative analysis* of metal ions is desired: detecting and distinguishing which metal ions might be present in a mixture. For other applications,

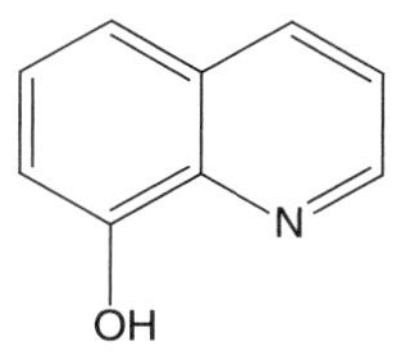

Figure 7.1   Structure of 8-hydroxyquinoline (HQ).

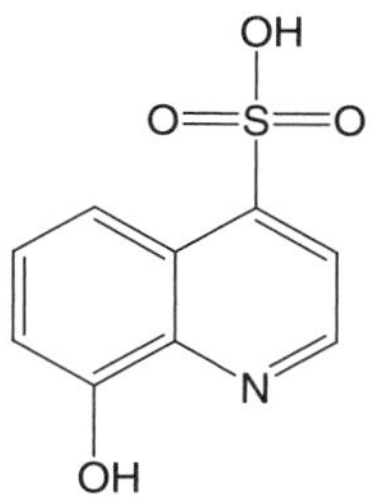

Figure 7.2   Structure of 8-hydroxyquinoline-5-sulfonic acid (HQS).

determining the concentration of a particular metal ion—a *quantitative analysis*—may be important. Using fluorescence spectroscopy, either of these determinations may be possible. Two objectives have emerged in the research literature for using HQ and related ligands for the qualitative analysis and detection of metal ions. The first is the goal of developing *specific reagents* and conditions for the selective analysis of a one-metal ion analyte in a solution that may have a mixture of species. This selectivity would be based on a large change in the fluorescent properties of HQ only in the presence of one particular metal ion in the mixture and avoiding interference from other species. The second kind of application relies on forming fluorescent HQ complexes with *all* metal ions present in a system (using *nonspecific reagents*), thereby allowing for a sensitive fluorescence detector to be used with some other instrumental method such as HPLC.

In order for HQ to be useful as an analytical reagent for the quantitative determination of a metal ion concentration (a *fluorimetric* reagent), the relationship between the concentration of the metal ion and the fluorescence signal from the metal complex would have to be understood. Upon coordination of the chelating HQ ligand to a metal ion, two types of changes in the fluorescence properties may occur. One effect is a shift in the wavelength (and, therefore, the color) of the fluorescence. By measuring the ratio of light emitted at two different wavelengths, the concentration of metal ion can be determined; this method is known as a ratiometric fluorescence sensor. Chelation enhanced fluorescence (CEF) is an analytical method that relies on a large enhancement of the fluorescence signal upon binding of a metal ion to a chelating ligand.

Among the derivative of HQ, 8-hydroxyquinoline-5-sulfonic acid (HQS) is attractive because of the enhanced water-solubility of the metal complexes it forms (Fig. 7.2). HQS is weakly fluorescent in solution. Investigating the effect of various metal ions and solution conditions on the fluorescence properties of HQS is the purpose of the experiment. By designing appropriate experiments, you can explore how HQS might be useful as either a specific or nonspecific fluorescent reagent.

Pre-lab 7.1.b: Introduction to Luminescence

Luminescence, the emission of light from an excited-state molecule, is a most dramatic example of the fascinating colors of inorganic chemistry. The beautiful colors of luminescent inorganic complexes are also a useful tool for important studies in inorganic chemistry. In this experiment, changes in luminescence intensity and color of an organic chelating ligand can be used in the detection of trace metal ions. In Experiment 8.4, changes in luminescence intensity of a luminescent Cr(III) complex are used to measure the rates of excited-state electron transfer reactions. A brief overview of excited state and luminescence spectroscopy is presented here.

Excited States The absorption of UV–visible light can result in the promotion of an electron from a lower energy orbital (often the highest occupied molecular orbital, or HOMO) to a higher energy empty orbital (often the lowest unoccupied molecular orbital, or LUMO). The resulting *excited state*, having a different electron configuration, can be thought of as an "electronic isomer" of the ground state. Although the metal ions we often deal with in inorganic chemistry can have partially filled d orbitals, let us pick an example of a metal complex having an even number of electrons, all paired. Therefore, in the excited state the HOMO and LUMO each have only one electron, and these two electrons may have opposite spin (total spin $S = 0$ and a multiplicity of "singlet") or the same spin (total spin $S = 1$ and a multiplicity of "triplet"). The triplet excited state is always lower energy than the corresponding singlet excited state. These configurations are summarized in Figure 7.3.[†]

After excitation by visible or UV radiation, molecules relax back to the ground state by dissipating the excitation energy through molecular vibrations and collisions with solvent molecules. This process is called *nonradiative decay* and is typically a very fast set of processes for molecules in fluid solution at room temperature. Alas, this process, which is simply the conversion of photonic energy to heat, is rarely useful.

However, some molecules are capable of relaxing back to their ground state by also emitting a photon. The energy of the emitted photon is lower than the energy of the photon used to excite the molecule, with the resulting difference in energy being lost as a small amount of heat. The luminescence can be classified as *fluorescence* if the excited state and the ground state have the same spin (e.g., both are singlet) or *phosphorescence* if the excited state and the ground state have the different spin (e.g., triplet excited state and singlet ground state). The fluorescence rate is much faster (with a corresponding shorter excited state lifetime) than the phosphorescence rate (with a corresponding longer

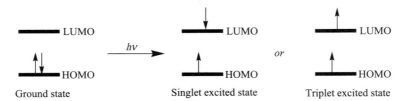

Figure 7.3 Ground state and singlet and triplet excited state electron configurations.

[†]Total spin is the vector sum of the individual electron spins. A molecule having equal numbers of $+\frac{1}{2}$ and $-\frac{1}{2}$ electrons has a total spin of $S = 0$. A molecule having two unpaired electrons, both with spin $+\frac{1}{2}$, has a total spin of $(+\frac{1}{2}) + (+\frac{1}{2}) = 1$.

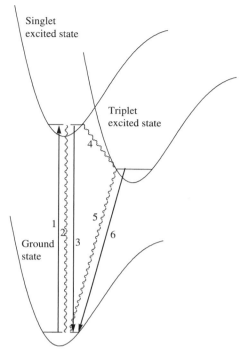

Figure 7.4   1 = absorption; 2 = nonradiative decay; 3 = fluorescence; 4 = intersystem crossing; 5 = nonradiative decay; 6 = phosphorescence.

excited state lifetime), because there is no "spin flip" required for fluorescence. Whether a molecule fluoresces or phosphoresces depends on another process: inter-system crossing from the singlet excited state to the triplet excited state. Typically, fast intersystem crossing is observed for transition metal complexes (especially those of the second and third row), and so only phosphorescence is observed. A summary of these states and rates is presented in Figure 7.4.

Quantum Efficiency of Luminescence   The intensity of luminescence can be expressed as the quantum efficiency (Φ), which is the ratio of the amount of light emitted to the amount of light absorbed. It can also be expressed as the ratio of the rate of luminescence to the sum of all rates that deactivate the excited state.

Luminescence Sensing   Some luminescent compounds can serve as probes of analytes such as metal ions. Upon coordination with a metal ion, the luminescence of a ligand may be enhanced, quenched, or shifted in color (the wavelength of maximum intensity). In addition, changes in the solution conditions (pH, ionic strength, temperature or the presence of macromolecular hosts capable of forming organized structures) often affect luminescence properties.

Two features of luminescence make it a powerful method for analytical detection of metal ions. First is *sensitivity*; fluorimetric reagents have been developed for the detection of even nanomolar metal ion concentrations. Indeed, even single molecule fluorescence is possible.

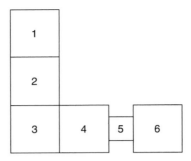

Figure 7.5 Schematic of the spectrofluorimeter. See text for description.

Second is *selectivity*; fluorimetric reagents may bind selectively to only particular metal ions, or only particular metal ions may lead to changes in luminescence of the ligand.

Luminescence Spectroscopy A *fluorescence spectrophotometer* ("fluorimeter") is used to measure luminescence intensity as a function of wavelength emitted (color), yielding a spectrum. A sample solution is placed in a cuvette similar to that used in UV–vis spectroscopy, except that all four windows are optically transparent and the glass is a special type of quartz, allowing for transmission of UV light. The sample cuvette is placed in the instrument, an excitation wavelength is selected and the intensity of the resulting luminescence is measured through a region of emission wavelengths.

The components of the fluorimeter include (Fig. 7.5): (1) *power supply*—powers the lamp; (2) *light source*—UV–vis lamp; (3) *excitation monochromator*—selects a particular wavelength of light from the lamp to excite the sample; (4) *sample*—solution in a cuvette in a holder; (5) *emission monochromator*—scans through a set of wavelengths where the sample emits light; (6) detector—a photomultiplier tube "PMT" detects the number of

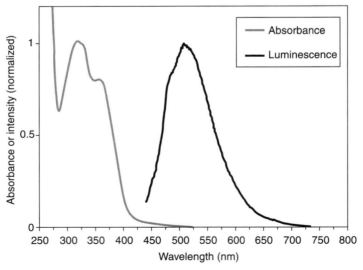

Figure 7.6 The absorption and emission spectra of HQS.

photons emitted at each wavelength scanned, transferring them into an electrical current signal.

Using a computer software program interfaced to the spectrophotometer, you can select: the wavelength of excitation; the wavelength range for scanning the emission; the increment of wavelengths to measure in this range; the amount of time spent measuring photons at each wavelength; and the slit width of the excitation and emission monochromators (which adjusts how much light passes through each monochromator).

Within the fluorimeter software or using an external spreadsheet or plotting program, you can calculate the area underneath an emission curve (integration), which is a measure of the fluorescence intensity.

The absorption and emission spectra of HQS are shown in Figure 7.6. Note that the compound absorbs at shorter wavelengths than it fluoresces.

Pre-lab Questions

Q7.1 Either HQ or HQS will chelate to metal ions, but most often alkaline solutions are employed. What happens to these organic ligands at high pH and why does this facilitate coordination of a metal ion? What would happen at low pH?

Q7.2 Why would you expect metal complexes of HQS to be more water-soluble than those of HQ?

Q7.3 Which occurs at longer wavelength: fluorescence or phosphorescence? Why?

Q7.4 What color is a solution of HQ? (Hint: consider Fig. 7.6.)

Q7.5 What color does HQ fluoresce?

Q7.6 If HQ emits light at longer wavelengths than it absorbs, what happens to the excess energy?

Q7.7 Could HQ be excited using light of a wavelength 500 nm? Why or why not?

Procedure 7.1.a: Fluorescence Spectrum of HQS

Clean glassware and cuvettes thoroughly before using. Interference from impurities can lead to a tremendous waste of time and money. Wash cuvettes with soap, de-ionized water, and acetone, and rinse with spectroscopic-grade MeOH. Keep your work area clean! Impurities that fluoresce or that quench fluorescence are the bane of fluorescence spectroscopy.

You will be provided with 8-hydroxyquinoline-5-sulfonic acid solid ($225.23 \, g \, mol^{-1}$) water-soluble nitrate, sulfate, or halide salts of: Ag(I), Al(III), Cd(II), Co(II), Cr(III), Cu(II), Hg(II), Mg(II), Ir(III), Ni(II), Pb(II), Pt(II), Zn(II), and hexadecyltrimethyl -ammonium chloride (HTAC); 1 cm pathlength quartz fluorescence cuvette; 100– 1000 µl auto-pipetter and tips; 10 and 25 ml volumetric flasks; 10 and 25 ml round-bottom flasks; ~40 four-dram sample vials with screw caps; gloves; wash acetone; and spectroscopic-grade methanol

1. Become familiar with the operation of the fluorescence spectrophotometer in your laboratory. In particular, you should understand how the following instrumental parameters affect the intensity and signal-to-noise ratio (S/N) of a luminescence spectrum:
   • excitation monochromator slit width;

- emission monochromator slit width;
- signal averaging through integration time or multiple averaged scans.

You should also explore the issue of background signal due to dark noise, scattered light or (possibly) luminescent impurities in the cuvette or water.

When comparing luminescence intensities of various solutions, the absorbance (A) at the excitation wavelength (λ_{ex}) is important. Therefore, absorption spectra (or at least A @ λ_{ex}) should be measured for all samples. The absorbance at the excitation wavelength should be less than 0.1 (known as being *optically dilute*) in order to minimize the inner filter effect.[†] Differences in absorbance at the excitation wavelength between samples will affect luminescence intensity, so a correction is needed (see Results/Summary page).

2. Prepare an aqueous solution of HQS, and determine what excitation wavelength would be optimum for minimizing background signal and maximizing fluorescence signal. Keep in mind that you may need to adjust the concentration so that the solution is optically dilute at the excitation wavelength. Vary instrumental parameters such as slit width, integration (signal averaging), and emission wavelength ranges to familiarize yourself with how these parameters affect the luminescence spectrum. Finally, determine the integrated area of the HQS fluorescence band and estimate the uncertainty of this measurement. Quantifying changes in fluorescence intensity among various samples requires calculating this integrated area (I) and correcting for differences in absorbance (see Results and Summary section, below). *Save all spectra* files in an appropriate subdirectory of the computer or storage device.

Procedure 7.1.b: Designing Experiments to Investigate the Use of HQS as a Fluorimetric Reagent for Metal Ion Detection

The objective of this exercise is to design meaningful experiments to investigate the potential use of HQS as a fluorimetric reagent for trace metal ions. Changes in luminescence properties that you can measure include quantum efficiency, wavelength maximum (λ_{max}), and/or band shape of the emission. Keep in mind that many factors can affect the luminescence intensity of a sample. In a well-designed experiment, the effect that a particular variable (metal ion, temperature, pH, etc.) has on the HQS fluorescence intensity needs to be separated from other effects.

Pre-lab Preparation An example of a sample solution appropriate for investigation would be 0.5 mM HQS(aq) at pH = 9 with 12 μM metal ion M^{n+}(aq). This could be prepared by:

1. Preparing a stock solution of 0.5 mM HQS (113 mg solid HQS dissolved in 1 l water) and adjusting the pH to 9 using KOH(aq).

[†]The inner filter effect refers to the fact the highly absorbing solutions ($A > 0.1$ in a 1 cm^2 cuvette) can act as filters of the excitation light; molecules of the front side of the cuvette absorb a significant amount of the excitation light and limits how much reaches the far side of cuvette, producing a gradient of excitation light intensity along the path length of the cuvette. Decreased excitation light intensity reaching the back side of the cuvette results in less luminescence from that region, and the total luminescence intensity, measured at a right angle to the path length, is decreased due to the inner filtering effect. Below an absorbance of 0.1, the excitation light intensity along the pathlength is essentially constant.

2. Preparing stock solutions of 3 mM M^{n+}(aq) (approximately 15 mg salt in 20 ml water).

3. Adding 80 μl of 3 mM M^{n+}(aq) to 20 ml of HQS stock and re-adjust pH, as needed to give 12 μM M^{n+} solution.

This procedure can be adjusted for other volumes or concentrations, accordingly. Your goal is to determine how various metal ions affect HQS luminescence. Consider interesting comparisons of different metal ions that may commonly be found together in biological/medical applications, in the environment or in industrial settings. See for example, Experiment 7.5.

Some possible aspects of HQS luminescence that you could design experiments to explore include the following:

1. How do various metal ions affect HQS luminescence? Consider interesting comparisons of different metal ions that may commonly be found together in biological/medical applications, in the environment or in industrial settings.

Some questions to investigate:

- Do all of the ions in your sample series form fluorescent complexes with HQS (an ideal property for a nonspecific fluorimetric reagent)?
- Or do only some metal ions affect fluorescence (a trait of a specific fluorimetric reagent)?
- Specifically, how are Φ and λ_{max} affected by the metal ions?

Some experimental conditions to consider:

- How sensitive is HQS luminescence to a change in metal ion concentration? Would a meaningful comparison involve solutions having the same metal ion concentration?
- Is absorbance affected by the metal ions? If so, how does this affect luminescence?

2. How do solution conditions affect fluorescence enhancement? For example:
   - What is the effect of solution pH?
   - What happens in solvent mixtures (try DMF−water mixtures, for example)?
   - What happens if a surfactant (0.5 mM HTAC, for example) is added?
3. Could HQS serve as a fluorimetric reagent for the quantitative analysis of a particular metal ion?
   - How could a calibration curve (intensity of fluorescence vs $[M^{n+}$(aq)]) be prepared?
   - Could this be used to determine the $[M^{n+}$(aq)] for a solution of unknown concentration?

Name _____   Section _____

Results Summary for Fluorescence Detection of Trace Metal Ions
in Aqueous Solution

1. Excitation wavelength for HQS _____
2. HQS concentration _____
3. Integrated area of the HQS fluorescence band (I) _____
   Include estimated uncertainty
4. A @ λ_{ex} _____
5. a. State the hypothesis of question you were attempting to address with your experiment.
   b. Are your results conclusive?
   c. Did you have a control experiment as part of your experimental design?
   d. What additional experiment(s) could help support or refute your conclusion?

When comparing luminescence intensity of two or more solutions, small variations in A @ λ_{ex} affect the emission intensity (I). The relative quantum efficiency can be calculated for two solutions (A and B) using this equation:

$$\phi_A = \phi_b \cdot (I_A / I_B)(A_B / A_A)$$

This assumes that A and I are linearly related, which should be the case for solutions that are optically dilute.

EXPERIMENT 7.2: SOLID-PHASE PEPTIDE SYNTHESIS AND HPLC PURIFICATION OF THE PHYTOCHELATIN, PENTAPEPTIDE, (γ-Glu-Cys)$_2$-Gly (PC$_2$)

Level 5

This multiweek laboratory introduces inorganic students to bioorganic synthetic techniques. While the preparation of the pentapeptide is relatively straightforward, the purification can be challenging. The exercise itself requires time and dedication. Students should work in groups and share responsibilities.

Pre-lab 7.2.a: Phytochelatins (γ-Glu-Cys)$_n$-Gly (PC$_n$)[8-10]

In response to heavy metal exposure, certain plants and yeasts produce PCs, analogous to the sulfur-rich protein metallothionein made by vertebrates. PCs are sulfur-rich peptides of the general chemical makeup (γ-Glu-Cys)$_n$-Gly where $n = 2$-11 (Fig. 7.7). PCs are

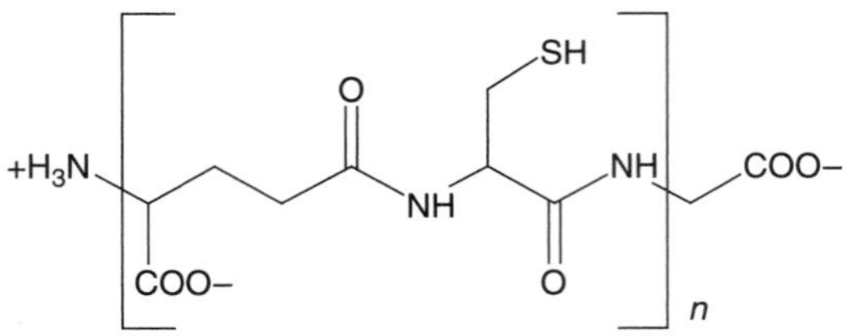

Figure 7.7   General chemical composition of phytochelatins (PCs).

therefore homologs of glutathione, an important antioxidant and biological redox agent. In the plant cell PC synthase (γ-glutamylcysteine dipeptidyl transpeptidase) is activated by heavy metal binding (Cd^{2+}, Cu^{2+}, Zn^{2+}, Pb^{2+}, Hg^{2+}, and Ag^{+}, but Cd^{2+} is most potent), catalyzing the transpeptidation reaction, equation (7.1), that results in PC formation. Along with varying (γ-Glu-Cys) chain lengths, peptide isoforms [e.g., (γ-Glu-Cys)$_n$-β-Ala and (γ-Glu-Cys)$_n$-Ser] in certain species are also known. The factors governing specific phytochelatin synthesis are the subject of current research. In this *three week* lab, we will use solid-state techniques to synthesize the smallest PC, where $n = 2$.

$$(\gamma\text{-Glu-Cys})_n\text{-Gly} + (\gamma\text{-Glu-Cys})\text{-Gly} \xrightarrow[\text{glutathione}]{} (\gamma\text{-Glu-Cys})_{n+1}\text{-Gly} + \text{Gly} \qquad (7.1)$$

Pre-lab 7.2.b: Introduction to Solid-Phase Peptide Synthesis (SPPS)[11,12]

Solid-phase synthesis of peptides was pioneered by Bruce Merrifield in the early 1960s. This work, for which he won the Nobel Prize in 1984, set in motion the modern approach to drug discovery called combinatorial chemistry. Through combinatorial chemistry, millions of compounds are generated by the synthesis of libraries on solid supports and screened for therapeutic activity by high-throughput assays. The importance of this work is attested by the numerous combinatorial research units that are now an integral part of most major pharmaceutical companies.

Solid-phase peptide synthesis involves the linear assembly of a sequence of amino acids on a polymeric support (resin beads) derivatized with a reactive functional group. When the synthesis of the peptide on the beads is complete, the peptide is then released into solution by cleaving the bond that attaches it to the resin. A key aspect of this strategy is the development of protecting groups that allows selective assembly of the peptide chain in the presence of reactive functional groups on the side chains of the amino acids. This solid-state method is much preferred over conventional solution approaches: along with its rapidity, ease of purification and high yield are benefits. Figure 7.8 depicts the building of the (γ-Glu-Cys)$_2$-Gly pentapeptide. Post-lab question 7.9 asks you to look up and detail the chemistry of each step.

Procedure 7.2: Synthesis of (γ-Glu-Cys)$_n$-Gly[13,14]

The general reagents and apparatus for SPPS are given in Truran *et al.*[13]

Preparation of Resin   The amounts used are based on the Fmoc–Gly–Wang resin, having a loading capacity of 0.75 meq g^{-1} and an average amino acid MM of 500 g mol^{-1}.

1. Add 250 mg of Fmoc–Gly–Wang resin to a 10 cm^3 syringe barrel that has been fitted with a polypropylene filter.

Fmoc-AA$_5$-AA$_4$-AA$_3$-AA$_2$-AA$_1$-linker $\{$wR$\}$

Figure 7.8   Structure of (γ-Glu-Cys)$_n$-Gly.

2. Wash the resin with dichloromethane (DCM, 3×5 ml) and then with DMF (3×5 ml). Allow each solvent to contact the resin for 2 min with occasional gentle stirring.

3. Remove each wash by aspiration leaving dry beads and perform a Kaiser test on the resin (see below). Your test should indicate an Fmoc-protected amine.

Kaiser Test

1. Transfer several resin beads with a Pasteur pipet to a 3 in. test tube.

2. Add three drops of each of the following solutions: ninhydrin–ethanol, phenol–ethanol, and potassium cyanide–pyridine.

3. Heat the mixture for 3 min at 100 °C in a heating block. If your test is *blue*, you have a resin-bound free amine. Either deptrotection has been successful, or you have incomplete coupling.

Coupling of the Amino Acids   For all washes, the contact time for each solvent aliquot is 2 min, unless noted otherwise, with occasional *gentle* stirring. Each wash aliquot is then removed completely by aspiration, leaving dry resin beads. A schematic of your peptide bound to the resin is shown in Figure 7.9. Note that your peptide is synthesized and is read from right (C-terminus) to left (N-terminus).

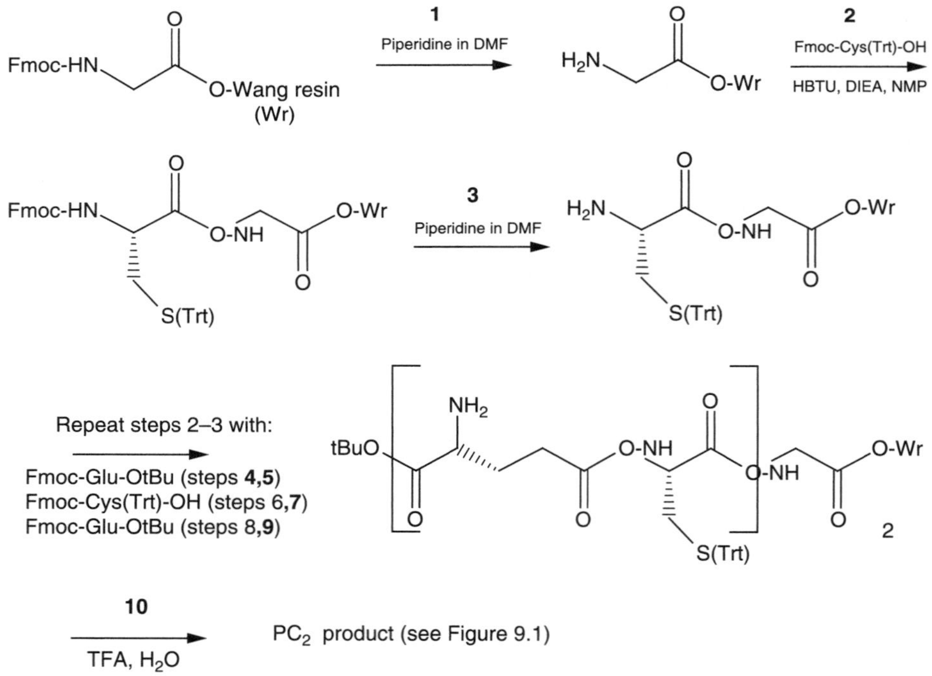

Figure 7.9   Synthesis of PC$_2$. DMF = dimethylformamide; Fmoc = 9-fluorenylmethoxycarbonyl; Trt = trityl; DIEA = diisopropylethylamine; NMP = N-methylpyrolidine. HBTU = O-(benzotriazol-1-yl)-N,N,N',N'-tetramethyluronium hexafluorophosphate.

Deblocking—Fmoc Removal

1. First, you need to remove the Fmoc protecting group on AA_1, Gly, by treating the resin with a solution of 20% piperidine in DMF. Carry out a 5 ml treatment with a contact time of 2 min followed by a second 5 ml treatment with contact time 8 min.

2. Wash the resin with DCM (4 × 5 ml).

Coupling and Washing

3. To a 4 ml glass vial add 0.439 g of AA_2, Fmoc-Cys(Trt)-OH (4 equiv.), 0.284 g HBTU (4 equiv.), 0.101 g HOBt (4 equiv.) and a solution of DIEA in NMP (1.0 ml containing 0.194 g or 8 equiv. DIEA).

4. Mix the solution by drawing up the solution up into a Pasteur pipet and ejecting the solution repeatedly until the solution is clear and colorless.

5. Quantitatively transfer the solution to the prepared resin.

6. Allow the coupling reaction to proceed for ~30 min with occasional stirring.

7. Remove the reaction solution by aspiration and wash the resin with DMF (5 × 5 ml).

8. Check if the reaction coupling is complete using the Kaiser test. If the reaction is finished, proceed to the next step. If incomplete coupling occurs, couple again with fresh reagents.

9. Remove the Fmoc protecting group from the dipeptidyl-resin as in steps 1–2. You will repeat the cycle of coupling, washing, deblocking, and washing (steps 1–8 of the coupling of amino acids) for each amino acid, AA_3–AA_5 using the following amounts: AA_3 (Fmoc-Glu-OtBu), 0.319 g; AA_4 [Fmoc-Cys(Trt)-OH], 0.439 g; AA_5 (Fmoc-Glu-OtBu), 0.319 g; however, after the final AA_5 is coupled, wash the resin with DMF (4 × 5 ml) and DCM (4 × 5 ml) and stop. *Do not* add 20% piperidine in DMF.

10. Stopper the syringe and add 4 ml of DCM, seal with parafin, and store your peptide in the refrigerator. (This step can be done after any coupling as a stopping point.)

Cleavage of the Peptide from the Resin

11. Remove the Fmoc group of AA_5 with 20% piperidine in DMF (as in step 1).

12. Wash the resin with DMF (3 × 4 ml), MeOH (1 × 4 ml), DMF (3 × 4 ml), and finally DCM (4 × 4 ml).

13. Cleave the pentapeptide from the resin by treating the resin with 4 ml of the 90% trifluoroacetic acid, TFA, solution (see Appendix 9.1) for 1.5–2 h.

14. Drain the solution into a 50 ml conical vial that has been cooled on ice.

15. Complete the transfer by washing the resin with 90% TFA solution (3 × 4 ml) and add the washes to the conical vial. (Use your PTFE stir rod to return any beads adhering to the syringe wall back into the solution.)

16. Precipitate the pentapeptide by adding 20 ml of ice cold *anhydrous* ether—fresh new reagent is highly recommended. If no precipitate is obtained, remove the ether by evaporation and repeat the precipitation step with fresh ether.

17. Centrifuge the 50 ml conical vial containing your precipitated peptide for 5 min at 2000*g*.

18. Pour off the ether supernatant and wash the pellet with ice cold anhydrous ether (3 × 5 ml) centrifuging and removing the ether each time.

19. Allow the pellet to dry and then dissolve the pellet in 10 ml ddH$_2$O and transfer to a tared, acid-washed, 50 ml round-bottom flask.

20. Lyophilize overnight, determine the crude yield (MM = 539.58 g/mol), and store the product as the lyophilized powder in the refrigerator.

The peptide may be purified by either HPLC or column chromatography. Both methods are given below. If HPLC is your method of choice, continue today's work with the analytical HPLC instructions. Precaution: if your peptide is dissolved in solution for any length of time, it is best to store it at $-20\,°C$ to prevent degradation.

Analytical HPLC

1. Prepare, filter, and degas HPLC buffers: buffer A, H$_2$O with 0.1% TFA; buffer B, 80:20 (v/v) CH$_3$OH–H$_2$O containing 0.1% TFA. Use a reversed-phase C$_{18}$ column with detection at 214–220 nm.

2. Make up a solution of the pentapeptide standard (1 mg ml^{-1} in 0.1% TFA HPLC solvent). You may add a reducing reagent such as TCEP (Tris[2-carboxyethyl]-phsophine) or dithiothreitol, DTT (∼30-fold excess), before HPLC if you want to ensure no disulfide cross-linking of your peptides occurs; however, the acidic conditions should prevent significant disulfide formation. Starting with an initial run with a linear gradient of 0–100% B in 50 min, determine the conditions for the purification you will do next week.

Peptide Purification
Semipreparative HPLC Purification

1. Purify your peptide according to the conditions determined on the analytical HPLC. You can test your fraction collected for the appropriate product by TLC (Kieselgel G) using the solvent system: nBuOH, pyridine, HOAc, and H$_2$O (1:1:1:1). Your product will have an R_f value of 0.84.

2. Lyophilize your solution to obtain your pentapeptide product (obtained as the PC$_2^+$ CF$_3$COO$^-$ salt). Record your yield.

Sephadex G-15 Column Chromatography

1. Prepare 500 ml of 3% aqueous acetic acid (HOAc).

2. Using your HOAc solution, pour a 1.5 × 28 cm Sephadex G-15 column (see Experiment 5.4 for preparing chromatography columns).

3. Dissolve peptide in a minimum amount of deoxygenated ddH$_2$O. You *may* add a reducing reagent such as TCEP (Tris[2-carboxyethyl]-phosphine) or dithiothreitol, DTT (∼30-fold excess), to ensure no significant disulfide cross linking of your peptides occurs prior to purification.

4. Carefully load your solution of peptide onto the column and elute your peptide with 3% HOAc.

5. Collect 1.5 ml fractions. Test your fractions for product by TLC (see step 2 under the previous section).

6. Pool and lyophilize your fractions to obtain your PC$_2$ product (obtained as the PC$_2$CH$_3$COO$^-$ salt). Record your yield.

At this point, there are several optional experiments you can do to confirm your product. *Note*: experiments (c) and (d) are standard biochemistry techniques. Coupling this lab with biochemistry lab may be an ideal partnership:[15-17†]

a. ^{1}H NMR spectroscopy (see Chapter 3).

b. Determine the $[\alpha]_D^\circ$ value in H_2O, $c = 0.2$. Theoretical optical rotation is $-37.0°$ (see Chapter 8 for optical rotation theory).

c. Determine the SH content (SH mol^{-1} peptide) by Ellman's (DTNB) method, DTNB = 5,5′-dithiobis(2-nitrobenzoic acid).

d. Perform an Edman degradation on your product to determine the amino acid sequence.

Name _____   Section _____

Results Summary for the Synthesis of (γ-Glu-Cys)$_2$-Gly (PC$_2$)

1. Yield of crude PC$_2$:_____

HPLC Purification

2. Determined conditions for HPLC purification: _____to _____% B in _____ min.
3. Retention time(s) for pooled fractions: _____
4. R_f value of product: _____
5. Yield of purified product: _____

Sephadex G-15 Column Purification

1. Fractions pooled for collection of product: _____
2. R_f value of product: _____
3. Yield of purified product: _____

Post-lab Questions

Q7.8 If you performed any of the optional tests for identification, describe your results.

Q7.9 Detail the chemistry involved at each step in Figure 7.8 for your peptide synthesis reaction.

EXPERIMENT 7.3: A QUALITATIVE UV SPECTROSCOPIC STUDY OF Cd^{2+}-(γ-Glu-Cys)$_2$-Gly (PC$_2$) BINDING[8-10,14,18]

Level 4

Pre-lab 7.3: Metal Complexes of Phytochelatins

Phytochelatins [γ-Glu-Cys)$_n$-Gly], PC$_n$, may be induced by a variety of heavy metals *in vivo* and *in vitro*. The response is complex and varies with each system of study. It is certain, however, that Cd^{2+} is by far the strongest inducer of PC synthesis and is the

---

†There are several protocols available on the web, e.g., http://www.interchim.com/interchim/bio/produits_uptima/tech_sheet /FT-UP01566(DTNB). pdf by Ultima.

primary target of the detoxification mechanism. PCs of varied chain length ($n > 2$) form cadmium complexes of substantial complexity.

The binding of Cd^{2+} to the smallest PC, (γ-Glu-Cys)$_2$-Gly (PC$_2$), has been investigated in some detail. PC$_2$ ligands exist in multiple protonated states over the pH range 0–12 [Fig. 7.10(a)] and a variety of $[Cd^{2+}]_n[PC_2]_m$ species exist depending on pH and reactant ratios [Fig. 7.10(b)]. Further, the atoms binding to the cadmium center vary depending on conditions. The specifics of Cd–PC$_2$ binding are discussed in greater detail in Experiment 7.4.

Why are PCs synthesized, when the PC precursor, glutathione (γ-Glu-Cys-Gly) itself is an effective metal ion chelator? In this experiment, we investigate the role of increasing (γ-Glu-Cys) chain length by studying the cadmium(II) binding capacity of PC$_2$ and glutathione. For comparison, we will look at copper(II) binding as well. If pure PC$_2$ is not available, students can still obtain a valuable learning experience from titrating glutathione alone and comparing the data with literature values for PC$_2$.

Procedure 7.3: Spectrophotometric Titration of PC$_2$ and Glutathione with Cadmium(II) and Copper(II)

Solution Preparation

1. Prepare 100 ml of 10 mM Tris HCl buffer (pH 7).

2. Prepare a 5 ml aqueous solution of 5 mM CdCl$_2$ and a 5 ml aqueous solution of 5 mM CuCl$_2$. Use acid-washed volumetric glassware.

3. Prepare 25 ml of 0.15 mM (as SH) peptide [MM(PC$_2$) = 539.58 g/mol not including triflate or acetate anion; MM(glutathione, reduced) = 307.33] in your Tris–HCl buffer. Prepare each just prior to analysis to minimize SH oxidation.

4. If necessary, the SH content of your peptide can be determined by a DTNB test.[15]

UV Analysis

1. To 5 ml of your PC$_2$ solution, add 5 μl Cd^{2+} solution. Allow the reaction to equilibrate with occasional stirring for ~2–5 min. Obtain the absorbance at 250 nm (Cd^{2+}-mercaptide).

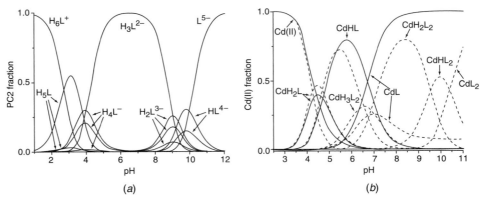

Figure 7.10 (a) pH distribution for PC$_2$; (b) species distribution of Cd^{2+}PC$_2$ complexes of ratios 2 mM PC$_2$:2 mM Cd^{2+} (solid line) and 1 mM PC$_2$:1 mM Cd^{2+} (dashed line). [Reproduced with permission from Dorcak and Krezel.[14]]

2. Continue to titrate Cd^{2+} in 5 μl increments until 50 μl is reached, equilibrating and reading the absorbance after each addition. Record your results in the Results Summary section.

3. Repeat steps 1–2 with PC_2 and your Cu^{2+} solution but take readings at 265 nm.

4. Repeat steps 1–2 with glutathione and Cd^{2+} and then glutathione and Cu^{2+}, recording at their respective wavelengths.

5. For each experiment, plot the absorbance (250 nm for Cd^{2+} and 265 nm for Cu^{2+}) vs $[M^{2+}]$.

6. As a control, obtain absorbance spectra of your metal ions at their highest concentration in buffer only.

Name _____   Section _____

Results Summary for the UV Spectroscopy Binding Study of Cd^{2+} and Cu^{2+}-$(\gamma$-Glu-Cys$)_2$-Gly (PC$_2$)

1. Complete Table 7.1 below.
2. Plot and describe your binding curves in each case.

Post-lab Questions

Q7.10   What can you conclude about Cd^{2+} vs Cu^{2+} PC_2 binding properties? Glutathione binding properties?

Q7.11   Predict with each metal ion what might happen when $n > 2$.

Q7.12   What is the significance of plants and organisms synthesizing PCs with $n > 2$?

Q7.13   Read Pre-lab 3.7.b. Why can you not treat the data by the Benesi–Hildebrand method?

TABLE 7.1   UV Spectroscopy Results

Cd^{2+}–PC$_2$			Cu^{2+}–PC$_2$		
Vol. Cd^{2+} (μl)	[Cd^{2+}] (mM)	A_{250}	Vol. Cu^{2+} (μl)	[Cu^{2+}] (mM)	A_{265}
5			5		
10			10		
15			15		
20			20		
25			25		
30			30		
35			35		
40			40		
45			45		
50			50		
60			60		
70			70		
80			80		
90			90		
100			100		

(Continued)

TABLE 7.1 *Continued*

Cd²⁺–Glutathione			Cu²⁺–Glutathione		
Vol. Cd²⁺ (μl)	[Cd²⁺] (mM)	A_{250}	Vol. Cu²⁺ (μl)	[Cu²⁺] (mM)	A_{265}
5			5		
10			10		
15			15		
20			20		
25			25		
30			30		
35			35		
40			40		
45			45		
50			50		
60			60		
70			70		
80			80		
90			90		
100			100		

EXPERIMENT 7.4: STABILITY CONSTANT DETERMINATION OF CD²⁺ AMINO ACID COMPLEXES

Level 4

Pre-lab 7.4.a: Cd²⁺–PC₂ Complexes[14,18]

The species distribution for (γ-Glu-Cys)₂-Gly, PC₂ and its cadmium(II) complex Cd²⁺–PC₂ in aqueous solution is complex, varying with pH and reactant ratios (see Fig. 7.10). The PC₂ ligand protonation macroconstants and the numbers of different species of each form (in parentheses) are summarized in Figure 7.11.

As we saw in Experiment 7.3, PC₂ ligands show enhanced binding of Cd²⁺ over lower molecular weight ligands such as glutathione. $H_3L^{2=}$, the predominant species over a wide pH range, is the form that most likely first binds the Cd²⁺ ion. At an L:M ratio = 1:1 and pH 4.0, CdH₂L predominates; with increasing pH, CdHL and then CdL forms take over. At

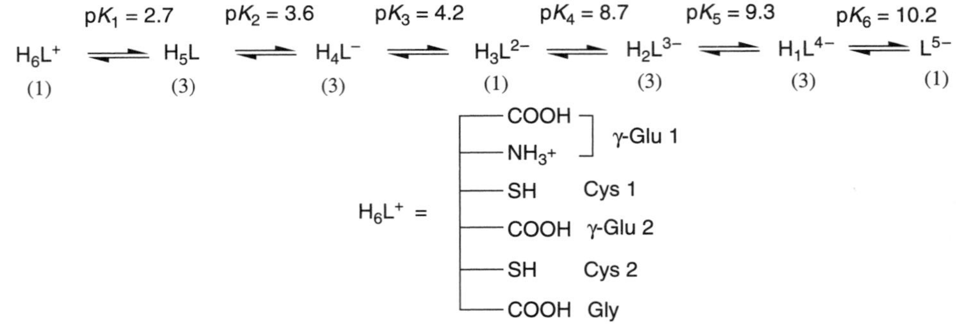

$$pK_1 = 2.7 \quad pK_2 = 3.6 \quad pK_3 = 4.2 \quad pK_4 = 8.7 \quad pK_5 = 9.3 \quad pK_6 = 10.2$$

$$H_6L^+ \rightleftharpoons H_5L \rightleftharpoons H_4L^- \rightleftharpoons H_3L^{2-} \rightleftharpoons H_2L^{3-} \rightleftharpoons H_1L^{4-} \rightleftharpoons L^{5-}$$

$$(1) \qquad (3) \qquad (3) \qquad (1) \qquad (3) \qquad (3) \qquad (1)$$

$$H_6L^+ = \begin{array}{ll} \text{COOH} & \\ \text{NH}_3^+ & \end{array} \right] \text{γ-Glu 1}$$

```
            ┌── COOH ┐
            │          } γ-Glu 1
            ├── NH₃⁺  ┘
            │
            ├── SH    Cys 1
H₆L⁺ =      │
            ├── COOH  γ-Glu 2
            │
            ├── SH    Cys 2
            │
            └── COOH  Gly
```

Figure 7.11 Protonation of the PC₂ ligand.

an L:M ratio = 2:1, the *bis* ligand complexes, CdH$_3$L$_2$ and CdH$_2$L$_2$ predominate. In these complexes, the Cd^{2+} shows 4 coordinate binding, adopting a tetrahedral molecular geometry.

In this experiment, you will determine the protonation constants of the amino acids glycine and cysteine as models for the PC$_2$ ligand and predict the structures of the PC$_2$ forms indicated in Figure 7.11. Since glutamate is bound through its γ-COO$^-$ in the PC$_2$ peptide, glycine is an appropriate model for the γ-glu N- and C-terminal binding sites shown in H$_6$L, Figure 7.11. You will then determine the Cd^{2+} binding constants of each amino acid and predict the most favorable Cd^{2+}–PC$_2$ molecular structures. An optional full pH range Cd–PC$_2$ analysis is outlined for those having the appropriate computational software.

Pre-lab 7.4.b: Potentiometric Titrations for Determining Ligand pK_a Values and Metal Ligand Stability Constants[19]

This experiment is adapted from Girolami *et al.,*[19] in which the theory behind the experiment is extensively detailed. We will cover only a portion of the binding analysis here.

The reaction that we will be looking at involves the binding of amino acid, HA, to the Cd^{2+} ion in aqueous solution, and potentiometric titration will be used to determine thermodynamic constants. The general equilibrium reactions and mass balance expressions for the Cd^{2+}–HA system are given in equatios (7.2)–(7.3). Here, HA represents an amino acid such as glycine or cysteine, in which the C-terminal carboxylate is already deprotonated (pK_a^{COOH} = 1.8–2.1). We will assume that *bis* complexation occurs; however, higher L:M ratios exist and can be analyzed.

$$\text{HA}_{(aq)} \leftrightharpoons \text{H}^+{}_{(aq)} + \text{A}^-{}_{(aq)} \quad K_a = [\text{H}^+][\text{A}^-]/[\text{HA}] \tag{7.2}$$

$$\text{Cd}^{2+}{}_{(aq)} + 2\text{A}^-{}_{(aq)} \leftrightharpoons \text{CdA}_{2(aq)} \quad \beta_2 = [\text{CdA}_2]/[\text{Cd}^{2+}][\text{A}^-]^2 \tag{7.3}$$

Ligand pK_a Determination

The structures of glycine and cysteine are given in Figure 7.12, with labile H$^+$ highlighted in bold. [We will not consider glutamic acid in this experiment. Because PC$_2$ has a linkage through the γ-COO group, glycine provides an appropriate model for Cd^{2+} binding (see Fig. 7.7)]. The pK_a for a weak acid such as the RNH$_3{}^+$ of glycine, HA, can be determined from the Henderson–Hasselbach equation (7.4). It is always desirable to have [HA] and [A$^-$] in terms of A_{total}, a known term, and therefore, by using both the laws of conservation of mass, A_{total} = [HA] + [A$^-$], and charge, [H$^+$] + [Na$^+$] = [A$^-$] + [OH$^-$], equation (7.4) can be rewritten as equation (7.5).

$$\text{pH} = \text{p}K_a + \log[\text{A}^-]/[\text{HA}] \tag{7.4}$$

$$\text{p}K_a = \text{pH} + \log\frac{[A_{total} - ([\text{Na}^+] + [\text{H}^+] - [\text{OH}^-])]}{[\text{Na}^+] + [\text{H}^+] - [\text{OH}^-]} \tag{7.5}$$

We will treat the successive pK_as of the diprotic cysteine similarly, as two separate deprotonation events. We still have one more correction, however. Because pH meters measure H$^+$ activity (a_{H+}) and not H$^+$ concentration, we must correct for this. The

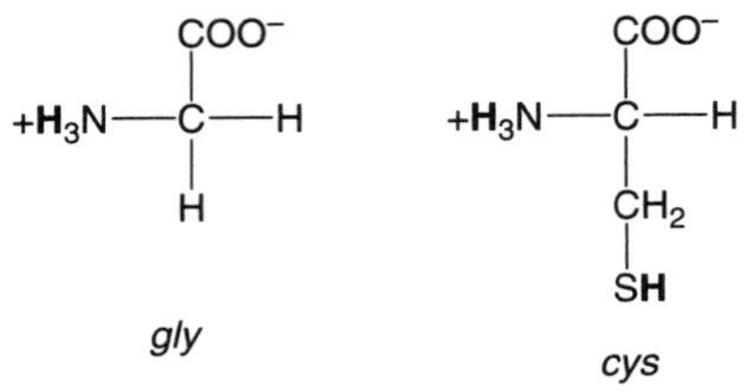

Figure 7.12   Structures of glycine and cysteine. The C-terminus is assumed deprotonated.

mean activity, $\gamma_{\pm}$, for a given ionic strength, μ, is given by the Davies equation (7.6), where z_1 and z_2 are the charges on the electrolyte used to maintain ionic strength as in the calculation of μ [see equation (5.12)].

$$- \log \gamma_{\pm} = \frac{0.50 \, z_1 z_2 \mu^{1/2}}{1 + \mu^{1/2}} - 0.10\mu \tag{7.6}$$

Since $a_{H+} = \gamma_{\pm}[H^+]$, the H^+ concentration (7.7). Similarly, the OH^- concentration is calculated from equation (7.9), using the autoionization constant of water, K_w, at 25 °C and corrected for ionic strength using equation (7.9) where $m_+ = m_- = 10^{-7}$ mol kg^{-1} and $m° = 1$ mol kg^{-1}.

$$\log[H^+] = -pH - \log(\gamma_{\pm}) \tag{7.7}$$

$$\log[OH^-] = pH - pK_w + \log(\gamma_{\pm}) \tag{7.8}$$

$$K_w° = \gamma_{\pm}^2 (m_+ m_- / m^{o2}) \tag{7.9}$$

Now, all quantities in equation (7.5) needed for calculating the pK_a are either known or can be calculated. The necessary data for determining the pK_a of HA from equation (7.5) are obtained by titrating NaOH into a solution of HA containing electrolyte, either 0.1 M KNO$_3$ (gly) or 3 M NaCl (cys). The pH after each measured addition is obtained and recorded. The pK_a is then calculated at each pH value in the linear (20–80%) portion of the typical titration plot (pH vs added OH$^-$ equivalents). The average pK_a for each region is then recorded. Note that two regions and therefore two pK_as will be observed for the diprotic cysteine.

Determination of Stepwise Stability Constants, K_1 and K_2

The numerous Cd–PC$_2$ species that exist in aqueous solution require specialized software for complete stability constant analysis. The basic principles behind the calculation, however, are well pointed out by considering a very simple reaction between the already doubly deprotonated amino acid, A$^-$, and Cd^{2+} aqueous ion, equations (7.10)–(7.11). [*Note*: this holds strictly only for glycine. In the cysteine system, the N-terminus is protonated (Cys should really be represented as HA).]

$$Cd^{2+} + A^- \leftrightharpoons CdA^+ \quad K_1 = [CdA^+]/[Cd^{2+}][A^-] \tag{7.10}$$

$$CdA^+ + A^- \leftrightharpoons CdA_2 \quad K_2 = [CdA_2]/CdA^+][A^-] \tag{7.11}$$

Considering the binding of two amino acids per Cd^{2+} metal ion, it can be shown that the average number of ligands per metal ion, n, is given by equation (7.12) and experimentally by equation (7.13). [Recall that $\beta_2 = K_1 K_2$.] Equation (7.12) can be expanded for higher order binding.

$$n = \frac{\text{moles } Cd^{2+} \text{ bound } A^-}{\text{total mol } Cd^{2+}} = \frac{K_1[A^-] + 2K_1K_2[A^-]^2}{1 + K_1[A^-] + 2K_1K_2[A^-]^2}$$

$$= \frac{\beta_1[A^-] + 2\beta_2[A^-]^2}{1 + \beta_1[A^-] + 2\beta_2[A^-]^2} \tag{7.12}$$

$$n = \frac{A_{\text{total}} - [HA] - [A^-]}{M_{\text{total}}} \tag{7.13}$$

Again free $[A^-]$ and $[HA]$ need to be put into terms that we can measure or that are known. Under our given assumptions, they become:

$$[HA] = H^+ \text{ (added)} + [OH^-] - [H^+] \tag{7.14}$$

$$[A^-] = (K_a/H^+)(H^+(\text{added}) + [OH^-] - [H^+]) \tag{7.15}$$

where H^+(added) is the concentration of strong acid (e.g., HNO_3) added at the start of the experiment. If you add inital base, OH^- (added) would need to be subtracted. With appropriate substitution into the β-form of equation (7.12) followed by rearrangement we obtain equation (7.16). Analysis of the linear part of a plot of $n/\{(1-n)[A^-]\}$ vs $\{(2-n)[A^-]\}/(1-n)$ (at low $[A^-]$) gives slope β_2 and intercept β_1.

$$\frac{n}{(1-n)[A^-]} = \beta_1 + \frac{(2-n)[A^-]}{(1-n)}\beta_2 \tag{7.16}$$

Potentiometric Titration of the PC_2–Cd^{2+} System and Analysis Using the BEST Program[20,21]

In the above example, we simplified our analysis by considering only the doubly deprotonated amino acid species. In a complete analysis, all possible species and their equilibria over a wide pH range must be considered. For the PC_2 ligand, three mass balance equations in terms of total PC_2 ligand (T_{PC2}), total Cd^{2+} metal ion (T_{Cd2+}), and total initial hydrogen ion concentration (T_{H+}) must be set up. Borrowing from the distribution diagrams in Figure 7.10, the species that need to be considered for the Cd^{2+}–PC_2 system are: $H_6L^+, \ldots, L^{5-}$, CdH_2L, $CdHL$, CdL, CdH_3L_2, CdH_2L_2, $CdHL_2$, CdL_2, Cd^{2+}, OH^-, H^+(added) and OH^- (added).[†] The computer program sets the equations up in terms of concentrations and β values. With experimentally determined ligand pK_as and initial

---

[†]A lot of work goes in to determining these species. In the Cd–PC_2 system, ^1H NMR data on PC_2 in the presence and absence of Cd^{2+} and UV titrations, along with information from other known and similar systems, helped define the species to be considered. Comparison of initial potentiometric titration curves for the PC_2 ligand, Cd^{2+}–PC_2 complex, and $6H^+$ (to represent the curve if all titratable protons were displaced by the Cd^{2+}) also provides some insight.

guesses for β values, the program solves the simultaneous equations for each component, comparing the calculated $[H^+]$ with the actual $[H^+]$ at each equilibrium point. The program then uses an iterative process to refine the β values until a reasonable fit in $[H^+]$ values is reached.

Procedure 7.4.a: Potentiometric Titration of Amino Acids, $\mu = 0.1$ M (KNO_3) for Gly and $\mu = 3.0$ M $(NaCl)$ for Cys, RT[19,22]

Solution Preparation

1. Carbonate-free ddH_2O is obtained by boiling ddH_2O and, after cooling to the touch, sealing the ddH_2O in an air-tight container. All solutions, especially $NaOH_{(aq)}$, should be prepared using this H_2O.

2. Prepare 100 ml of 0.5 M aqueous NaOH using carbonate-free ddH_2O and standardize using the primary standard, potassium hydrogen phthalate ($MM = 204.22$ g mol^{-1}):

    a. Weigh out three 0.6–0.7 g portions of potassium hydrogen phthalate into a 250 ml Erlenmeyer flask. Record actual amounts.

    b. Add 75 ml of carbonate-free ddH_2O to each portion, stopper the flask and swirl to dissolve the salt. Add a couple drops of phenolphthalein indicator.

    c. Titrate each solution with your prepared NaOH to a slight pink color change.

    d. Determine, average, and record your NaOH concentration.

3. Prepare either 500 ml of aqueous 0.2 M KNO_3 (gly) or 500 ml of aqueous 6 M NaCl (cys).

4. Prepare 25 ml of aqueous 0.4 M glycine (free base, zwitterionic form— $^+H_3NCH_2COO^-$), or 0.4 M cysteine (hydrochloride salt, $Cl^{-.+}H_3NCHCH_2SHCOOH$).

Amino Acid pH Titration

5. Into a 400 ml beaker, place 100 ml of 0.2 M KNO_3 (gly) or 6 M NaCl (cys), 10 ml amino acid, and 90 ml H_2O. Note that the pH for glycine should be ~6. For cysteine, neutralize the C-terminus by adding 1 equiv. of NaOH. Record initial pH and the room temperature.

6. Using a burette and with stirring, titrate your amino acid solution with the standardized 0.5 M NaOH solution in ~0.5 ml aliquots until pH ≈ 12. After each addition, record the exact amount of NaOH added and the corresponding pH.

7. Plot pH vs equivalents of OH^- added to view your titration. Estimate your $pK_a(s)$ from the plot.

8. Use equations (7.6)–(7.8) to calculate $[OH^-]$ and $[H^+]$.

9. Using equation (7.5) and data points in the 20–80% range, calculate the pK_a at each titration point and from these determine the average pK_a. Remember that you will need to determine two pK_as for cysteine.

10. Repeat steps 5–9 for the other amino acid.

Procedure 7.4.b: Determination of Cd^{2+}–Amino Acid Stability Constants, $\mu = 0.1$ M (KNO_3) for Gly and $\mu = 3.0$ M (NaCl) for Cys, RT[14,18]

This procedure assumes *bis* complex formation. It is recommended that a Job's plot be performed with cysteine (A_{250}; see Experiment 3.6).

Solution Preparation Carbonate-free water should be used in the preparation of all solutions (see Procedure 7.4.a, step 1).

1. Prepare and standardize a 1.0 M $NaOH_{(aq)}$ solution as in Procedure 7.4.a, step 2.

2. Prepare a 0.1 M solution of $HNO_{3(aq)}$ and standardize this with a 0.1 M NaOH solution.

3. Prepare a 500 ml solution of aqueous 0.2 M KNO_3.

3a. *Glycine*: prepare 20 ml of 0.4 M deprotonated glycine by weighing out the determined amount of glycine and neutralizing *one* equivalent of H^+ with your standardized NaOH solution. Record the pH of your solution.

3b. Into a 400 ml beaker, place 0.3085 g (1 mmol) $Cd(NO_3)_2 \cdot 4 H_2O$ (MM = 308.47) and add to this, 10 ml (1 mmol) of the 0.1 M HNO_3 solution, 100 ml of the 0.2 M KNO_3 solution and 90 ml ddH_2O.

4a. *Cysteine*: weigh out 1.261 g (0.008 mol) cysteine $\cdot$ HCl (MM = 157.6 g mol^{-1}) and dissolve in 20 ml aqueous NaOH solution (16 ml 1 M NaOH:4 ml H_2O), neutralizing two equivalents of H^+ on cysteine.

4b. Into a 400 ml beaker, place 0.2284 g (1 mmol) $CdCl_2 \cdot 2.5 \; H_2O$ (MM = 228.36 g mol^{-1}), 35 g NaCl and 10 ml (1 mmol) of the 0.1 M HNO_3 solution. Bring the total volume to 200 ml with ddH_2O.

5. Charge a burette with the deprotonated amino acid solution prepared in step 3. This is your titrant.

6. With continual stirring, titrate your Cd^{2+} solution with your amino acid solution in $\sim$0.2 ml increments until you add $\sim$10 ml of your titrant.

7. Calculate $[A^-]$ and n from your titration data and determine β_1 and β_2 from the plot of equation (7.16) using n values from 0.2–0.8. For each amino acid, determine K_1 and K_2.

Note on Cysteine

Your beginning pH will be $\sim$1.7. Titrate your solution with your cysteine solution until pH $\approx$ 2.5 in 0.5 ml increments. Note how much volume of titrant you use, but do not start recording data in Table 7.2 until after pH 2.5.

Procedure 7.4.c: Determination of Cd^{2+}–PC_2 (1:1) Stability Constants, $\mu = 0.1$ M (KNO_3), RT[20,21]

You may want to first test your technique and analysis on a simple, known system (e.g., Cd–glycinate).

Solution Preparation Carbonate-free water should be used in the preparation of all solutions (see Procedure 7.4.a, step 1).

1. Prepare and standardize a 0.5 M $NaOH_{(aq)}$ solution as in Procedure 7.4.a, step 2.

2. Prepare a 0.1 M solution of $HNO_{3(aq)}$ and standardize this with your 0.1 M NaOH solution.

3. Prepare a 5 ml solution 2 mM $Cd(NO_3)_2 \cdot 4H_2O$ (MM = 308.47), 2 mM PC_2 ligand, at pH = 2.5 and $\mu = 0.1$ M (KNO_3). Be sure to record the exact amount of HNO_3 added to bring the pH to 2.5.

4. With constant stirring, titrate your $Cd-PC_2$ solution with your standardized NaOH solution in 5–10 μl increments, recording the pH after each aliquot addition.

5. Using the ligand pK_a values given in Figure 7.11, the $Cd-PC_2$ species of Figure 7.10, and a computer analysis program such as BEST (see Reference section), compute the stability constants for the 1:1 $Cd-PC_2$ complexes. Compare your answers to the literature.[14,18]

Name _____   Section _____

Results Summary for the Potentiometric Titration Studies of Cd^{2+}–Amino Acid Complexes

Make copies of this sheet

Amino acid system:_____

	Trial 1	Trial 2	Trial 3	Average
1. Concentration of 0.5 M NaOH solution:	_____	_____	_____	_____

2. Room temperature (°C): _____

3. log $\gamma_\pm$ _____

4. K_w _____

Amino Acid pK_a Determinations

5. Complete Table 7.2.

6. Average pK_{a1} _____   average pK_{a2} (if applicable) _____

7. What atom(s) are being deprotonated in the titration of your amino acid?

Cd^{2+}–Amino Acid Stability Constant Determination

8. [H^+] added (M) _____

9. Amino acid K_a _____

10. [Cd]$_{total}$ (M), M_T _____

11. Cysteine only—volume of titrant used to reach pH $\approx$ 2.5 _____

12. Complete Table 7.3—*you will need to continue this table in your notebook.*

13. β_1 _____ β_2 _____

14. K_1 _____ K_2 _____

15. Predict and draw structures of your Cd–amino acid complexes.

16. (Optional experiment) Summarize your results for the $Cd-PC_2$ potentiometric titration.

Post-lab Questions

Q7.14 Compare the Cd–glycine and Cd–cysteine formation constants. Which amino acids (gly, cys, glu) do you expect to play a significant role in $Cd-PC_2$ binding?

Q7.15 Predict the protonation state of the PC_2 forms in Figure 7.11.

Q7.16 Predict and draw the binding modes for the CdH_2L, $CdHL$, CdL, CdH_3L_2, and CdH_2L_2 species.

TABLE 7.2 Titration Data for pK_a Determinations

NaOH added (ml)	[Na$^+$] (M)	pH	[H$^+$] (M)	[OH$^-$] (M)	pK_a	NaOH added (ml)	[Na$^+$] (M)	pH	[H$^+$] (M)	[OH$^-$] (M)	pK_a

TABLE 7.3 Data Work-up for Cd^{2+}-Amino Acid Titration for Stability Constant Determination

Titrant added (ml)	pH	[H$^+$] (M)	[OH$^-$] (M)	[A$^-$] (M)	[HA] (M)	A$_{total}$ (M)	n value	x value[a]	y value[a]

[a]See equation (7.16)—calculate for n = 0.2–0.8.

EXPERIMENT 7.5: BIOSYNTHESIS OF PHYTOCHELATINS BY THE FISSION YEAST (*SCHIZOSACCHAROMYCES POMBE*) IN THE PRESENCE OF Cd^{2+} AND ANALYSIS BY ATOMIC ABSORPTION SPECTROSCOPY

Level 5

This experiment involves cell biology techniques and theory.

Pre-lab 7.5.a: Fission Yeast, *Schizosaccharomyces pombe*, and its Role in the Study of PC$_n$ Biosynthesis [8,23,24]

Both plants and yeast are known to produce phytochelatins (PC_n), peptide metal-binding ligands, in response to heavy metal (especially Cd^{2+}) toxicity: heavy metal ions activate the enzyme, PC synthase (PCS), which produces PC_ns from glutathione (GSH); see equation (7.1).

Yeasts, a type of fungi, are unicellular microorganisms with cells that are structurally similar to plants and animals. Yeasts have a small genome ($\sim 200\times$ smaller than the human genome) and a rapid doubling time of approximately 2 h. For these reasons yeasts are often used as model systems for studying biological processes. (Because of their relatively sophisticated cellular machinery, yeasts are also used in genetic engineering to produce proteins such as the hepatitis B vaccine.) The fission yeast, *Schizosaccharomyces pombe*, has been used extensively in the study of PC biosynthesis.

Activation of PC synthase leads to the production of low molecular weight (LMW) PC_ns that sequester Cd^{2+} and transport the metal into the vacuole of the cell. High molecular weight (HMW) PC_ns and sulfide are known to play a role in Cd^{2+} storage in the vacuole. A schematic of this process is shown in Figure 7.13.

In this experiment, we will culture fission yeast cells from the *S. pombe* strain in the presence of Cd^{2+}. In response, the yeast will express PCS that will begin PC_n synthesis and uptake of the Cd^{2+} into the vacuole of the cell. We will then extract the $Cd-PC_2$ complexes and analyze the Cd^{2+} content by atomic absorption spectroscopy. Appendix 5 describes the apparatus and gives reagents needed for yeast culture.

Pre-lab 7.5.5.b: Atomic Absorption Spectroscopy[25,26]

Atomic absorption spectroscopy (AAS) will be used to qualitatively visualize and to quantify the amount of Cd^{2+} taken up by the yeast cells. A schematic of the AA spectrometer is shown below. The layout is very similar to a UV–visible spectrophotometer, Figure 7.14. In AAS we detect the resonance absorption of a gaseous atom, in our case cadmium, with unique electronic transition energies. The source is usually either a hollow cathode lamp containing a mixture of several metals or an electrodeless discharge lamp. Each contains the metal of interest and emits the appropriate atomic line spectrum for absorption by the element. Another key difference is the sample holder in which the sample is vaporized. In general, either a flame or furnace technique is used. An air–acetylene flame with a lamp current of 4 mA, slit width of 0.5 nm, and wavelength 228.8 nm is used for cadmium. An advantage of the flameless electrothermal atomizer, however, is that impurities are removed prior to atomization of the sample; high sensitivity with small sample volumes is achieved. You should familiarize yourself with the type of spectrometer

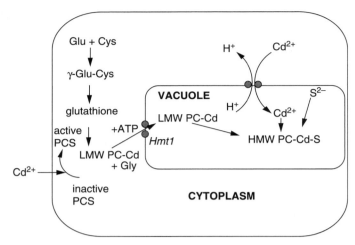

Figure 7.13   Schematic of PC_n biosynthesis in yeast cells. LMW = low molecular weight; HMW = high molecular weight, *Hmt1* is a gene found in fission yeast for the LMW–PC–Cd transporter protein. [Adapted from Zenk[8] and Cobbett[24].]

you will be using. A good source for this is Skoog, Holler, and Nieman in *Principles of Instrumental Analysis*.[25]

Procedure 7.5: Culturing Yeast for Optimal Growth in Liquid[27,28]

See Appendix 5 for information on yeast culture technique and apparatus. Aseptic technique must be practiced throughout the procedure.

Yeast Culture

1. Streak out yeast onto YPD plates (see Appendix 5) and allow to grow for 3–4 days at RT or for 2 days at 30 °C. Note that the plate may be stored for short periods at 4 °C.
2. Using sterile technique, inoculate 5–10 ml of liquid YPD media in a test tube with yeast from this plate.
   a. Flame the opening of your testtube and pipet 5–10 ml of YPD media into the test tube. Reflame the opening of your testtube.
   b. Using a sterilized pipet tip or the flat end of a sterile toothpick, add a colony of yeast to the media. Reflame your test tube and cover with foil.
3. Grow overnight at 30 °C with shaking (200–250 rpm).

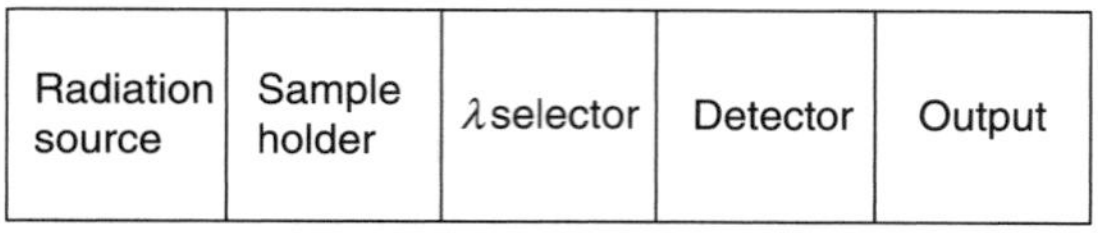

Radiation source	Sample holder	λ selector	Detector	Output

Figure 7.14   Diagram of an atomic absorption spectrophotometer.

4. The next day, measure the optical density (OD) at 600 nm of the overnight culture. Remember to use good sterile technique!

   a. Dilute 0.1 ml of the overnight culture by adding 0.9 ml of YPD media (a 1:10 dilution).

   b. Dilute 0.05 ml of the overnight culutre by adding 0.95 ml of YPD media (a 1:20 dilution).

   c. After blanking with YPD, measure the OD_{600} of these dilutions.

   d. Calculate the OD_{600}/ml value for the undiluted overnight culture.

5. Use the overnight culture to inoculate a new culture to which you will add your Cd^{2+}: Let us say that you will be growing up 100 ml of yeast in YPD containing Cd^{2+}; you will want to harvest the culture at mid-log phase (i.e., $OD_{600} \approx 0.4-0.8$), and you want the culutre to go through seven to eight doublings. How do you inoculate your 100 ml of YPD using your overnight culture so that after seven to eight doublings, the OD_{600} of the culture is $\sim 0.4-0.8$?

If the final desired OD_{600} is 0.4, then back calculate to the beginning OD by:

$$0.4 \to 0.2 \to 0.1 \to 0.05 \to 0.025 \to 0.0125 \to 0.00625 \to 0.00313 \to 0.00156$$

Thus, if you inoculate 100 ml of YPD so that you have 0.00156 OD_{600}, then after eight doublings of the yeast, the OD_{600} should be ~ 0.4. Use your typical $OD_1 V_1 = OD_2 V_2$ to determine the volume (V_1) of your original overnight stock at OD_1 to obtain an $OD_2 = 0.00156$ in a final volume (V_2) of 100 ml.

Cd^{2+} Incubation

6. Prepare two 100 ml cultures. To one, add 1 ml of your 10 mM Cd^{2+} salt solution and nothing to the other as a control and grow the cultures at 30 °C with shaking (200–250 rpm). *Schizosaccharomyces pombe* has a doubling time of 2–3 h depending on conditions. Taking the doubling time as 2 h, eight doublings will be 16 h. Add an extra 2 h recovery period. Incubate a total 18 h.

7. To harvest the cells from the liquid culture, centrifuge at $\sim 3000-5000$ rpm for 10–20 min. You should record the mass of your centrifuge tube(s) to obtain a yield for your pellet. Pour off the supernatant and save. Wash the pellet with ddH_2O and centrifuge. Repeat the wash. Obtain a mass for the washed pellet and store at -80 °C.

Isolation of the Crude $Cd^{2+}-PC_2$ Extract

8. Add 750–1000 μl of Tris buffer (pH 7.8) that contains 1 mM PMSF (see Appendix 5) to your thawed yeast pellet and transfer to a 1.5 ml centrifuge tube.

9. Break open the cells by vortexing the pellet–buffer mixture in the presence of acid washed beads for 30 s, followed by 1 min rests on ice. Repeat at least five times.

10. Quantitatively transfer your solution to a weighed centrifuge vial and centrifuge at 15,000 g for 5 min.

TABLE 7.4   Calibration Curve for Cadmium

Solution	Concentration (μg/ml)	Absorbance
1		
2		
3		
4		
5		
6		

11. Remove and save (note the exact volume) the supernatant for atomic absorption analysis. Note exact weight and also save also the pellet in case the extraction needs to be performed again.

Optional Experiments

1. Determine the total protein content of the supernatant (crude extract) by performing a Bradford protein assay.
2. Use nondenaturing polyacrylamide gel electrophoresis (PAGE) to separate and visualize $LMW-Cd^{2+}-PC_2$ and $HMW-Cd^{2+}-PC_2$ complexes contained in the crude extract. A detailed protocol for doing this is given by Abrahamson et al.[27]. The peptides should be visualized by silver staining but referenced to the [109]Cd result since proteins other than PCs may be present.

Procedure 7.5.b: Determination of Cd^{2+} Uptake by Atomic Absorption Spectroscopy[29,30]

1. Prepare a standard stock solution (1000 μg/ml of Cd) by dissolving 1 g Cd metal (record exact mass) in a minimum volume of 1:1 $HCl:ddH_2O$. Dilute to 1000 ml with ddH_2O.
2. Using the optimum concentration range for Cd (0.05–2 mg l^{-1}) prepare a series of six concentrations to be used in constructing a calibration curve for Cd. Record your concentrations and absorbance readings in the table in the Results Summary section. Plot the calibration curve for cadmium (absorbance vs concentration).
3. Analyze the supernatant from your peptide isolation procedure using atomic absorption spectroscopy. Follow the directions for your particular instrument. Assuming 100% extraction, determine the amount of Cd taken up by the yeast. Make dilutions of your supernatant as needed using ddH_2O.

Name _____   Section _____

Results Summary for the Biosynthesis of PC_n and Cd^{2+} Binding in a Yeast Model

1. Complete Table 7.4.
2. Amount of Cd in sample.

3. Read up on the biochemistry of Cd^{2+} as a heavy metal. Why might yeast and plants biosynthesize phytochelatins to sequester this ion? What happens to this when it enters the cell? What would happen if it were released from storage?

REFERENCES

1. Spiro, T. G., Stigliani, W. M. *Chemistry of the Environment*, 2nd edn. Prentice Hall: Upper Saddle River, NJ, 2003.

2. Soroka, K., Vithanage, R. S., Phillips. D. A., Walker, B., Dasgupta, P. K. *Anal. Chem.* **1987**, *59*, 629–636.

3. Hollingshead, R. G. W. *Oxine and its Derivatives*, Vols I–IV. Buttersworth: London, 1954–1956.

4. Hung, L. S., Chen, C. H. *Materials Science and Engineering R—Reports* **2002**, *39(5–6)*, 143–222.

5. Pearce, D. A., Jotterand, N., Carrico, I. S., Imperial, B. *J. Am. Chem. Soc.* **2001**, *123(21)*, 5160–5161.

6. Lucy, C. A., Ye, L. W. *J. Chromatogr. A* **1994**, *671(1–2)*, 121–129.

7. Czarnik, A. W. *Topics in Fluorescence Spectroscopy*, Vol. 4 *Probe Design and Chemical Sensing*, Lakowicz, J. R. (ed.). Kluwer Academic: New York, 2002, Chapters 3 and 4.

8. Zenk, M. H. *Gene* **1996**, *179*, 21–30.

9. Grill, E., Loffler, S., Winnacker, E. L., Zenk, M. H. *Proc. Natl Acad. Sci. USA* **1989**, *86*, 6838–6842.

10. Klapheck, S., Fliegner, W., Kimmer, I. *Plant Physiol.* **1994**, *104*, 1325–1332.

11. Chan, W. C., White, P. D. (eds). *Fmoc Solid-Phase Peptide Synthesis: a Practical Approach*. Oxford University Press: Oxford, 2000, pp. 9–76.

12. Dunn, B. M., Pennington, M. W. (eds). *Peptide Analysis Protocols*. Humana Press: Totowa, NJ, 1994, pp. 23–53.

13. Truran, G. A., Aiken, K. S., Fleming, T. R., Webb, P. J., Markgraf, J. H. *J. Chem. Educ.* **2002**, *79*, 85–86.

14. Dorcak, V., Krezel, A. *Dalton Trans.* **2003**, 2253–2259.

15. Ellman, G. L. *Arch. Biochem. Biophys.* **1959**, *82*, 70–77.

16. P. Edman, *Acta Chem. Scand.* **1950**, *4*, 283.

17. Walker, J. M. (ed.). *Basic Protein and Peptide Protocols*. Methods in Molecular Biology Vol. 32. Humana Press: Totawa, NJ, 1994, pp. 329–334.

18. Johanning, J., Strasdeit, H. *Angew. Chem. Int. Edn.* **1998**, *37*, 2464–2466.

19. Girolami, G. S., Rauchfuss, T. B., Angelici, R. J. *Synthesis and Technique in Inorganic Chemistry—a Laboratory Manual*, 3rd edn. University Science Books: Sausalito, CA, 1999.

20. Martell, A. E., Hancock, R. D. *Metal Complexes in Aqueous Solution*. Plenum Press: New York, 1996, pp. 223–226.

21. Martell, A. E., Motekaitis, R. J. *Determination and Use of Stability Constants*, 2nd edn. VCH: Weinheim, 1992.

22. Bottari, E., Festa, M. R. *Talanta* **1997**, *44*, 1705–1718.

23. Atlas, R. M. *Principles of Microbiology*, 2nd edn. Brown: Boston, MA, 1997, Chap. 19.

24. Cobbett, C. S. *Plant Physiol.* **2000**, *123*, 825–832, Figure 8.5.2.

25. Skoog, D. A., Holler, F. J., Nieman, T. A. *Principles of Instrumental Analysis*, 5th edn. Saunders: Philadelphia, PA, 1998, Chap. 9.

26. Schwedt, G. *The Essential Guide to Analytical Chemistry*. Wiley: New York, 1997.

27. Atlas, R. M., Brown, A. E. *Experimental Microbiology: Fundamentals and Applications*. Prentice Hall: Upper Saddle River, NJ, 1997.

28. Abrahamson, S. L., Speiser, D. M., Ow, D. W. *Anal. Biochem.* **1992**, *200*, 239–243.

29. Sawyer, D. T., Heineman, W. R., Beebe, J. M. *Chemistry Experiments for Instrumental Methods*. Wiley: New York, 1984, Chap. 9.

30. Pappenhagen, J. M. *Standard Methods for the Examination of Water and Waste Water*, 16th edn. American Public Health Association (AWWA, WPCF): Washington, DC, 1985, p. 154.

Metals in Molecular Biology—Synthesis, Photophysical and Chiral Properties of Tris(1,10-Phenanthroline)Chromium(III): Metal Complex DNA Interactions and Reactivity

Metals use such unique methods of binding to and reacting with biological molecules; they broaden our understanding of the chemical principles underlying living processes, as well as our ability to detect and treat the diseases that adversely affect them.

(Elizabeth Boon, Stony Brook University)

PROJECT OVERVIEW[1,2]

The discovery of the anticancer properties of *cis*-platin fueled the imaginations of inorganic chemists, leading to new therapies to fight this disease. In the case of *cis*-platin, the generally toxic platinum ion was turned into a beneficial therapy. Not long afterwards, the discovery of the anticancer properties of the natural antibiotic iron-bleomycin showed how damaging reactive oxygen species, when delivered in a controlled manner, made for a powerful medicinal treatment. Both *cis*-platin and iron-bleomycin are examples of the ligand environment controlling chemical and physical properties about a metal center, through which important biochemical problems can be addressed. Both *cis*-platin and iron-bleomycin bind to and act on DNA and it was not long before research chemists and biologists realized the potential for such DNA-binding complexes to probe biomolecular structure and reactivity. In particular, metal complexes that activate dioxygen to species that react with DNA and proteins have been used to map DNA–protein, DNA–drug, and protein–protein binding sites. These complexes are called chemical footprinting agents. In addition, chiral metal complexes are used to probe biomolecular conformation, and luminescent transition metal complexes are valuable tools in the study of electron transfer reactions of nucleic acids and DNA.

In this laboratory series you will investigate the chemistry and biological reactivity of the Tris(1,10-phenanthroline) chromium(III) ion, $[Cr(phen)_3]^{3+}$. First, $[Cr(phen)_3]^{3+}$ will

Integrated Approach to Coordination Chemistry: An Inorganic Laboratory Guide. By Rosemary A. Marusak, Kate Doan, and Scott D. Cummings

be resolved into its optical isomers and its optical activity measured. Its stereospecific binding to DNA will then be explored. Because $[Cr(phen)_3]^{3+}$ also has luminescent properties, it can be used in investigations of electron-transfer reactions involving DNA. The quenching of the luminescence of $[Cr(phen)_3]^{3+}$ using guanosine, guanosine monophosphate, and DNA will be investigated and the electron transfer quenching rate constants will be measured using steady-state and time-resolved luminescence spectroscopy.

EXPERIMENT 8.1: SYNTHESIS OF Tris(1,10-PHENANTHROLINE)CHROMIUM(III) HEXAFLUOROPHOSPHATE, [Cr(phen)₃] (PF₆)₃

Level 5

Pre-lab 8.1: On the Synthesis of $[Cr(phen)_3]^{3+}$

The traditional preparation of chromium(III) complexes involves reacting the brilliant blue $Cr_{(aq)}^{2+}$ ion with phenanthroline ligand. In this experiment we will prepare $[Cr(phen)_3]^{3+}$, Figure 8.1, using a revised method in which the precipitation of $AgCl_{(s)}$ drives the complex formation and eliminates the need for inert atmosphere conditions.

Procedure 8.1: Synthesis of Tris(1,10-Phenanthroline)Chromium(III) Hexafluorophosphate, [Cr(phen)₃] (PF₆)₃[3,4]

All manipulations must be carried out in the dark (use aluminum foil on reaction flasks as indicated) and in the hood.

$$CrCl_{3(aq)} + 3AgNO_{3(aq)} \rightarrow Cr(NO_3)_3 + 3AgCl_{(S)} \tag{8.1}$$

1. Dissolve 0.800 g $CrCl_3 \cdot 6H_2O$ (3.0 mmol) in 15 ml dd H_2O.
2. Dissolve 1.530 g $AgNO_3$ (9.00 mmol) in a separate 15 ml dd H_2O.
3. Combine the two solutions in a 100 ml round-bottom flask wrapped in foil and containing a magnetic stir bar. Silver chloride will precipitate from the resulting green-blue solution.
4. Stir the mixture for 15 min.

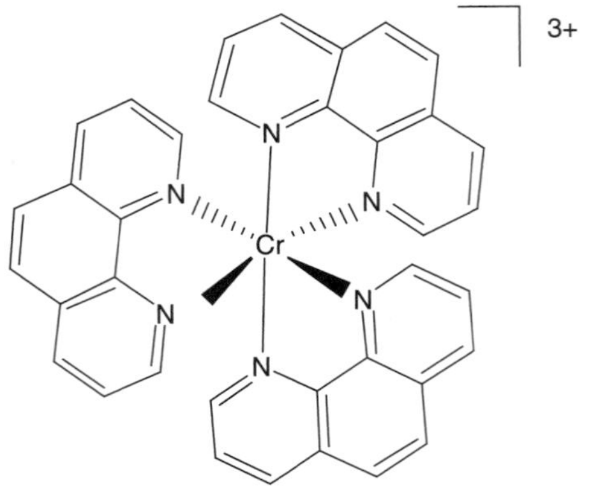

Figure 8.1 The $[Cr(phen)_3]^{3+}$ complex ion.

5. While the above mixture is stirring, dissolve 1.620 g of solid 1,10-phenanthroline (9.00 mmol) in 30 ml dd H_2O with mild heating. Before the temperature reaches 50 °C add 1.5 ml of 3 M HCl. When the temperature reaches 50 °C add about four to five more drops of HCl or until the solid is fully dissolved.

6. Vacuum filter the AgCl suspension and return the bright blue filtrate to the foil-wrapped round-bottom flask.

$$Cr(NO_3)_3 + 3\ phen_{(aq)} \longrightarrow [Cr(phen)_3](NO_3)_{3(aq)} \qquad (8.2)$$

7. Add the fully dissolved 1,10-phenanthroline to the above blue solution, turning the solution yellow-green with precipitate formation.

8. Gently heat the solution to 50 °C using a heating mantle and variac. After 30 min the solution should begin to turn reddish-purple with precipitate.

9. Stir the reaction mixture at 50 °C for 12 h (at least 5 h is required).

10. Cool the solution to room temperature and then in an ice bath for 15 min.

$$[Cr(phen)_3](NO_3)_{3(aq)} + 3NH_4PF_6 \rightarrow [Cr(phen)_3](PF_6)_{3(s)} + 3NH_4NO_3 \quad (8.3)$$

11. Prepare a solution of 2.45 g (15.0 mmol) NH_4PF_6 in 15 ml dd H_2O and add this to the brick red filtrate. A light red precipitate should form immediately.

12. Filter the solid and wash with cold dd H_2O and then with several aliquots of diethyl ether–isopropanol (5:1).

13. Dry in a vacuum dessicator and record yield.

Determination of Sample Purity

14. Prepare $\sim 5 \times 10^{-4}$ M (~ 5 mg/10 ml), $\sim 5 \times 10^{-5}$ M, and $\sim 5 \times 10^{-6}$ M solutions of $[Cr(phen)_3](PF_6)_3$. Determine the purity of your complex by obtaining a UV–visible spectrum from 200 to 700 nm and comparing the ratios of the experimental $A_{430}{:}A_{320}$ with $\varepsilon_{430}{:}\varepsilon_{320}$. Excited state absorbances and molar absorptivities for $[Cr(phen)_3]^{3+}$ are: ε_{430} (shoulder) $= 642$ M^{-1}cm^{-1}; ε_{320} (shoulder) $= 13200$ M^{-1}cm^{-1}; $\varepsilon_{266} = 65700$ M^{-1}cm^{-1}.

Name _____   Section _____

Results Summary for the Synthesis of Tris(1,10-phenanthroline)chromium(III) Hexafluorophosphate

1. Limiting reagent for synthesis: _____

2. Theoretical yield: _____

   Actual yield: _____

   Percentage yield: _____

3. Compare the ratios of the experimental $A_{430}{:}A_{320}$ and comment on the purity of your product.

EXPERIMENT 8.2: RESOLUTION OF TRIS(1,10-PHENANTHROLINE)-CHROMIUM(III) HEXAFLUOROPHOSPHATE, [Cr(phen)₃] (PF₆)₃ BY FRACTIONAL PRECIPITATION USING ANTIMONYL-D-TARTRATE

Level 5

Pre-lab 8.2.a: Optical Isomers of Metal Coordination Complexes[5,6]

From organic chemistry, you are familiar with chiral centers and optical activity in carbon-containing compounds. While metal complexes can also be optically active, this is not always due to the presence of an asymmetric atom. Metal complexes are chiral if they have only proper rotation axes. The most cited examples include the Tris-bidentate chelate complexes [e.g., $[Co(en)_3]^{3+}$] of D_3 symmetry and the *cis-bis*-bidentate chelate complexes [e.g., $Co(en)_2(NO_2)_2$], of C_2 symmetry. Schematics of these are shown in Figure 8.2. If the chelating ligands (L^L) form a left-handed screw about the C_3 axis of rotation, the isomer is of the Λ absolute configuration. A right-handed screw about this axis defines the Δ absolute configuration.

   Chiral complexes hold a special place in the history of coordination chemistry. Definitive proof of optical isomerism in metal coordination compounds secured acceptance of Werner's coordination theory. Werner's theory predicted that optical isomerism would exist for octahedral complexes and, in 1911, his American student, Victor L. King, successfully resolved *cis*-amminechloro-*bis*(ethylenediamine)cobalt(III) chloride, $[Co(en)_2Cl(NH_3)]Cl_2$ using D-bromocamphorsulfonic acid. Still, there were skeptics who believed that it was the carbon in the ethylenediamine (en) ligand that was responsible for this property. In 1914 Werner's lab synthesized and resolved the noncarbon-containing $\{Co[(OH)_2Co(NH_3)_4]_3\}Br_6$,[†] silencing the critics. Optically active metal complexes continue to play major roles in inorganic chemistry today. Such complexes are used as chiral reagents in organic synthesis, probes of biomolecular structure, and probes of outer-sphere electron transfer mechanisms.

Pre-lab 8.2.b: Optical Isomer Resolution Using Tartrate Salts[5,6]

Two primary resolving agents for metal complexes are D-tartrate and antimonyl D-tartrate ions. Optical resolution is achieved using either chromatography (e.g., ion exchange with D-tartrate salt as the eluent) or by fractional precipitation. We will use this latter technique for resolutions. When a racemic mixture of a metal complex is combined with an optically

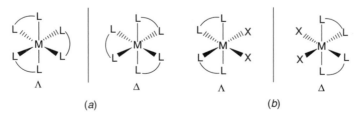

Λ $\qquad$ Δ $\qquad$ Λ $\qquad$ Δ

(a) $\qquad\qquad$ (b)

Figure 8.2   The enantiomeric or nonsuperimposible mirror images of (a) the *Tris*-bidentate chelate $[M(L^L)_3]$ and (b) the *cis-bis*-bidentate chelate, $[M(L^L)_2X_2]$.

[†]The chelating ligands about the central cobalt(III) center are *cis*-$[Co(NH_3)_4(OH)_2]^+$ ions.

active substance (e.g., D-tartrate), one combined configuration (e.g., $\{\Lambda\text{-}[M(L^{\wedge}L)_3]^{n+}\}_2-$ D-tartrate$_n$) will be less soluble than its respective combined antipode. This allows the step-wise precipitation and removal of one ion-pair precipitate. The most optically pure substance of the $\{\Lambda\text{-}[M(L^{\wedge}L)_3]^{n+}\}_2-$D-tartrate$_n$ diastereomer is expected to be in the initial fraction(s) and the most optically pure antipode in the last fractions. Yoneda and coworkers[5] have examined the origin of chiral discrimination during the resolution process. These often involve unique face-to-face (either C$_2$ or C$_3$) ion-pair structures governed by hydrogen bonding.

Procedure 8.2: Resolution of Tris(1,10-Phenanthroline) Chromium(III) Hexafluorophosphate, [Cr(phen)$_3$] (PF$_6$)$_3$, by Fractional Precipitation[7,8]

All manipulations must be carried out in the dark (keep aluminum foil on reaction flasks).

Preparation of Silver Antimonyl D-Tartrate
1. Dissolve 2.27 g (7 mmol) potassium antimonyl-D-tartrate hydrate [K(SbO) C$_4$H$_4$O$_6$ · xH$_2$O] in 15 ml dd H$_2$O in a 50 ml beaker equipped with a stir bar.
2. Dissolve 1.19 g (7 mmol) AgNO$_3$ in 10 ml dd H$_2$O and add this with stirring to the K(SbO)C$_4$H$_4$O$_6$ solution.
3. Vacuum filter the silver antimonyl D-tartrate product, wash with dd H$_2$O and air dry.

Conversion of [Cr(phen)$_3$] (PF$_6$)$_3$ to the Highly Soluble [Cr(phen)$_3$]Cl$_3$ Salt
4. Suspend 1.54 g (1.5 mmol) [Cr(phen)$_3$] (PF$_6$)$_3$ in 19 ml dd H$_2$O.
5. Add small increments of moist Dowex 1X8 ion-exchange resin (Cl$^-$ form).
6. After the complex dissolves, vacuum filter the solution to remove the resin.

Isolation of Impure L-[Cr(phen)$_3$] (SbO-D-tart)$_3$
7. Over a period of ~2–3 min, and with continual stirring, add 1.77 g (4.5 mmol) of silver antimonyl-D-tartrate. Filter the AgCl precipitate.
8. Warm the solution (filtrate) to ~50 °C and add absolute ethanol slowly until the solution becomes cloudy.
9. After chilling in an ice bath, filter fraction 1.
10. Wash with cold absolute ethanol, ether, and air dry. Record yield.
11. Repeat steps 8–10 but with warming to only 40 °C to collect fractions 2–4.

Isolation of D-[Cr(phen)$_3$] (SbO-D-tart)$_3$
12. After collecting the last fraction of L-[Cr(phen)$_3$] (SbO-D-tart)$_3$, cool the filtrate in an ice-bath and add dropwise an aqueous solution of NH$_4$PF$_6$ until incipient precipitation.
13. Filter when approximately one-half of the complex has precipitated (estimate based on previous L-[Cr(phen)$_3$] (SbO-D-tart)$_3$ yield).
14. To the remaining filtrate solution, add an excess of NH$_4$PF$_6$ to obtain your purest fraction of D-[Cr(phen)$_3$] (PF$_6$)$_3$. Record yield.

Optical purities will be assessed in Experiment 8.3.

Name _____   Section _____

Results Summary for the Resolution of Tris(1,10-phenanthroline)chromium(III)
Hexafluorophosphate

L-[Cr(phen)₃]–(Antimonyl-D-Tartrate)₃ Diastereomer

1. Theoretical yield: _____

<table>
<tr><td></td><td colspan="4" align="center">Fraction number</td></tr>
<tr><td></td><td align="center">1</td><td align="center">2</td><td align="center">3</td><td align="center">4</td></tr>
<tr><td>2. Actual yield:</td><td>_____</td><td>_____</td><td>_____</td><td>_____</td></tr>
<tr><td>Percentage yield:</td><td>_____</td><td>_____</td><td>_____</td><td>_____</td></tr>
</table>

D-[Cr(phen)₃] (PF₆)₃ (Final Fraction)

Theoretical yield: _____   Actual yield: _____
Percentage yield: _____

Post-lab Questions

Q8.1   Draw the optical isomers of $[Co(1,2\text{-pn})_3]^{3+}$ (pn = propanediamine). Discuss in terms of enantiomers and diastereomers.

Q8.2   Why is Ag(SbO) tartrate used instead of K(SbO) tartrate for these resolutions?

EXPERIMENT 8.3: OPTICAL ANALYSIS OF Δ AND Λ [Cr(phen)₃] (PF₆)₃

Level 5

Pre-lab 8.3: Methods for Measuring Optical Activity: Polarimetry, Optical Rototary Dispersion, and Circular Dichroism Spectrophotometry[5,6]

Once a metal complex has been resolved into its optical isomers, several optical techniques can be used to probe its chiral properties: polarimetry, optical rotatory dispersion (ORD), and circular dichroism (CD) spectrophotometry. The basis of these techniques lies in the fact that plane polarized light (ppl) can be resolved into two circularly polarized components of equal magnitude and opposite (right- and left-handed) rotation. The molecular asymmetry of a chiral metal complex in solution will cause a difference in indices of refraction ($n_l - n_r$), causing a rotation of the ppl. The angle of rotation, α (rad unit length^{-1}) and its wavelength (λ) dependence is described by equation (8.4) and can be measured using a *polarimeter*. The specific $[\alpha]$ and molar $[\alpha]_M$ rotations are given by equations (8.5) and (8.6), respectively, where c is concentration in mg cm^{-3}, d is the pathlength in decimeters, and MM is the molar mass of the substance being measured.

$$\alpha = (n_l - n_r)/\lambda \tag{8.4}$$

$$[\alpha] = \alpha/c \cdot d \tag{8.5}$$

$$[\alpha]_M = MM[\alpha]10^{-2} \tag{8.6}$$

If the rotation is in a clockwise direction, the substance in solution is said to be dextrorotatory [D or (+)]. Rotation in the counter clockwise direction infers a levorotatory substance

[L or (−)]. It should be noted that D (+) and L (−) refer only to the sign of rotation at a particular wavelength (for *polarimetry*, this is often the sodium D line, 589.5 nm). These signs cannot be used to determine absolute configuration (D and L). Absolute configuration can only be assigned by careful analogy to a known substance or by X-ray crystallography. Note that the convention for $[Co(en)_3]^{3+}$ is $\Delta(-)_{589}$ and $\Lambda(+)_{589.}$

Because a rotation might be zero at a particular wavelength, it is advisable to obtain rotations at more than one wavelength. A plot of magnitude of specific or molar rotation as a function of wavelength is called a *rotatory dispersion curve*. In the narrow region of maximum absorption of light for the complex, the rotatory dispersion curve displays a characteristic maximum, followed by a sharp decrease that passes through zero rotation and continues to a maximum trough of opposite rotation. This anomalous effect is due to the differential absorption of the right and left circularly polarized components of light, and this particular region of the curve is known as the *optical rotatory dispersion* curve. The total differential interactions of left and right circularly polarized light are termed the Cotton effect. The differential absorption ($\Delta\varepsilon = \varepsilon_l - \varepsilon_r$) is *circular dichroism* and the measurement of $\Delta\varepsilon$ is called CD spectroscopy. A *circular dichroism spectrophotometer* actually measures ellipticity, φ (mdeg), as defined in Figure 8.3 and equations (8.7) and (8.8); molar ellipticity, $[\theta]$, is related to $\Delta\varepsilon$ by equation (8.9). The parameters are as defined for equations (8.5) and (8.6). CD and ORD effects are interrelated by the Kronig–Kramer relationship (not discussed further here).

$$specific \ ellipticity \quad [\varphi] = \alpha/c \cdot d \tag{8.7}$$

$$molar \ ellipticity \quad [\theta] = MM[\varphi]10^{-2} \tag{8.8}$$

$$\Delta\varepsilon = 0.3032 \times 10^{-3}[\theta] \tag{8.9}$$

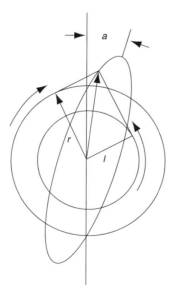

Figure 8.3 Elliptically polarized radiation after passing through a solution in which $\varepsilon_d > \varepsilon_l$ and $n_l > n_d$ for the chiral complex [reproduced with permission from Cotton, F. A., Wilkinson, G. *Advanced Inorganic Chemistry*, 5th edn. Wiley-Interscience: New York, 1988.]

Although polarimetry and ORD were used during the time of Werner it was not until the1960s that CD and ORD instruments became commercially available and these techniques were commonly employed. Polarimetry, revealing only the sign and magnitude of rotation at one wavelength, is the most limiting of all three techniques and can be used only for identifying optical isomers and quantifying optical purity. ORD and CD spectroscopy, on the other hand, can also be used for studying electronic transitions. Absorption bands for $d-d$ transitions are broadened due to vibronic coupling. The lowering of symmetry (from O_h symmetry, for example) due to varying ligand types predicts multiple absorption bands that are often not resolved in the absorption spectrum. CD and ORD offer attractive alternatives for gathering information on electronic transitions. Each technique has advantages and disadvantages: CD spectra are accumulated only within electronic absorption bands while ORD can be measured outside these regions as well. On the other hand, CD provides better resolution of peaks and, hence, more and clearer information about optical transitions.

Procedure 8.3: Analysis of the D- and L-Tris(1,10-Phenanthroline) Chromium(III) Hexafluorophosphate, [Cr(phen)₃] (PF₆)₃, Complexes[7,8]

All manipulations must be carried out in the dark (keep aluminum foil on reaction flasks).

Initial analysis of the L-[Cr(phen)₃]–(Antimonyl-D-Tartrate)₃ Diastereomer Fractions (See Experiment 8.2) using Polarimetry/ORD

1. Prepare 0.2% solutions (2 mg/ml ddH₂O) of the L-[Cr(phen)₃]–(antimonyl-D-tartrate)₃ fractions obtained in Experiment 8.2, and measure the optical rotation, $(g\ cm^{-3})^{-1}dm^{-1}$, at the sodium d-line (589.5 nm).

2. Recrystallize fraction 1 by dissolving the solid in 25 ml of warm water ($\sim$40 °C), and adding ethanol to precipitate the least soluble fraction. Check optical rotation as previously described. Ideally, recrystallizations should continue until there is no change in the optical rotation.

Conversion of L-[Cr(phen)₃]–(Antimonyl-D-Tartrate)₃ to L-[Cr(phen)₃] (PF₆)₃

3. Dissolve the recrystallized solid of fraction 1 in a small amount of ddH₂O.

4. Add NaPF₆ precipitating the L-[Cr(phen)₃](PF₆)₃ salt.

Polarimetry/ORD: Optical Rotation of D- and L-[Cr(phen)₃] Cl₃

5a. Measure the optical rotation using a polarimeter or ORD instrument (589.5 nm) of a 0.2% aqueous solution: sonicate a slurry of 10 mg [Cr(phen)₃] (PF₆)₃ in 5 ml ddH₂O. Then, with stirring, add a small amount of moistened Dowex 1X8 ion-exchange resin (Cl⁻ form) as in Procedure 8.2, steps 4–6. When the solution is clear, filter off the resin.

6a. Repeat step 5 for the D-[Cr(phen)₃] (PF₆)₃ complex isolated in Procedure 8.2 steps 12–14. Proceed to step 7.

CD Analysis of D- and L-[Cr(phen)₃] Cl₃

5b. Prepare 25 ml of each D- and L-[Cr(phen)₃] Cl₃ at 5×10^{-4} M (see step 5a for conversion to the Cl⁻ salt).

TABLE 8.1   ORD^a and CD^b Values for the D-and L-[Cr(phen)₃]$^{3+}$ Optical Isomers

[Cr(phen)₃]$^{3+}$ Isomer	CD/ORD Values for Pure Enantiomers
Δ(L)-(−)₅₄₆	CD: $\Delta\varepsilon_{457} = -2.48$ M^{-1} cm^{-1}
	ORD: $[\alpha]_D = -1320°$ (g cm^{-3})$^{-1}$ dm^{-1}
Λ(D)-(+)₅₄₆	CD: $\Delta\varepsilon_{457} = +2.48$ M^{-1} cm^{-1}
	ORD: $[\alpha]_D = +1320°$ (g cm^{-3})$^{-1}$ dm^{-1}

^{a} See Lee et al.[7] ^{b} Miyoshi, K., Matsumoto, Y., Yoneda, H. *Inorg. Chem.* **1982**, *21*, 790–793; Kaizaki, S., Hidaka, J., Shimura, Y. *Bull. Chem. Soc. Jpn*, **1969**, *42*, 988; **1970**, *43*, 1100; Mason, S. F., Peart, B. J. *J. Chem. Soc. Dalton Trans.* **1973**, 949.

6b. Obtain a CD spectrum from 200–650 nm. Note, you will need to make dilutions such as those described in Procedure 8.1 step 14 for UV analysis. Proceed to step 7.

7. Determine the percentage optical purity from literature values, Table 8.1.

Name _____   Section _____

Results Summary for the Optical Analysis of the
Tris(1,10-phenanthroline)chromium(III) Ion, D- and L-[Cr(phen)₃]$^{3+}$

Polarimetry/ORD

	Δ(L)-(−)₅₄₆ isomer	Λ(D)-(+)₅₄₆ isomer
1. α	_____	_____
2. $[\alpha]_D$	_____	_____
3. Percentage optical purity	_____	_____

CD Spectrophotometry

	Δ(L)-(−)₅₄₆ isomer	Λ(D)-(+)₅₄₆ isomer
1. $[\theta]$	_____	_____
2. $\Delta\varepsilon$	_____	_____
3. Percentage optical purity	_____	_____

Post-lab Questions

Q8.3   Why is it important to convert your metal complex from the tartrate salt to a halide salt for determining optical purity?

Q8.4   Show pictorially how plane polarized light is made up of two circular components of equal but opposite magnitude.

EXPERIMENT 8.4: INVESTIGATION OF THE PHOTO-OXIDATION OF NUCLEIC ACIDS BY [Cr(phen)$_3$]$^{3+}$ USING SPECTROFLUORIMETRY[9-14]

Level 5

Pre-level 8.4: Photochemistry of Metal Complexes

The study of biochemical electron transfer reactions has been driven by a quest for a better understanding of intracellular phenomena such as oxidative damage to DNA and the respiratory electron transport chain. Recent advances in the nascent field of nanotechnology have demonstrated that biopolymers can serve as molecular wires, and the conductivity of DNA has become an issue of vigorous debate and active research. Changes in luminescence can be used to measure the rates of electron transfer reactions involving DNA bases, and excited-state metal complexes can serve as photo-cleavage agents of the DNA polymer. Metal complexes of polypyridines, such as the luminescent complex [Ru(bpy)$_3$]$^{3+}$ (bpy = 2,2′-bipyridine), are often used in this capacity because variation in the metal ion and the polypyridine ligands can offer a wide range of photophysical and redox properties. [Cr(phen)$_3$]$^{3+}$ (see Fig. 8.1) is another example of a luminescent transition metal complex used in investigations of electron-transfer reactions involving DNA; its photophysical properties and reaction with nucleic acids will be investigated in this experiment. A general introduction to luminescence is given in Pre-lab 7.1.b.

Photo-induced Electron Transfer

The ground-state d^3 Cr^{3+} complex is a quartet state (given the term symbol ^{4}A) owing to its three unpaired electrons. Upon excitation with visible light ($\lambda_{ex} = 450$ nm), an electron is promoted from a partially filled d orbital (d_{xy}, d_{xz} or d_{yx}) to an empty d orbital ($d_{x^2-y^2}$ or d_{z^2}) yielding a metal-centered excited-state complex *[Cr(phen)$_3$]$^{3+}$ also having a quartet spin state (^{4}T). Boletta *et al.* and Jamieson give detailed discussions of the nature of these orbitals and other transitions. This quartet state relaxes quickly to a set of lower-energy metal-centered doublet states (^{2}T and ^{2}E) by a process known as inter-system crossing. These doublet excited states then relax back to the ground state by two major pathways: phosphorescence and a set of nonradiative decay pathways that involve intramolecular vibrations and collisions with solvent that result in the formation of heat. The phosphorescence is long-lived due to the required electron spin flip involved. For [Cr(phen)$_3$]$^{3+}$, the phosphorescence consists of two bands—one at 700 nm originating from the higher energy ^{2}T state and one at 730 nm originating at the lower energy ^{2}E state. The absorption and luminescence spectra of [Cr(phen)$_3$]$^{3+}$ in water are shown in Figure 8.4. In addition to relaxing back to the ground state, the energy-rich excited-state complex is capable of oxidizing a suitable substrate, if present, and thereby being reduced to [Cr(phen)$_3$]$^{2+}$.

Thermodynamics of Electron-transfer Quenching

One important aspect of photochemistry is that an excited-state species is a stronger oxidant than the corresponding ground-state species. Therefore, the absorption of light can drive electron-transfer reactions that may be thermodynamically unfavorable in the ground state. The driving force for electron transfer can be expressed as a change in Gibbs energy (ΔG^0) or in terms of a redox potential (E^0). The two parameters are

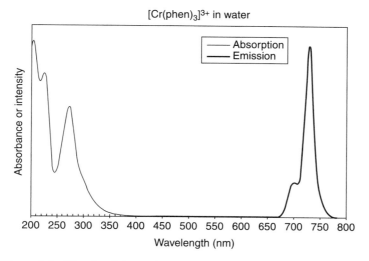

Figure 8.4 The absorbtion and emission spectra for [Cr (phen)$_3$] in water.

related, equation (8.10), by the number of electrons involved in the electron transfer process (*n*) and Faraday's constant (*F*). The ground-state reduction potential of [Cr(phen)$_3$]$^{3+}$ can be determined using cyclic voltammetry, equation (8.11).

$$\Delta G^0 = -nFE^0 \tag{8.10}$$

$$[Cr(phen)_3]^{3+} + e^- \rightarrow [Cr(phen)_3]^{2+} \quad E^0(Cr^{3+}/Cr^{2+}) = -0.28 \text{ V vs NHE}$$

$$\Delta G^0 = +27 \text{ kJ mol}^{-1} \tag{8.11}$$

The excited-state reduction potential, $E^0(*Cr^{3+}/Cr^{2+})$, can be estimated using an analysis similar to Hess's law of heat summation (Fig. 8.5). Using the emission maximum (730 nm) in the luminescence spectrum and converting units yields an excited-state energy of 164 kJ mol^{-1} for *[Cr(phen)$_3$]$^{3+}$. That means that relaxation of the ^2E excited state to the ground state involves $\Delta G^0 = -164$ kJ mol^{-1} or a one-electron electrochemical potential of $E_{00} = 1.70$ V. Combining these two potentials or ΔG values (Fig. 8.5) yields the excited-state reduction potential of +1.42 V vs NHE and a corresponding change in Gibbs energy of $\Delta G = -137$ kJ mol^{-1} for the reduction of the excited state complex *[Cr(phen)$_3$]$^{3+}$ to [Cr(phen)$_3$]$^{2+}$.

	Process	Rate	Description
(i)	$Cr(phen)_3^{3+} + h\nu \rightarrow$ *$Cr(phen)_3^{3+}$	I_{abs}	Absorption
(ii)	*$Cr(phen)_3^{3+} \rightarrow Cr(phen)_3^{3+} + h\nu$	$k_r[*Cr(phen)_3^{3+}]$	Luminescence
(iii)	*$Cr(phen)_3^{3+} \rightarrow Cr(phen)_3^{3+} +$ heat	$k_{nr}[*Cr(phen)_3^{3+}]$	Nonradiative decay
(iv)	*$Cr(phen)_3^{3+} + Q \rightarrow Cr(phen)_3^{2+} + Q^+$	$k_q[*Cr(phen)_3^{3+}][Q]$	Dynamic quenching

Figure 8.5 Series of reactions used to estimate the excited-state reduction potential, $E^0(*Cr^{3+}/Cr)^{2+}$.

The excited-state and ground-state reduction potentials indicate that a substrate (Q) having an oxidation potential (for the process $Q \rightarrow Q^+ + e^-$) of E^0 $(Q/Q^+) = -1.29$ V vs NHE, for example, could be oxidized by the excited-state complex $*[Cr(phen)_3]^{3+}$ but not by the ground state complex $[Cr(phen)_3]^{3+}$:

$$*[Cr(phen)_3]^{3+} + Q \rightarrow [Cr(phen)_3]^{2+} + Q^+$$

$$E^0 = (+1.42 \text{ V}) + (-1.29 \text{ V}) = +0.13 \text{ V} \tag{8.12}$$

$$[Cr(phen)_3]^{3+} + Q \rightarrow [Cr(phen)_3]^{2+} + Q^+$$

$$E^0 = (-0.28 \text{ V}) + (-1.29 \text{ V}) = -1.57 \text{ V} \tag{8.13}$$

The excited-state redox reaction, equation (8.12), is thermodynamically favorable ($E^0 > 0$) while ground-state reaction, equation (8.13) is not ($E^0 < 0$). Therefore, a mixture of $[Cr(phen)_3]^{3+}$ and such a substrate will only undergo a redox reaction after the chromium complex has been excited. This is the process of *photo-induced electron transfer*: light initiates an electron-transfer reaction. This experiment will explore how substrates such as DNA may be oxidized by the excited-state $*[Cr(phen)_3]^{3+}$ complex. Because the electron-transfer reaction competes kinetically with luminescence, the presence of such a suitable substrate leads to a decrease in the intensity of luminescence. For this reason, the substrate is termed a *quencher*.

Kinetics of Electron-Transfer Quenching Figure 8.6 shows that the excited-state complex $*[Cr(phen)_3]^{3+}$ can have either two or three reaction pathways, depending whether or not a quencher Q is present in solution. As the concentration of Q increases, the rate of the bimolecular electron-transfer pathway increases. We can take advantage of this to determine the rate constant of the electron-transfer pathway.

The rate of absorption (step i) can be expressed as the "absorption intensity" I_{abs} (which depends on the concentration of the ground-state complex and the intensity of light used in excitation), while the rate of luminescence (step ii) and the rate of all the nonradiative decay paths (step iii) follow first-order rate laws that depend only on the concentration of the excited-state complex (nonradiative collisional quenching by solvent is not unimolecular, but follows first-order kinetics because of the large excess of solvent molecules). *Dynamic quenching* (step iv) results from an electron-transfer reaction during a collision between a quencher molecule and an excited-state complex. This bimolecular reaction has a

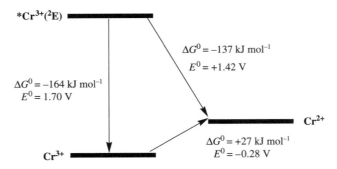

Figure 8.6 Estimating the excited state redox potential.

second-order rate law depending on the concentrations of both the excited-state complex and the quencher. In this model, the quencher does not interact with the ground-state complex.

Experimentally, the intensity of luminescence can be measured using a fluorescence spectrometer and expressed as the luminescence *quantum efficiency* (ϕ), defined as the ratio of the rate of luminescence to the rate of absorption (rate of light emitted/rate of light absorbed):

$$\phi = k_r[^*Cr(phen)_3{}^{3+}]/I_a \tag{8.14}$$

Additionally, under continuous excitation, the excited-state complex concentration is essentially constant, so the rate of formation and the rates of deactivation are equal. If there is no quencher in solution, step (iv) does not occur, so this steady-state expression becomes:

$$I_a = k_r[^*Cr(phen)_3{}^{3+}] + k_{nr}[^*Cr(phen)_3{}^{3+}] \tag{8.15}$$

Combining equations (8.14) and (8.15) yields an expression for the quantum efficiency in the absence of quencher (ϕ_0):

$$\phi_0 = k_r/(k_r + k_{nr}) \tag{8.16}$$

By adding a known concentration of quencher to the solution, we can increase the rate of step (iv). Because there is now an additional deactivation pathway for the excited state complex, the steady-state expression and the quantum efficiency expression become:

$$I_a = k_r[^*Cr(phen)_3{}^{3+}] + k_{nr}[^*Cr(phen)_3{}^{3+}] + k_q[^*Cr(phen)_3{}^{3+}][Q] \tag{8.17}$$

and

$$\phi_q = k_r/(k_r + k_{nr} + k_q[Q]) \tag{8.18}$$

The luminescence quantum efficiency when quencher is present is termed ϕ_q and the bimolecular quenching rate constant is termed k_q.

The luminescence lifetime (τ) can be measured using a laser system. The lifetime without quencher present (τ_0) and the lifetime with quencher present (τ_q) can be expressed in terms of the same rate constants from Figure 8.6 in a similar manner:

$$\tau_0 = 1/(k_r + k_{nr}) \tag{8.19}$$

and

$$\tau_q = 1/(k_r + k_{nr} + k_q[Q]) \tag{8.20}$$

That is, the lifetime is the inverse of the sum of the decay rates.

The Stern–Volmer Equation

In this experiment to determine the electron-transfer quenching rate constant, you will measure the luminescence quantum efficiency for a series of solutions having a constant concentration of $Cr(phen)_3{}^{3+}$ and varying concentrations of an oxidizable quencher. By combining equation (8.16) with equation (8.18) and equation (8.19) with equation (8.20), we find that the ratio of quantum efficiencies for solutions without and with quencher is:

$$\phi_0/\phi_q = 1 + k_q\tau_0[Q] \tag{8.21}$$

and

$$\tau_0/\tau_q = 1 + k_q\tau_0[Q] \tag{8.22}$$

These two forms of the Stern–Volmer (SV) equation indicate that plots of ϕ_0/ϕ_q vs [Q] and τ_0/τ_q vs [Q] should be linear with a slope of $k_q\tau_0$ and an intercept of unity. This is how the electron-transfer quenching rate constant can be determined by measuring changes in luminescence quantum efficiency and/or lifetimes for solutions having varying quencher concentration. Several research groups have used the SV kinetics analysis of luminescence quenching as an experimental method for determining the rate constants of electron-transfer reactions involving $[Cr(phen)_3]^{3+}$ and a wide range of organic electron donors.

Pre-lab 8.4: Questions (See Procedure 8.4 and Table 8.2)

Q8.5 Based on the precision of an analytical balance and volumetric glassware, what is the precision of the calculated stock concentration?

Q8.6 Using the given molar absorptivity for $Cr(phen)_3]^{3+}$, calculate the expected A_{354} value for your Cr stock solution. Is this absorbance within the measurable range of the UV–vis spectrophotometer? Based on the precision of the spectrophotometer, what is the precision of your calculated concentration?

Q8.7 Based on the precision of an analytical balance and volumetric glassware, what is the relative error in the guanosine stock solution calculated concentration?

Q8.8 How can you confirm that the volumetric pipettes you will be using are operating correctly?

Q8.9 What could happen if either the stock solutions or sample solutions were left in unsealed containers? How might this affect your experiment?

Q8.10 Why should instrumental parameters on the fluorescence spectrophotometer *not* be changed when measuring luminescence intensities for a given series of solutions?

TABLE 8.2 Final Concentrations and Amounts of Reagents in 3 ml Total Volume[a] Reaction Mixture

Solution	V_{Cr} (ml)	V_Q (ml)	V_{buffer} (ml)	[Cr] (M)[b]	[Q] (M)
1	1.00	0.00	2.00	5.00×10^{-5}	0
2	1.00	0.50	1.50	5.00×10^{-5}	4.00×10^{-5}
3	1.00	1.00	1.00	5.00×10^{-5}	8.00×10^{-5}
4	1.00	1.50	0.50	5.00×10^{-5}	1.20×10^{-4}
5	1.00	2.00	0.00	5.00×10^{-5}	1.60×10^{-4}
Total	5.00	5.00	5.00		

[a]A standard 1 cm^2 fluorescence cuvette holds $\sim$3.5 ml of solution, but a total volume of 3 ml often works fine.
[b]The solution should be optically dilute ($A < 0.10$) at the excitation wavelength used.

Q8.11 When measuring a series of solutions having an increasing concentration of quencher, would it make better sense to begin with the solution having no quencher or the solution having the largest concentration of quencher?

Q8.12 If only 10% quenching ($\phi_0/\phi_q > 1.11$) is observed through a series of five solutions, how should the quencher concentration be adjusted for a repeated trial?

Q8.13 Can you derive the SV equation from the model mechanism presented in the Pre-lab?

Q8.14 Can you hypothesize as to what happens to the products of the electron-transfer reaction $[Cr(phen)_3{}^{2+} + Q^+]$?

Procedure 8.4: Luminescence Studies of $[Cr(phen)_3]^{3+}$

Solution Preparation *Note*: clean glassware and cuvettes thoroughly before using. Interference from impurities can lead to a tremendous waste of time and money. Wash cuvettes with soap, de-ionized water and wash acetone, and rinse with spectroscopic-grade MeOH. Keep your work area clean! Impurities that fluoresce or that quench fluorescence are the bane of fluorescence spectroscopy.

A general approach to preparing a series of solutions for a quenching experiment involves adding a set aliquot of a stock solution of the luminescent compound ("lumophore") to a series of vials, then adding various aliquots of a stock solution of the quencher and adding various aliquots of the solvent so as to have the constant total volume for each solution. The result is a series of solutions having the same concentration of the lumophore and a range of concentrations of the quencher. In this experiment, $[Cr(phen)_3](PF_6)_3$ is the lumophore (hereafter abbreviated "Cr") and either guanosine (G), GMP, or DNA is the quencher "Q". All solutions are to be prepared in 50 mM Tris–HCl buffer (pH 7). The synthesis of $[Cr(phen)_3](PF_6)_3$ is given in Experiment 8.1.

1. Prepare 500 ml of a 50 mM Tris–HCl (*aq*) solution (pH 7).
2. Prepare 25 ml of approximately 1.5×10^{-4} M Cr aqueous stock solution (MM = 339 g mol^{-1}).
3. The molar absorptivity at 354 nm for $[Cr(phen)_3]^{3+}$ is $\varepsilon_{354} = 4200$ M^{-1} cm^{-1}. Determine the precise concentration of your stock solution by UV–visible analysis.
4. Prepare 10 ml of a 2.4×10^{-4} M guanosine (MM = 283.2 g mol^{-1}) stock solution. The mixtures of guanosine in water may require sonication to dissolve the solid.
5. Prepare a solution having [G] the same as the last in your series, but with no chromium complex. This will serve as a control sample.
6. Prepare reaction solutions according to Table 8.2 in 4 dram sample vials with screw caps.

Luminescence Spectrum of [Cr(phen)$_3$]$^{3+}$

1. Check the background signal for your Tris–HCl buffer solution using large slit-widths and 440 nm excitation. You should observe the Raman scattering, but no other signal.
2. Using your Cr aqueous stock solution prepare a solution of approximately 5×10^{-5} M $[Cr(phen)_3]^{3+}$ (*aq*) in 50 mM Tris–HCl buffer, determining the actual concentration as precisely as possible. Obtain a UV–vis spectrum and a luminescence spectrum of this solution.

Consult your instructor on the proper use of the fluorescence spectrophotometer and the choice of appropriate collection parameters. You should understand how to optimize the following instrumental parameters in order to maximize the signal-to-noise ratio (S/N) of a luminescence spectrum:

- excitation monochromator slit width;
- emission monochromator slit width;
- signal averaging through integration time or multiple scans.

The choice of an excitation wavelength takes into consideration:

- the wavelength region of absorption by $[Cr(phen)_3]^{3+}$;
- the wavelength region of absorption by the quencher;
- the wavelength region of luminescence by $^*[Cr(phen)_3]^{3+}$;
- light scattering from the cuvette and solvent.

An excitation wavelength of approximately 450 nm is appropriate to selectively excite $[Cr(phen)_3]^{3+}$ but not the quencher, and scattering from this excitation wavelength should not interfere with collecting the full luminescence spectrum. Other excitation wavelengths may be used, though. Once instrumental parameters have been set for this solution, they should not be changed for a given series of solutions.

3. Determine how to calculate the integrated emission intensity (I) of the luminescence band.

Steady-state Luminescence Quenching

1. For each solution in the $Cr + G$ series, Table 8.2: (1) measure the UV–vis spectrum to obtain the absorbance A at λ_{ex}, and (2) measure an emission spectrum and obtain the integrated emission intensity, I. Also, prepare a solution having $[G]$ the same as the last in your series (Table 8.2), but with no chromium complex. This will serve as a control sample. Save your sample solutions for *time-resolved luminescence quenching* if you are conducting these experiments (see below). Save all UV–vis and luminescence spectra files that you acquire—they may be useful for data analysis.

2. Repeat the entire experiment, but prepare a new stock solution of guanosine; the same stock solution of $[Cr(phen)_3]^{3+}$ (aq) could be used, although fresh solutions should be prepared each day. Before beginning this duplicate experiment you should analyze the data from your first trial, as you may need to adjust the quencher concentration so that the total amount of quenching (comparing solutions 1 and 5) is at least 20% ($\phi_0/\phi_q > 1.25$), but not more than 90% ($\phi_0/\phi_q < 10$). Divide the responsibilities among group members to make efficient use of your time for solution preparation, spectroscopic measurements and data analysis.

Data Analysis

Calculating Quantum Yield Absorbance values should be corrected for scattering by inspecting the baseline of each UV–vis spectrum in the long-wavelength region. The spectrum, and importantly the absorbance at the excitation wavelength used in the luminescence measurements, should be adjusted to yield a baseline of zero absorbance. This small correction can have a large impact on the calculated quantum yield.

Variations in the absorbance (A) of each solution affect the emission intensity (I). If the luminescent compound absorbs more light, it will emit more light. Therefore, the ratio of quantum efficiencies needed in the SV equation can be obtained using equation (8.23), which corrects for absorbance differences:

$$\phi_0/\phi_q = [I_0/I_q][A_q/A_0] \qquad (8.23)$$

This assumes that A and I are linearly related, which should be the case for solutions that are optically dilute ($A < 0.1$). If the absorbance values increase steadily through the series, check that the quencher is not absorbing at the excitation wavelength.

Time-resolved Luminescence Quenching

1. Using a solution of approximately 5×10^{-5} M $[Cr(phen)_3]^{3+}$ (aq), obtain a luminescence lifetime decay profile. Consult your instructor on the proper use of the laser system. In particular, you should consider:
   - the selection of excitation and emission wavelengths;
   - the choice of monochromator slit width;
   - other instrumental parameters (such as PMT voltage) that affect the detector.

2. Determine how to fit a single exponential decay to the data to determine the luminescence lifetime, τ.

3. For each solution in the Cr + G series (Table 8.2), measure the luminescence decay to obtain a lifetime.

Optional: Variations in the Nature of the Quencher

One objective of this experiment is to explore how the nature of the nucleic acid quencher affects the rate of electron-transfer quenching. Once acceptable results (good fit to the SV equation and good agreement between different trials) are obtained for the Cr+ G series, conduct a similar experiment (in duplicate, if possible) using guanosine monophosphate (GMP) as a quencher with $0 \leq$ [GMP] $\leq \sim 0.12$ mM (again, prepare two separate stock solutions of GMP). Conduct a third experiment (in duplicate, if possible) using DNA as the quencher with $0 \leq$ [DNA] $\leq \sim 1.0$ mM (again, prepare two separate stock solutions of DNA). Adjust concentrations of the quenchers between trials, as needed, to obtain appropriate amounts of quenching. Other quenchers that could be explored include adenosine, cytidine, and thymidine, or the related mononucleosides.

Name _____   Section _____

Results Summary for the Luminescence Studies of the Tris(1,10-phenanthroline)chromium(III) Ion, $Cr(phen)_3]^{3+}$

Luminescence Spectrum of $[Cr(phen)_3]^{3+}$

1. $[Cr(phen)_3]^{3+}$ determined by UV–vis spectroscopy   _____ M

TABLE 8.3   Results of Steady-state Luminescence Quenching and Emission Lifetime of the $[Cr(phen)_3]^{3+}$–Guanosine Reaction

Solution No.	Absorbance at λ_{ex}	Integrated Intensity, I_q	ϕ_0/ϕ_q	τ	τ_0/τ
1					
2					
3					
4					
5					

Steady-state Luminescence Quenching and Emission Lifetime Data of $[Cr(phen)_3]^{3+}$ in Reaction with Guanosine

2. Complete Table 8.3   Quencher:_____

3. From the slope of the plot of ϕ_0/ϕ_q vs $[G]$, $k_q\tau_0 =$ _____; $k_q =$ _____

4. From the slope of the plot of τ_0/τ_q vs $[G]$, $k_q\tau_0 =$ _____; $k_q =$ _____

5. If you have obtained steady-state emission intensity data and/or emission lifetime data for other quenchers, list and compare $k_q\tau_0$ values obtained with guanosine. Discuss.

Post-lab Questions

General Technique Issues

Q8.15   What could happen if either the stock solutions or sample solutions were left in unsealed containers? How might this affect your experiment?

Q8.16   Why should instrumental parameters (excitation wavelength, slit width, etc.) on the fluorescence spectrophotometer *not* be changed when measuring luminescence intensities for a given series of solutions?

Q8.17   When measuring luminescence spectra for a series of solutions having an increasing concentration of quencher, would it make better sense to begin with the solution having no quencher or the solution having the largest concentration of quencher? Why?

Q8.18   If only 10% quenching ($\phi_0/\phi_q > 1.11$) is observed through a series of five solutions, how should the quencher concentration be adjusted for a repeated trial?

Theory

Q8.19   Show that a wavelength of 730 nm corresponds to an energy of 164 kJ mol^{-1}.

Q8.20   Can you derive the two forms of the SV, equations (8.21) and (8.22), from the model mechanism presented in the Introduction?

Q8.21   Can you hypothesize as to what happens to the products of the electron-transfer reaction $[Cr(phen)_3{}^{2+} + Q^+]$.

EXPERIMENT 8.5: STEREOSELECTIVE BINDING OF $[Cr(phen)_3]^{3+}$ TO DNA: AN EQUILIBRIUM DIALYSIS STUDY

Level 5

Pre-lab 8.5.a: Chiral Discrimination in Metal Complex–DNA binding[15,16]

In Experiment 8.4, we found that $[Cr(phen)_3]^{3+}$ exhibits strong luminescence in aqueous solution and that this luminescence is effectively quenched by an electron transfer mechanism in the presence of guanosine and B-DNA. In both cases, the excited state $[^*Cr(phen)_3]^{3+}$ lumophore directly oxidizes the guanosine acceptor. Because of its luminescence properties and chiral nature, $[Cr(phen)_3]^{3+}$ has the potential to be a conformational probe and a probe of DNA binding interactions (i.e., a chemical nuclease).

Transition metal complexes can interact with the DNA biomolecule either covalently as with *cis*-platin (see Chapter 6), or noncovalently, as we see with the coordinatively saturated, inert $[Cr(phen)_3]^{3+}$ complex. Noncovalent interactions can be one of three types: electrostatic binding of a positively charged metal complex to the negatively charged DNA backbone, surface binding involving primarily hydrogen bonds in either the major or minor groove of the DNA, and intercalative binding, where planar aromatic ligands insert themselves between stacked base pairs. Only the latter two binding modes are expected to show chiral discrimination, and it is thought that discrimination is greater for intercalation. For example, the $[Cr(bpy)]^{3+}$ (bpy = bipyridine) complex, which binds to DNA predominantly through electrostatic interactions, exhibits no stereoselective binding, while the $[Ru(phen)_3]^{2+}$ complex, which exhibits significant non-electrostatic binding, does show chiral descrimination. Three different binding modes for $[Ru(phen)_3]^{2+}$ in the minor groove of DNA are reported: two surface interactions (one preferring Δ and the other preferring Λ) and one partial intercalation, (a strong Δ preference), resulting in an overall preference for the Δ-isomer. From luminescence data, the $[Cr(phen)_3]^{3+}$–DNA binding constant, K_{DNA}, is determined to be 3500 M^{-1}. Given the structure of $[Cr(phen)_3]^{3+}$ (Fig. 8.1), it is expected that stereoselective binding may be possible should $[Cr(phen)_3]^{3+}$ bind by either a surface or intercalative interaction. In this experiment, you will analyze, by equilibrium dialysis, chiral discrimination in the binding of $[Cr(phen)_3]^{3+}$ to DNA.

Pre-lab 8.5.b: Equilibrium Dialysis[17]

Equilibrium dialysis is a convenient and simple method for studying the binding of small molecules and ions to macromolecules. Generally, equal volumes of specified amounts of macromolecule and small binding molecule (ML) are placed on either side of a membrane or dialysis tubing that is permeable only to ML, Figure 8.7(a). The system reaches equilibrium when the *free* concentrations of ML on both sides of the membrane become equal [Fig. 8.7(b)]. The total concentration of small molecule, however, is greater on the side containing the macromolecule after binding occurs. The ML dialysate from analysis chamber B, for example, can then be removed and the free ML concentration measured by spectroscopy. In doing so, we can determine the amount of ML bound to the macromolecule in A. By varying the initial concentration of ML, the association constant (K), number of binding sites, and binding capacity can be determined. From K, thermodynamic parameters ΔG, ΔH, and ΔS can be obtained.

In this experiment, we will simply be looking for chiral descrimination in ML binding to DNA. The two enantiomers are designated as Δ-ML and Λ-ML in Figure 8.7. After equilibrium, an excess of the enantiomer not preferentially bound (Δ-ML) to the DNA is found in analysis chamber B, which can be detected by circular dichroism spectroscopy. To carry out this experiment, you will use small volumes: 1 ml in each A and B chamber. Bel-Art products (Pequannock, NJ, USA) sells a conveniently small dialysis cell for a price that is reasonable for a small inorganic lab section. Alternatively, SPECTRA/ POR® 1 regenerated cellulose dialysis tubing is commercially available in 6.4 mm diameter.

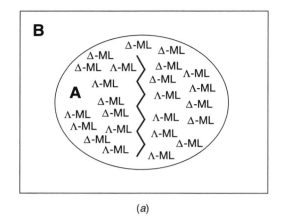

(a)

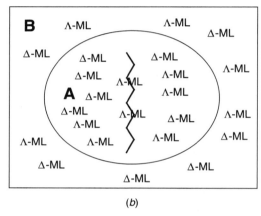

(b)

Figure 8.7   (a) Initial equilibrium dialysis conditions showing 12 Δ-ML and 12 Λ-ML in chamber A with a DNA molecule. (b) After equilibrium is reached, if 2 Λ-ML bind to the DNA, the free ML statistically distributed across the two chambers shows an excess of Δ-ML to Λ-ML (6:5) exists in the dialysate of analysis chamber B.

Procedure 8.5: Stereoselective Binding of $Cr(phen)_3]^{3+}$ to DNA Studied by Equilibrium Dialysis and CD Spectroscopy[15]

Pre-lab Work: DNA Stock Standardization   Prepare a 1 ml stock solution of calf thymus DNA type I from Sigma (1 mg/ml) in ddH_2O (this may take gentle agitation overnight at 4 °C). The A_{260} units mg^{-1} for the DNA is given on the bottle. If this number is, for example, 16.9 A_{260} units mg^{-1}, then dissolving 1 mg DNA in 1 ml should give an absorbance of 16.9.

To determine the actual concentration per nucleotide base, prepare a 1:20 dilution: remove 50 μl from your DNA stock, dilute to 1000 μl and obtain a UV spectrum. The molar absorptivity per nucleotide base is $\varepsilon_{260} = 6600$ M^{-1} cm^{-1}. In this example, the A_{260} should be ~0.845. Using Beer's law, the concentration of nucleotide base is $0.845 \cdot 1$ $cm^{-1} \cdot 6600$ M^{-1} $cm^{-1} = 1.28 \times 10^{-4}$ mol nucleotide l^{-1}. In terms of mol *base-pair* l^{-1}, $(1.28 \times 10^{-4}/2) = 0.64 \times 10^{-4}$ mol base-pair l^{-1}. Back-calculating, the molarity (in terms of base-pairs) in your original 1 mg ml^{-1} stock is 1.28×10^{-3} M.

Solution Preparation In this experiment, we will start with both DNA and metal complex in the same dialysis chamber. You will want to prepare a 1 ml solution containing both $Cr(phen)_3{}^{3+}$ and DNA such that the mole ratio of metal complex to DNA base pair is $1:4$. The total *rac*-$[Cr(phen)_3]^{3+}$ concentration should be approximately 1.25×10^{-4} M and the DNA base-pair, therefore, 5×10^{-4} M *after* equilibrium is reached (i.e., in 2 ml total volume). Given this, the initial concentrations of $[Cr(phen)_3]^{3+}$ and DNA in chamber A should be approximately 2.5×10^{-4} and 1×10^{-3} M, respectively.

1. Prepare 25 ml of 50 mM Tris–HCl–100 mM NaCl buffer (pH 7.4). Place 0.151 g Trizma base (MM $= 121.14$ g mol^{-1}) and 0.146 g NaCl in a 50 ml beaker. Add approximately 15 ml ddH$_2$O and titrate to pH 7.4 with 3 M HCl. Quantitatively transfer the contents to a 25 ml volumetric flask and add ddH$_2$O to the mark.

2. From the Pre-lab work, determine the amount of DNA to be dissolved in 1 ml to obtain a 1×10^{-3} M (in base-pair) solution.

3. Prepare 5 ml of a 2.5×10^{-4} M solution of *rac*-$[Cr(phen)_3]^{3+}$ by dissolving 1.28 mg of *rac*-$[Cr(phen)_3](PF_6)_3$ (MM $= 1027.3$ g mol^{-1}) in 5 ml of Tris–HCl buffer. Obtain a visible spectrum (use buffer as blank) and determine the actual concentration using $\varepsilon_{354} = 4200$ M^{-1}cm^{-1}.

Equilibrium Dialysis

1. Prepare either your two chambered dialysis cell or dialysis tubing (molecular weight cut-off for the membrane or tubing should be 6000 Da).

2. Dissolve the predetermined amount of calf thymus DNA in 1 ml of your 2.5×10^{-4} M solution of *rac*-$[Cr(phen)_3]^{3+}$. Add this to chamber A of your dialysis set up.

3. In chamber B, place 1 ml Tris–HCl buffer.

4. Dialyze at $4\,^\circ$C, in the dark, for 48 h.

Circular Dichroism Spectroscopic Analysis

1. Using a 3 ml circular dichroism cell, obtain a blank of your Tris–HCl buffer from 300 to 500 nm.

2. Remove your dialysate (not containing the DNA) from chamber B and dilute this to 3 ml with buffer. Obtain a CD spectrum.

3. Using the information in Experiment 8.3, determine the isomer that preferentially binds to DNA and to what extent (i.e., determine the enantiomeric excess in the dialysate).

Name _____   Section _____

Results Summary for the Equilibrium Dialysis Study of the
Tris(1,10-phenanthroline)chromium(III) Ion, $Cr(phen)_3]^{3+}$,
Binding to Calf-thymus DNA

1. What isomer is found in the dialysate? _____

2. Determine the percentage enantiomeric excess: $\Delta\varepsilon_{457}$ (actual)$/\Delta\varepsilon_{457}$ (theoretical) $\times$ 100

3. Propose the binding interaction of $[Cr(phen)_3]^{3+}$ to DNA. Discuss.

REFERENCES

1. Cowan, J. A. *Inorganic Biochemistry—an Introduction*, 2nd edn. Wiley-VCH: New York, 1997.

2. Marusak, R. A., Meares, C. F. *Active Oxygen in Biochemistry—Series 3*, Valentine, J. S., Foote, C. S., Greenberg, A., Liebman, J. F. (eds). Chapman and Hall: New York, 1995, pp. 336–400.

3. Kane-Maguire, N. A. P., Hallock, J. S. *Inorg. Chim. Acta* **1979**, *35*, L309–L311.

4. Lappin, A. G., Segal, M. G., Weatherurn, D. C., Sykes, A. G. *J. Am. Chem. Soc.* **1979**, *101*, 2297–2301.

5. Yoneda, H., Miyoshi, K. *Coordination Chemistry—a Century of Progress*, Kauffman, G. B. (ed.). ACS Symposium Series no. 565. ACS: Washington, DC, 1994.

6. Mizuta, T., Tada, T., Kushi, Y., Yoneda, H. *Inorg. Chem.* **1988**, *27*, 3836.

7. Lee, C. S., Gorton, E. M., Neumann, H. M., Hunt, H. R. Jr. *Inorg. Chem.* **1966**, *5*, 1397–1399.

8. Gillard, R. D., Hill, R. E. E. *J. Chem. Soc. Dalton Trans.* **1974**, 1217.

9. Balzani, V., Juris, A., Venturi, M., Campagna, S., Serroni, S. *Chem. Rev.* **1996**, *96*(2), 759–833.

10. Watson, R. T., Desai, N., Wildsmith, J., Wheeler, J. F., Kane-Maquire, N. A. P. *Inorg. Chem.* **1999**, *38*, 2683–2687.

11. Bolletta, F., Maestri, M., Moggi, L., Jamieson, M. A., Serpone, N., Henry, M. S., Hoffman, M. Z. *Inorg. Chem.* **1983**, *22*, 2502–2509.

12. Jamieson, M. A. *Coord. Chem. Rev.* **1981**, *39*, 121.

13. Lakowicz, J. R. *Principles of Fluorescence Spectroscopy*, 2nd edn, Chapter 8, Plenum Publishers Corporation: New York, 1999.

14. Ballardini, R., Varani, G., Indelli, M. T., Scandola, F., Balzani, V. *J. Am. Chem. Soc.* **1978**, *100*(23), 7219.

15. Barker, K. D., Benoit, B. R., Bordelon, J. A., Davis, R. J., Delmas, A. S., Mytykh, O. V., Petty, J. T., Wheeler, J. F., Kane-Maguire, N. A. P. *Inorg. Chim. Acta*, **2001**, *322*, 74–78.

16. Kane-Maguire, N. A. P., Wheeler, J. F., *Coord. Chem. Rev.* **2001**, *211*, 145–162.

17. www.nestgrp.com/pdf/Ap1/EqDialManual.pdf (accessed August 2004).

Oxidation of a Natural Product by a Vanadium Catalyst: Synthesis and Catalytic Activity of Vanadyl-*bis*(2,4-pentanedione), VO(acac)$_2$

What the ocean was to the child, the periodic table is to the chemist; new catalytic reactivity is of course, my personal coelacanth (K. Barry Sharpless, 2001)

PROJECT OVERVIEW[1–3]

In 2001 K. Barry Sharpless, along with William Knowles and Ryoji Noyori, won the Nobel Prize in chemistry for the development of asymmetric catalytic oxidation reactions. The ability to induce chirality in a nonchiral organic substrate was truly a great achievement in organic chemistry. However, this achievement would not have been possible without an understanding of the periodic table and inorganic chemistry. Organic chemistry became interwined with inorganic chemistry when Perkin discovered that the oxidation of aniline with potassium dichromate produced a mauve color; this led to the modern day chemical industry and industrial synthetic organic chemistry. Today's industry commonly uses colorful organic compounds as dyes.

Sharpless compares the discovery of a new and a more effective catalyst to a fishing expedition for a rare fish. If one looks at the number of metals that are capable of catalyzing olefin epoxidation (these include Sc, Ti, V, Cr, Y, Zr, Nb, Mo, La, Hf, Ta, W, Ce, Pr, Nd, Sm, Eu, Gd, Tb, Dy, Ho, Er, Tm, Yb, Lu, and U), it is indeed a fishing expedition. The coelacanth[†] for Sharpless turned out to be the metal titanium; his distinction, in part, involved titanium-catalyzed oxidation of olefins using the organic oxidant, *t*-butylhydroperoxide. We will look at a smaller catch—perhaps a cod—VO(acac)$_2$, which Sharpless and his co-worker Michaelson discovered on an earlier fishing

[†]This 400 million year old prehistoric fish was believed extinct until one was caught by an unsuspecting (and very lucky) fisherman off the coast of South Africa in 1938; http://www.austmus.gov.au/fishes/fishfacts/fish/coela.htm.

Integrated Approach to Coordination Chemistry: An Inorganic Laboratory Guide. By Rosemary A. Marusak, Kate Doan, and Scott D. Cummings

215

trip. VO(acac)$_2$, a precursor and actually a wrong turning on the way to the Nobel prize-winning work, also catalyzes the epoxidation of olefins. The vanadium center is formally vanadium(IV), with a d^1 electron configuration, offering a spectroscopic advantage for study over the titanium catalyst, which is generated *in situ* from a spectroscopically silent titanium(IV) starting material. Our studies will focus on how visible and infrared spectroscopy can reveal some hints about how to perform the catalysis, how catalysis may take place at the metal center, and how to modify the catalyst to carry out hetero-geneous reactions. Further, ^{51}V has a spin of 7/2 and VO(acac)$_2$ is excellent for introducing students to EPR spectroscopy and other magnetic measurements. EPR spectroscopy is a highly recommended addition to experiments in this chapter; details are provided by Butera, R. A., Waldeck, D. H. *J. Chem. Educ.*, **2000**, *77(11)*, 1489–1490.

EXPERIMENT 9.1: SYNTHESIS OF VANADYL-*bis*(2,4-PENTANEDIONE), VO(acac)$_2$

Level 5

Pre-lab 9.1.a: The Acetylacetonate (acac) Ligand[4-6]

One of the most versatile classes of ligands in coordination chemistry is that of the β-diketonates, of which the most common is the acetylacetonate, (acac), Figure 9.1. The coordination chemistry of this ligand first appears in the literature in work by Combes in 1887–1894. Alfred Werner also published on the chemistry of the acac ligand in 1901. The acac ligand is remarkable in that it forms complexes with virtually any metal, including beryllium, lead, aluminum, chromium, platinum, and gadolinium.

The acac ligand has one acidic hydrogen with a $pK_a \approx 9$. Removal of the acidic hydrogen results in a negatively charged species with considerable delocalization of the negative charge (Fig. 9.2). The versatility of the ligand itself, then, lies in part in the keto-enol tautomerism possible for the parent ligand (Fig. 9.1), and also the stability of the six-membered aromatic chelate ring formed upon metal complexation.

The acac ligand can bind to a metal atom through either the oxygens or the carbons. Binding of the metal through the oxygen atoms leads to formation of a stable six-membered chelate ring with considerable aromatic character. For example, some +3 metal complexes, such as Co(acac)$_3$ and Cr(acac)$_3$, are reactive to electrophilic aromatic substitution by bromine forming the 3-bromoacetylacetonate ligand while bound to the metal. The coordination chemistry of β-diketonates continues to be studied as the field of materials chemistry grows; these compounds make excellent precursors for metal

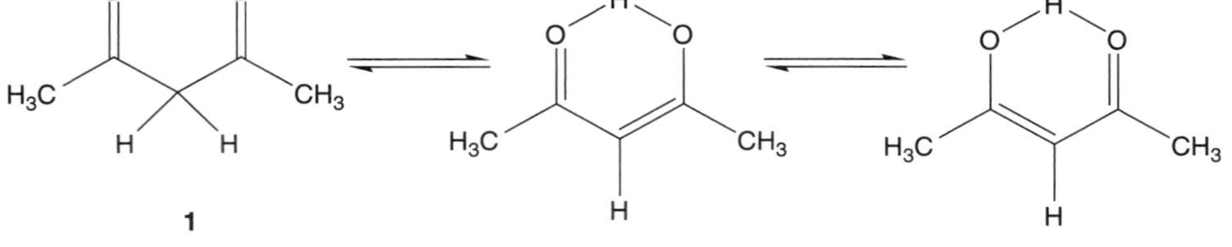

Figure 9.1 Tautomers of 2,4-pentanedione, acac.

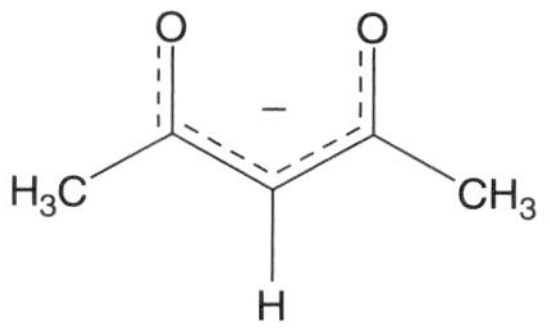

Figure 9.2   Delocalized of 2,4-pentanedione, acac.

organic chemical vapor deposition (MOCVD), and also can be used in extraction of radio-active metal salts from waste solutions.

Pre-lab 9.1.b: The VO(acac)₂ Complex[7–10]

In this chapter we explore the chemistry of the vanadium catalyst, VO(acac)₂. First synthesized in 1876 by Guyard, new uses for this compound continue to be discovered, including analogs as insulin mimetic drugs for type (II) diabetes and also as catalyst precursors for asymmetric oxidation.

Vanadium can bind either two or three acac ligands. The Tris-bidentate chelate complex, [V(acac)₃], shows the expected octahedral geometry, but the *bis*-acac complex of vanadium, in either the +3 or +4 oxidation state, is 5 coordinate with an approximate square pyramidal geometry. As such this complex has an open site for ligand binding (Fig. 9.3).

The VO bond in VO(acac)₂ is quite short (1.59 Å) and distinct from other metal oxygen bonds. Since this bond will be *trans* to a sixth added ligand, one expects that the IR stretching frequency of the VO bond will shift if the metal coordinates to another ligand. When a complex takes on a sixth ligand, such as a coordinating solvent, a substantial change in *d*-orbital splitting and corresponding change in the visible spectrum of the complex result. Both of these phenomena will be investigated in Experiments 9.2 and 9.3.

Pre-lab 9.1.c: Synthesis of the VO(acac)₂ Complex[11–13]

The most direct way to prepare VO(acac)₂ is by the reaction of vanadyl sulfate with a source of the ligand. Vanadium(V), such as V₂O₅, can be reduced to vanadium(IV) by ethanol solvent in the presence of sulfuric acid. Reaction with acetylacetone in sodium carbonate yields the desired product. The synthesis we will use produces the complex in high yield directly in a system that can visually shed light on the active catalyst species in the epoxidation of olefins, Figure 9.4.

Procedure 9.1: Synthesis of Vanadyl-bis(2,4 pentanedionate)[11–13]

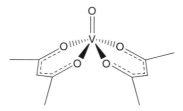

Figure 9.3   Structure of vanadyl-*bis*(2,4-pentanedionate).

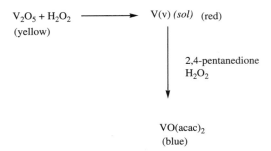

Figure 9.4 Direct synthesis of VO(acac)$_2$.

1. In a 250 ml Erlenmeyer equipped with a magnetic stir bar mix 6.0 ml of hydrogen peroxide (H$_2$O$_2$) with 10.0 ml ddH$_2$O.
2. Add 1.0 g vanadium(V) oxide slowly, while stirring the slurry at room temperature. The vanadium solution will initially dissolve as a yellow solution and then will change to a deep red color.
3. Once the mixture is deep red, place the flask in an ice-bath and cool for approximately 5 min.
4. Add 10.0 ml 2,4-pentanedione drop-wise over a 10–15 min period, as the reaction mixture is continually stirred. Make sure to add more ice as it melts.
5. Add an additional 2-4 ml H$_2$O$_2$ drop-wise until a blue precipitate begins to form.
6. Warm the reaction mixture to room temperature, and then place the flask in a boiling water bath for approximately 15 min until the precipitation is complete. During the heating wash any red material down with water or the reaction solution.
7. Collect the product by suction filtration and wash with cold water.
8. Dry the product in a dessicator over CaCl$_2$. Weigh the product and record the yield, and percentage yield.

Name _____   Section _____

Results Summary for the Synthesis of Vanadyl-*bis*(2,4-pentanedione), VO(acac)$_2$

1. Moles of V$_2$O$_5$ _____
2. Moles of 2,4 pentanedione   _____
3. Limiting reagent in synthesis   _____
4. Yield of VO(acac)$_2$ _____
5. Theoretical yield of VO(acac)$_2$ _____
6. Percentage yield of VO(acac)$_2$ _____

Post-lab Questions

There are numerous methods to make VO(acac)$_2$. Since vanadium is in the +4 oxidation state in this complex, one would expect that synthesis from a vanadium(IV) starting material would be the easiest approach. However, most syntheses start with V$_2$O$_5$ as the starting material followed by reduction of the metal to the $+4$ oxidation state using either ethanol as a reducing agent or sodium sulfite.[14]

Q9.1   What is the reducing agent in our synthesis?

Q9.2   What would one expect the product of the oxidation of our reducing agent to be?

Q9.3   What would be the product of the oxidation of the other reducing agents?

Q9.4   Why might our synthesis yield $VO(acac)_2$ in higher yield than when other conditions are used?

EXPERIMENT 9.2: INVESTIGATING SOLVATOCHROMATIC PROPERTIES OF VANADYL-*bis*(2,4-PENTANEDIONE), VO(acac)$_2$ USING VISIBLE SPECTROSCOPY

Level 5

Pre-lab 9.2: Solvatochromism in the VO(acac)$_2$ System[6,12]

The complex $VO(acac)_2$ adopts a square pyramidal structure with a short $V{=}O$ bond. Since the vanadium lies outside the plane of the acac oxygen atoms, the interactions between both the d_{xy} and d_{x2-y2} orbitals of the vanadium and the acac ligand are weakened, and the π interactions between the d_{xz} orbital and d_{yz} orbitals of the vanadium and the oxygen p orbitals become stronger. These interactions lift the degeneracy of the d_{xz}, d_{yz}, and d_{xy} orbitals as well as that of the d_{z2} and d_{x2-y2} orbitals that would normally be present in an octahedral complex. The splitting diagram for the parent complex, $VO(H_2O)_5{}^{2+}$, is shown in Figure 9.5. Note this is for an octahedral complex, so one might expect that removal of the sixth ligand, *trans* to the oxygen, will change the splitting slightly.

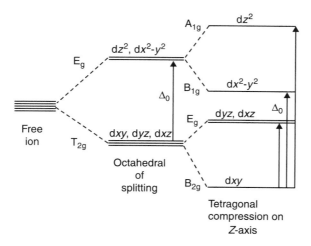

Figure 9.5   Crystal field splitting diagram for $VO(H_2O)_5{}^{2+}$. [Reproduced with permission from Ophardt and Stupgia.[12]]

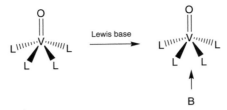

Figure 9.6   The 5 coordinate vanadyl complex acting as a Lewis acid.

When the 5 coordinate VO(acac)$_2$ complex dissolves in a coordinating solvent it acts as a Lewis acid, accepting electrons from a sixth ligand, Figure 9.6. The complex becomes more octahedral as a result of this binding interaction and this is reflected in the spectroscopy. Such changes constitute *solvatochromism*. In this experiment you will obtain spectra of VO(acac)$_2$ in various solvents of different coordinating ability and analyze the spectral changes.

Procedure 9.2: Visible Spectroscopic Analysis

1. Prepare solutions of 0.25 M VO(acac)$_2$ in the following solvents: methanol (CH$_3$OH), methylene chloride (CH$_2$Cl$_2$), acetonitrile (CH$_3$CN) and pyridine.
2. Record spectra in the visible range.
3. Note the color and any changes you observe in the spectrum of the complex in the Results Summary section.

Name _____   Section _____

Results Summary for the Visible Spectroscopic Analysis of
Vanadyl-*bis*(2,4-pentanedione), VO(acac)$_2$

1. Record your observations below.

   Solvent   CH$_3$CN   Color _____   Peaks in spectrum _____
                CH$_2$Cl$_2$ Color _____   Peaks in spectrum _____
                CH$_3$OH   Color _____   Peaks in spectrum _____
                Pyridine   Color _____   Peaks in spectrum _____

2. Sketch an approximate *d*-orbital splitting diagram for the parent complex and a 6 coordinate complex.
3. How would one expect the *d*−*d* transitions to change upon addition of a new ligand? Does this fit with what you see in the visible spectrum of the complex in different solvents?
4. Cashin *et al.*'s explored the Lewis base properties of several vanadyl complexes, in which a Lewis acid interacts with the lone pairs on the V=O oxygen, as shown in Figure 9.7. How would the addition of a Lewis acid capable of coordinating the oxygen change the *d*-orbital splitting pattern of the complex, and also the visible spectrum.

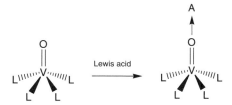

Figure 9.7   The vanadyl complex acting as a Lewis base.

EXPERIMENT 9.3: INFRARED STUDIES OF VANADYL-*bis*(2,4-PENTANEDIONE), VO(acac)₂

Level 5

Pre-lab 9.3: Infrared Spectroscopy of Metal Carbonyls[16,17]

In Experiment 9.2 we looked at how electronic structure of the metal complex can change the energy of electronic transitions, as manifested in spectral changes in the visible region. The electronic structure of a metal complex can also change the energy of infrared bands of complexes. For instance in metal carbonyls, the charge on the metal ion can cause considerable change in the energy of the carbonyl band. $Ti(CO)_6^{2-}$, for example, has a CO stretching frequency of 1748 cm^{-1}, whereas $Fe(CO)_6^{2+}$ has a CO stretching frequency of 2204 cm^{-1}. The more electron-rich titanium is a better π base than the positively charged iron, and thus less electron density is donated by titanium into the anti-bonding orbitals of the CO ligand. Furthermore, smaller changes in CO stretching frequency are observed in molybdenum carbonyl phosphane complexes by changing the basicity of the phosphane ligand.

In VO(acac)₂ we can directly probe how changing the electronic nature of the metal can change a metal–ligand stretch, as all vanadyl complexes have a distinct V=O stretch near 1000 cm^{-1}. The oxygen ligand acts as a π donor ligand; thus changing the electron density at the metal center should change the ability of the vanadium to act as a π-Lewis acid to the oxo ligand. This effect should be most pronounced if an additional donor is added *trans* to the oxo ligand, as the new ligand will interact with the same *d*-orbitals as the oxygen on the metal center. For our 5 coordinate complex it is rather simple to explore the effect of an added ligand by obtaining solution infrared spectra in different solvents. The solvents in this experiment have been chosen such that they have no distinctive infrared bands near the V=O stretching frequency. In addition, we will obtain infrared spectra of the VO(acac)₂ complex as a KBr pellet and of neat 2,4-pentanedione to see how metal complexation of the ligand changes the carbonyl stretch in the ligand.

Procedure 9.3: Infrared Spectroscopic Analysis of VO(acac)₂

Note: pyridine is toxic, and must be used only in the hood. Avoid contact with your skin.

1. Record the IR spectrum (KBr pellet) of the VO(acac)₂ complex (see Procedure 3.4).
2. Compare the spectrum of VO(acac)₂ with that of 2,4-pentanedione, prepared neat with salt plates.

3. Take solution IR spectra of VO(acac)$_2$ in methylene chloride (CH$_2$Cl$_2$), acetonitrile (CH$_3$CN), and pyridine using 0.2 mm NaCl solution cells.

4. Answer the questions and fill out the Results Summary sheet.

Name _____   Section _____

Results Summary for the Infrared Spectroscopic Analysis of
Vanadyl-*bis*(2,4-pentanedione), VO(acac)$_2$

1. Record your observations below.

   Position of the C=O stretching band(s) in 2,4-pentanedione: _____cm^{-1}

   Position of the C=O stretching band(s) in VO(acac)$_2$:   _____cm^{-1}

   What happens to carbonyl band of the ligand upon metal complexation? Explain.

   Position of the V=O stretching band in VO(acac)$_2$ (KBr pellet):   _____cm^{-1}

   Position of V=O stretching band in CH$_2$Cl$_2$:   _____cm^{-1}

   Position of V=O stretching band in CH$_3$CN: _____cm^{-1}

   Position of V=O stretching band in pyridine: _____cm^{-1}

   What happens to the V=O band as one changes solvents? Explain.

Post-lab Questions

Q9.5   How might a Lewis acid added to the oxo ligand of the complex (as described in Experiment 9.2, Results Summary section question 4) change the V=O stretching frequency?

EXPERIMENT 9.4: VO(acac)$_2$ CATALYZED EPOXIDATION OF GERANIOL USING *t*-BuOOH

Level 5

Pre-lab 9.4.a: Epoxidation Reactions[5,19,20]

Epoxidation of alkenes is an important transformation in synthetic organic chemistry as chemists use epoxides as highly reactive intermediates in the formation of diols. The formation of the *anti*-diol product using an epoxide intermediate gives a different stereochemical outcome, Figure 9.8, than one would obtain using the oxidant KMnO$_4$, which gives exclusively a *cis*-diol.

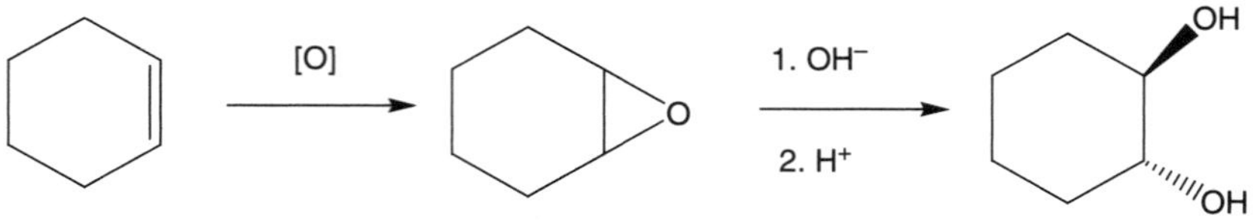

Figure 9.8   Synthesis of an *anti* diol from an alkene.

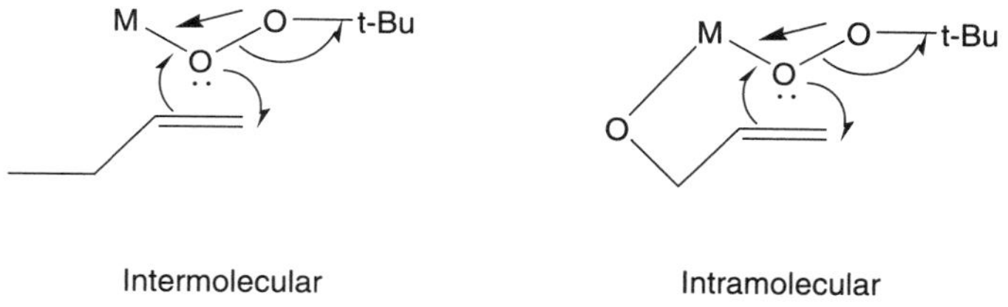

Intermolecular   Intramolecular

Figure 9.9   Epoxidation of olefins.

In 1909 it was discovered that organic peracids, such as *meta*-chloroperoxybenzoic acid, MCPBA, are capable of stoichiometrically epoxidizing olefins. In the 1960s it was shown in two industrial laboratories that transition metals were capable of catalyzing the epoxidation of olefins (Fig. 9.9) using alkyl hydroperoxides. Later work showed that transition metals are highly effective in catalyzing the epoxidation of allylic alcohols, as opposed to isolated double bonds. The epoxidation is believed to be the result of an intramolecular oxygen atom transfer as opposed to an intermolecular oxygen atom transfer. Such a reaction mechanism, in which the catalyst and the substrate can be brought together on a metal center, is an example of *inner-sphere* catalysis. Further evidence for the importance of the alcohol group on the substrate is that, in kinetic studies of the epoxidation of cyclohexene with *t*-BuOOH using the VO(acac)₂ catalyst, *t*-BuOH inhibits the reaction by binding to the metal center. In light of this, it is important to note that *t*-BuOH is a product of the reaction. In addition, although the precursor in this reaction is vanadium(IV), it is believed that an activation process occurs in which the active catalyst is actually vanadium(V).

Pre-lab 9.4.b: Epoxidation of Geraniol[1,2,20–23]

In this experiment we will study the epoxidation of the olefin geraniol with *t*-BuOOH using VO(acac)₂ as a catalyst. This work mimics that of Sharpless and Michaelson, and is a forerunner to the Sharpless asymmetric epoxidation reported 7 years later.

The natural product geraniol provides an excellent substrate to test selectivity in the epoxidation of olefins. There are two olefin sites that can be oxidized, and therefore using geraniol allows for a competitive study using one substrate for the rate of epoxidation of an isolated olefin vs that of an allylic olefin. Both olefin sites for epoxidation are *prochiral*, such that if one can distinguish between the two faces of the olefin one can attain enantioselectivity. With our achiral catalyst and epoxidizing agent we can have regioselectivity, but not enantioselectivity. Later work by Sharpless with the titanium catalyst gave one product specifically.

In considering the mono-epoxidation of geraniol there are four possible products, as shown in Figure 9.10. Because the olefinic protons on C6 and C2 of the products have different chemical shifts, the two sets of regioisomers are easily distinguished using NMR spectroscopy. The hydrogen on C2 has a chemical shift of 5.404 ppm while that on C6 is 5.097 ppm, Figure 9.11. Upon formation of the 2,3-epoxide, the proton on C2 shifts significantly up-field (between 2.5 and 3.0) and appears as a double doublet. The signal from the proton on C6, however, shifts only slightly. A similar scenario occurs with the formation of the 6,7-epoxide. The peak at 5.404 ppm remains largely unchanged while the peak at 5.097 ppm shifts into the 2.5–3.0 ppm range. If the diepoxide is formed, one can expect the loss of both olefinic hydrogen signals as well as two peaks in the range 2.5–3.0.

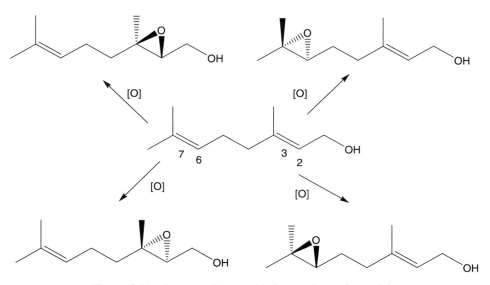

Figure 9.10   Four possible epoxidation products of geraniol.

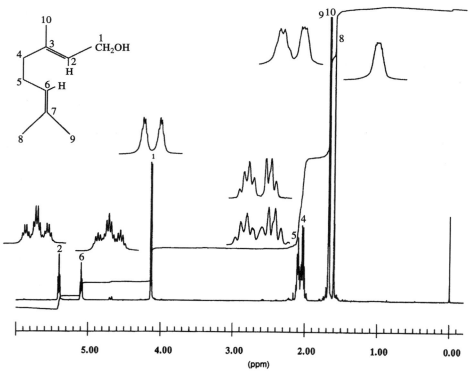

Figure 9.11   NMR spectrum of geraniol. [Reproduced with permission from Silverstein and Webster.[21]]

With this experiment, you will compare the VO(acac)$_2$ catalyzed epoxidation of geraniol using *t*-BuOOH with the epoxidation using a peracid as the oxidizing agent (results provided by instructor). In addition, you will test whether or not the alcohol group is indeed essential for this epoxidation to occur rapidly and selectively by using an alternative substrate, geranylacetate. One of the more pleasant aspects of this experiment is that both the starting material and the products smell nice!

Procedure 9.4.a: Epoxidation of Geraniol

1. Weigh 0.300 g of geraniol into a 10 ml round-bottom flask equipped with a stir bar.
2. Add 5 ml CH$_2$Cl$_2$ and then add 0.028 g VO(acac)$_2$ and stir.
3. When the mixture becomes homogeneous, add dropwise over 1 min 0.45 ml of 5.0 M *t*-BuOOH in decane.
4. Allow the reaction to proceed for 5 min and then take a TLC using 30% ethylacetate and 70% hexane. Use either a KMnO$_4$ dip or a phosphomolybdic acid dip to visualize.
5. When the reaction is complete concentrate the reaction mixture under a stream of air.
6. Purify the product by passing the product through a 7 cm plug of silica gel (70–230 mesh), using 5 ml of CH$_2$Cl$_2$ as the eluent.
7. Concentrate the semi-crude product again and pass the product through a second 7 cm silica gel plug again using 5 ml CH$_2$Cl$_2$ as the eluent.
8. Concentrate with an air stream again until a constant weight is reached.
9. Report the yield and obtain IR (neat) and NMR (CD$_2$Cl$_2$) spectra of the product.

Procedure 9.4.b: Epoxidation of Geranyl Acetate

To determine the importance of the alcohol group on the epoxidation reaction, carry out the Procedure 9.4.a using geranyl acetate. Begin with 0.382 g reaction by TLC using 5% ethyl acetate and 95% hexane. You do not need to isolate the product. Note the outcome of the reaction.

Procedure 9.4.c: Reaction of VO(acac)$_2$ with *t*-BuOOH

1. Weigh out 0.03 g of VO(acac)$_2$ and dissolve in 5.0 ml CH$_2$Cl$_2$.
2. Add one drop of *t*-BuOOH to the solution. Note observations in the Results Summary section.

An optional experiment—Sharpless epoxidation of geraniol—is given in the instructor notes.

Name _____   Section _____

Results Summary for the VO(acac)$_2$-catalyzed Epoxidation of Geraniol
Attach *labeled* NMR and IR spectra of your product to this page.

1. Yield: _____   2. Percentage yield: _____
Record your observations below:

3. What product was formed from the $VO(acac)_2$ epoxidation of geraniol?

4. What does the NMR spectrum indicate about the regioselectivity of the epoxidation? Compare this with the result obtained using MCPBA as the oxidant [Bradley, L. M., Springer, J. S., Delate, G. M., Goodman, A. *J. Chem. Educ.*, **1997**, *74(11)*, 1336–1337].

5. Researchers have suggested the hydroperoxide oxidizes V(IV) to V(V). What evidence do you have that might suggest this may be true?

6. What did you observe with geranylacetate as the substrate? Explain your results.

Post-lab Questions

Q9.6   Homo-allylic alcohols, where the alcohol group is removed one extra carbon from the double bond, also show increased reactivity toward metal catalyzed epoxidations by t-BuOOH.[3] Explain why this might be so.

Q9.7   What would you expect the product to be for the reaction of homogeraniol with t-BuOOH catalyzed by $VO(acac)_2$?

Q9.8   What would happen to the selectivity as one continued to add carbons between the olefin and the alcohol group?

EXPERIMENT 9.5: ENCAPSULATION OF VO(acac)$_2$ AND CATALYTIC CHEMISTRY OF A HETEROGENEOUS CATALYST[24,25]

Level 5

Pre-lab 9.5.a: Microencapsulation of Metal Complex Catalysts

In the catalytic epoxidation of geraniol using $VO(acac)_2$ (Experiment 10.4), both catalyst and substrate were uniformly solvated in the reaction solution. One of the disadvantages of using such a homogeneous catalyst is that the catalyst cannot be easily separated and reused for another reaction. This then, might actually reduce the overall yield in a reaction. In industry, therefore, heterogeneous catalysts are often used because of their ease of separation. Recently, methods have been developed for encapsulating oxidation catalysts such as $VO(acac)_2$ in a thin film of polystyrene. The $VO(acac)_2$ complex can be microencapsulated in polystyrene and then used as a catalyst in a less polar and more environmentally friendly solvent than CH_2Cl_2, as in the homogeneous example. The principles of this idea are illustrated in Figure 9.12. The catalyst is co-dissolved into a solution with polystyrene and then precipitated out. Because of the aromatic group on the polymer, however, they do not phase separate and the catalyst is encapsulated into the polymer. A homogeneous inorganic catalyst can then be rendered as a heterogeneous catalyst if the catalysis can take place in a solvent in which the polymer is insoluble.

Procedure 9.5.a: Encapsulation of the VO(acac)$_2$ Catalyst

1. In a 200 ml beaker dissolve 1.000 g of polystyrene (MW 280,000) in 50 ml warm, (70 °C) cyclohexane.

2. Add 0.200 g $VO(acac)_2$ and stir at 70 °C for 1 h.

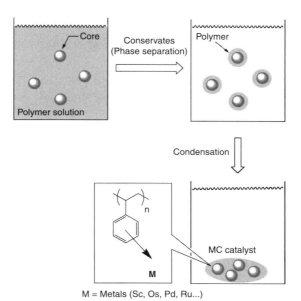

Figure 9.12 Microencapsulation technique. [Reproduced with permission from Kobayashi and Akiyama.[24]]

3. Slowly allow the mixture to cool to 0 °C while maintaining vigorous stirring. The polystyrene will solidify around the catalyst as it cools.

4. Add 30 ml hexane—this will harden the capsule walls—and stir for 1 h at room temperature.

5. Wash the polymer–catalyst mixture several times with hexane to ensure there is no "free" catalyst not encapsulated in the polymer.

6. Dry your polymer–catalyst mixture under vacuum overnight.

7. Measure the mass of the polymer–catalyst mixture to determine how much VO(acac)$_2$ is encapsulated into the polymer.

Procedure 9.5.b: Catalytic Chemistry with the Encapsulated Catalyst

8. In 8 ml hexane add 120 mg encapsulated catalyst, 0.300 g geraniol, and 0.45 ml *t*-BuOOH (5–6 M in decane).

9. Allow the reaction to stir at room temperature and moniter by TLC (see Experiment 9.4.a, step 4), until the reaction is complete. This may take 2.5 h.

10. Remove the catalyst by filtration. Concentrate the product and analyze by NMR spectroscopy (see Experiment 9.4).

Procedure 9.5.c: Catalytic Chemistry—Solvent Dependence

11. Use a similar procedure to that given Procedure 9.4.a for the homogeneous reaction mixture but change the catalyst solvent from methylene chloride to hexane. Allow the reaction to stir for 2.5 h and monitor by TLC.

12. Repeat Procedure 9.5.b above, but now do the reaction in methylene chloride.

Procedure 9.5.d: Catalytic Chemistry—System Recyclability

13. Repeat Procedures 9.5.b and c to see if the catalyst is recyclable.

Name _____   Section _____

Results Summary for the Encapsulated VO(acac)₂-catalyzed Epoxidation of Geraniol

1. Attach *labeled* NMR and IR spectra.
2. Epoxidation of geraniol conditions:
3. Complete the following:

Microencapsulated, ME (hexane)	Yield	_____
Microencapsulated, ME (CH₂Cl₂)	Yield	_____
Homogeneous hexane	Yield	_____
Homogeneous CH₂Cl₂	Yield	_____
Repeat ME (hexane)	Yield	_____
Repeat ME (CH₂Cl₂)	Yield	_____

4. How does microencapsulating the catalyst change the reactivity of the catalyst and the reaction, in terms of the choice of the solvent one should use for the chemistry?
5. Why might it be advantageous to have an easy way to render a catalyst heterogeneous?
6. Explain why it might be advantageous in any setting to be able to use hexane as a solvent for a reaction rather than CH_2Cl_2?

REFERENCES

1. Sharpless, K. B. http://nobelprize.org/chemistry/laureates/2001/sharpless-lecture.pdf (Last accessed December 3, 2006).
2. Ball, P. *Bright Earth*. Farrar, Straus, Giroux: New York, 2001.
3. Sharpless, K. B., Michaelson, R. C. *J. Am. Chem. Soc.* **1973**, *95(18)*, 6136–6137.
4. Mehrotra, R. C., Bohra, R., Gaur, D. P. *Metal β-Diketonates and Allied Derivatives.* Academic Press: New York, 1978.
5. Streitweiser, A., Heathcock, C. H., Kosower, E. *Introduction to Organic Chemistry*, 4th edn., Macmillan: New York, 1992, pp. 859–860.
6. Ballhausen, C. J., Gray, H. B. *Inorg. Chem.* **1962**, *1*, 111.
7. Selbin, H. *Chem. Rev.* **1965**, *65(2)*, 153.
8. Maurya, M. R. *Coord, Chem. Rev.* **2003**, *237*, 163–218.
9. Amin, S. S., Cryer, K., Zhang, B., Dutta, S. K., Eaton, S. S., Anderson, O. P., Miller, S. M., Reul, B. A., Brichard, S. M., Crans, D. C. *Inorg. Chem.* **2000**, *39*, 406–416.
10. Bolm, C. *Coord. Chem. Rev.* **2003**, *237*, 245–256.
11. Moeller, T. (ed.), *Inorg. Syn.* **1957**, *5*, 114.
12. Ophardt, C. E., Stupgia, S. *J. Chem. Educ.* **1984**, *61(12)*, 1102–1103.
13. Battarcharjee, M. *J. Chem. Res. (S)*, **1992**, 142.

14. Morley, C. *Inorganic Experiments*, Woollins, J. D. (ed.). VCH: New York, 1994.

15. Cashin, B., Cunningham, D., Daly, P., McArdle, P., Munroe, M., Chonchubhair, N. N. *Inorg. Chem.* **2002**, *41(4)*, 773–782.

16. Miessler, G. L., Tarr, D. A. *Inorganic Chemistry*, 3rd edn. Prentice Hall: Upper Saddle River, NJ, 2004.

17. Cotton, F. A. *Inorg. Chem.* **1964**, *3*, 702.

18. Sharpless, K. B. *CHEMTECH* **1985**, 692–700.

19. Gould, E. S., Hiatt, R. R., Irwin, K. C. *J. Am. Chem. Soc.* **1968**, *90(17)*, 4573.

20. Katsuki, T., Sharpless, K. B. *J. Am. Chem. Soc.* **1980**, *102*, 5976–5978.

21. Silverstein, R. M., Webster, F. X. *Spectrometric Identification of Organic Compounds*, 6th edn. Wiley: New York, 1998.

22. Martins, R. L., Neves, M. G. P. M. S., Silvestre, A. J. D., Simões, M. M. Q., Silva, A. M. S., Tomé, A. C., Cavaleiro, J. A. S., Tagliatesta, P., Crestini, C. *J. Mol., Catal. A: Chem.* **2001**, *172*, 33–42.

23. Emura, M. E., Nohara, I., Takaaki, T., Kanisawa, T. *Flav. Frag J.* **1997**, *12*, 9–13.

24. Kobayashi, S., Akiyama, R., *Chem. Commun.* **2003**, 449–460.

25. Lattanzi, A., Leadbetter, N. E. *Org. Lett.* **2002**, *4(9)*, 1519–1521.

Introduction to Pulsed NMR Spectroscopy of Metal Complexes[1-3]

ENERGY ABSORPTION BY A NUCLEUS

In 1946, both Purcell and Bloch and their coworkers independently reported the first ^{1}H NMR spectra of paraffin and water, respectively. They were awarded the Nobel Prize for physics in 1952. Twenty years later, Ernst and Anderson applied Fourier-transform mathematics to this technique, increasing instrument sensitivity and spectral resolution and opening the door to many possible applications. Today, NMR analysis of compounds not only reveals chemical structure and conformation, but also molecular mobility and internal dynamics of systems.

Like an electron, a nucleus has the fundamental property of spin. The value of the nuclear spin, I, depends on the structure of the nucleus. For both ^{1}H and ^{13}C nuclei, $I = 1/2$. When placed in a magnetic field (B_o), the nuclear spin, I, can take on $2I+1$ orientations given by the magnetic quantum number, m_I, which takes on the values $I, I-1, I-2, \ldots, -I$ in units of $h/2\pi$. A positive m_I value indicates that the spin is aligned by convention along the z-axis with the magnetic field and a negative value, aligned against the field.

The spinning nucleus, Figure A.1.1, possesses angular momentum, p, which from quantum mechanics, is given by equation (A.1.1). The nuclear magnetic moment, μ, for nuclei with nonzero I is given by equation (A.1.2), where the gyromagnetic ratio, γ, is a constant for the nucleus. The energy of each m_I state is proportional to the magnetic field strength, B_0 (in tesla, where $1 \text{ T} = 10^4$ Gauss), equation (A.1.3).

$$p = m_I \cdot h/2\pi \qquad (A.1.1)$$

$$\mu = \gamma \cdot p = \gamma \cdot I \cdot h/2\pi \qquad (A.1.2)$$

$$E = -\gamma \cdot h/2\pi(m_I B_o) \qquad (A.1.3)$$

A very small spin population difference between the lower (aligned with B_o) and upper (opposed to B_o) enables the nucleus to absorb energy and experience magnetic resonance. The energy of transition from one m_I state to another ($\Delta m_I = \pm 1$) is given by equation

Integrated Approach to Coordination Chemistry: An Inorganic Laboratory Guide. By Rosemary A. Marusak, Kate Doan, and Scott D. Cummings
Copyright © 2007 John Wiley & Sons, Inc.

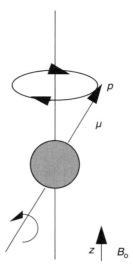

Figure A.1.1.   Angular momentum, p, and nuclear magnetic moment, μ, generated by a spinning nucleus in a magnetic field, B_o.

(A.1.4). Since $\Delta E = h\nu$, the frequency, ν, for the transition is given by eq. (A.1.5).

$$\Delta E = (\gamma \cdot h \cdot B_o)/2\pi \qquad\qquad\qquad (A.1.4)$$
$$\nu = (\gamma \cdot B_o)/2\pi \qquad\qquad\qquad (A.1.5)$$

For the nucleus to absorb energy, the frequency of incoming radiation (in the radio range for nuclear transitions) must match that described by equation (A.1.5).

FT NMR SPECTROSCOPY

To describe Fourier transform (FT) NMR spectroscopy and pulse techniques, it is best to picture the nuclear spins aligned with and against the field as in Figure A.1.2(a). The nuclear spin precesses about the z-axis with the frequency, ν, given in equation (A.1.5).

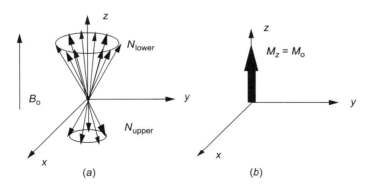

Figure A.1.2.   Description of the nuclear spin alignment (a) and net magnetization, M_z, when a sample is placed in a magnetic field.

This frequency is called the Larmor frequency and to distinguish this from that of incoming radiation the Larmor frequency is labeled ν_o. Sometimes the Larmor *angular* frequency, $\omega_o = 2\pi\nu_o$ is used. To simplify the picture, the observer places him/herself in a rotating reference frame, rotating along the z-axis at the rate of the Larmor frequency of the nuclear spins, such that the spins appear static. To further simplify, only the net magnetization, M_z, along the z-axis is considered, Figure A.1.2(b).

A typical one-dimensional pulsed experiment is shown in Figure A.1.3. At point (*a*), the initial net magnetization (M_o) lies along the z-axis in a positive direction ($M_o = M_z$). At (*b*) the nuclei are subjected to a pulse of RF radiation, B_{appld}, along the x-axis. B_{appld} encompasses a range of frequencies that interact with individual nuclear spins, causing a tilt of the net magnetization into the x–y plane. The time, t, of the RF pulse determines the tilt angle, θ ($\theta = \gamma B_{appld}t$). M_z at any time is described as $M_z = M_o\cos\theta$, along the y-axis, $M_y = M_o\sin\theta$ and along the x-axis, $M_x = 0$. If the detector is placed on the y-axis, a maximum signal will be reached when $\theta = \pi/2$, or with a 90° pulse (*c*). At (*c*), B_{appld} is turned off and the spins are allowed to relax back to equilibrium, along the z-axis. Spin relaxation occurs by two mechanisms: T_1 (spin-lattice or longitudinal) along the z-axis and T_2 (spin–spin, transverse) along the x–y plane. T_1 and T_2 are assumed to be exponential processes, and for small molecules, MM < 1000, $1/T_1 = 1/T_2$. T_2 is related to line broadening as given in equation (A.1.6), where $W_{1/2}$ is the linewidth at half height.

$$T_2 = 1/\pi W_{1/2} \tag{A.1.6}$$

The detector records this loss of spin generating a free induction decay (FID) (*d*), an intensity vs time plot. The FID is a measure of energy emitted by the spin system as it

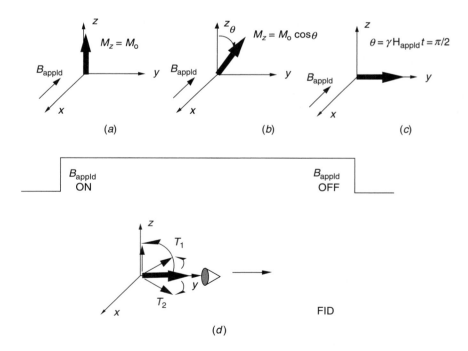

Figure A.1.3. Schematic of a typical NMR experiment.

returns to thermal equilibrium, reflecting the buildup of net magnetizaiton along the z-axis, given by equations (A.1.7) and (A.1.8).

$$\Delta M_z/\mathrm{d}t = -(M_z - M_o)/T_1 \qquad (A.1.7)$$

$$\Delta M_{x(y)}/\mathrm{d}t = -(M_{x(y)})/T_2 \qquad (A.1.8)$$

Fourier transformation converts the output from the time domain into the frequency domain (intensity vs frequency). Note that if the pulse angle exceeds $90°$, the intensity of the signal will decrease until $\theta = 180°$ when no net magnetization lies along the y-axis, resulting in no signal. Between 180 and $360°$ net magnetization lies in the $-yz$ plane and inverted signals are obtained.

ONE- AND TWO-DIMENSIONAL PULSE EXPERIMENTS

The design of a pulsed experiment is shown in Figure A.1.4. In a one-dimensional experiment, t_1, the amount of time B_{appld} is left on may be constant (as in obtaining a simple ^{1}H spectrum) or it may be varied (e.g., in an inversion recovery experiment that measures the T_1 relaxation time). In experiments where t_1 is varied, each FID acquired over time t_2 is processed and treated as an individual data point in a manner dependent on the experiment.

Similar to a one-dimensional experiment where t_1 is varied, in a two-dimensional experiment, a series of FIDs from experiments carried out at varying t_1 are processed. However, in the two-dimension experiment, the amplitudes of the signals at particular frequencies are read and Fourier-transformed a second time. The resulting signals are then a function of the two times t_1 and t_2 and after Fourier transformation, frequencies, v_1 and v_2. The plot is usually given as a contour.

A two-dimensional correlation spectroscopy (COSY) experiment is carried out as described above. Referring to Figure A.1.4, at the start of the evolution phase, a $90°$ pulse is applied along the x-axis ($90°_x$) followed by t_1 and then an extra $90°_x$ is applied at the end of t_1 prior to FID acquisition, t_2. This pulse mixes the population states of the transverse (xy) magnetization, at which time energy is transferred between spins that are coupled through bonds in the molecule (coherence transfer). Although a more rigorous mathematical description is needed to fully understand this process, we can say for now that the through bond coupling information that is transferred manifests itself in the off-diagonal signals in the v_1 vs v_2 contour plot. The signals along the diagonal are simply the one-dimensional spectrum.

An example of a COSY spectrum is shown in Figure A.1.5 for the gly−ala dipeptide. By performing a random walk in off-diagonal assignment, the peaks of the ^{1}H NMR

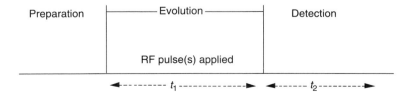

Figure A.1.4. Diagram of a pulsed experiment sequence.

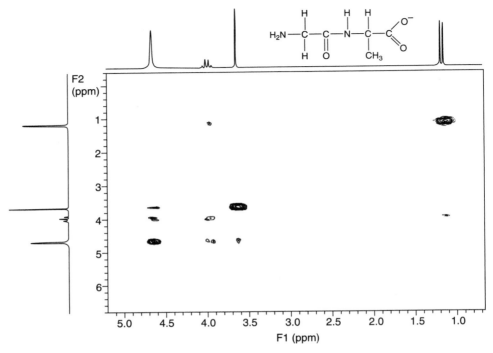

Figure A.1.5. The COSY spectrum of the gly–ala dipeptide in D_2O (ref. DSS). The HOD resonance is at 4.7 ppm.

spectrum are readily assigned. A variety of pulses may be incorporated into both the preparation and evolution stages and, by manipulating pulse sequences, a great deal of information about molecules in solution can be obtained.

REFERENCES

1. Derome, A. *Modern NMR Techniques for Chemistry Research.* Pergamon Press: New York, 1987.
2. Abraham, R. J., Fisher, J., Loftus, P. *Introduction to NMR Spectroscopy.* Wiley: New York, 1991.
3. James, T. J. *Nuclear Magnetic Resonance in Biochemistry.* Academic Press: New York, 1975.
4. Purcell, E. M., Torrey, H., and Pound, R. V. *Phys. Rev.* **1946**, *69*, 37.
5. Bloch, F., Hansen, W. W., and Packard, M. *Phys. Rev.* **1946**, *70*, 474.
6. Ernst, R. and Anderson, E. A. *Rev. Sci.* **1966**, *37*, 93.

Introduction to Cyclic Voltammetry[1–6]

The Nernst Equation

In Experiment 5.2, we saw that the mathematical model describing electronic transitions in solution is Beer's law ($A = \varepsilon bc$). Using visible spectroscopy we were able to determine ε, the molar absorptivity, which gives us information about the probability of electronic transitions occurring in coordination complexes. Similarly, E^0 from the Nernst equation (A.2.1) gives us information about redox activity of species in solution, where n is the number of electrons transferred.

$$E = E^0 - 0.0591/n \log Q \qquad \text{(A.2.1)}$$

Recall that the Nernst equation is the mathematical model describing the relationship between cell potential and concentrations and is readily derived from the fact that cell potential shows a concentration dependence due to its relationship to free energy, equations (A.2.2) and (A.2.3), where Q is the concentration ratio of oxidized (e.g., $[Fe^{3+}L_n]$) to reduced (e.g., $[Fe^{2+}L_n]$) species. In our system, E^0 is the reduction potential for the one electron transfer half reaction [equation (5.4.4)].

$$\Delta G = \Delta G^0 + RT \ln Q \qquad \text{(A.2.2)}$$

$$\Delta G = -nFE^0 + RT \ln Q \qquad \text{(A.2.3)}$$

$$[Fe^{3+}(CN)_5 L]^{n-} + e^- = [Fe^{2+}(CN)_5 L]^{(n+1)-} \qquad \text{(A.2.4)}$$

Current, *i*, Substituted for Concentration Ratio, *Q*

Your first inclination may be to set up an electrochemical cell, as taught in your general chemistry text, where one compartment contains a known concentration of your oxidized species and the other your reduced species. The potential difference between the two cells can then be measured and E^0 readily obtained. However, such an electrochemical cell is often not convenient for study. Instead, we will use *cyclic voltammetry*, a technique that observes redox behavior over a wide potential range. This technique can also be used to obtain kinetic and mechanistic information, giving valuable insight into the solution chemistry taking place during redox processes.

Integrated Approach to Coordination Chemistry: An Inorganic Laboratory Guide. By Rosemary A. Marusak, Kate Doan, and Scott D. Cummings

With cyclic voltammetry, we look at changes in the concentration ratio, Q, in response to changes in E, our applied voltage (now called E_{appl}). This is analogous to looking at changes in absorbance, A, as a function of concentration, c, in visible spectroscopy. However, unlike changes in absorbance readings, changes in Q are not readily measured. We can, however, take advantage of the fact that a change in Q at an electrode surface is directly proportional to a change in current, i, at that surface. Current change is something we can detect; the mathematical expression describing this is equation (A.2.5), where A = area of electrode; D = diffusion coefficient for either oxidized or reduced form of the analyte ($cm^2 s^{-1}$); c = concentration of oxidized or reduced species; x = distance from the electrode (cm).

$$i = nFAD(\delta c / \delta x)_{x=0} \tag{A.2.5}$$

With this substitution of i for Q, the Nernst equation becomes more complex. For one, i depends on the diffusion coefficient of the species. Indeed, the *cyclic voltammogram* (a plot of current vs applied potential) does not result in a straight forward linear plot as did Beer's law, but rather results in a plot such as that for $Fe(CN)_6^{4-/3-}$ shown in Figure A.2.1. Understanding how E^0 is obtained from this plot and the experimental

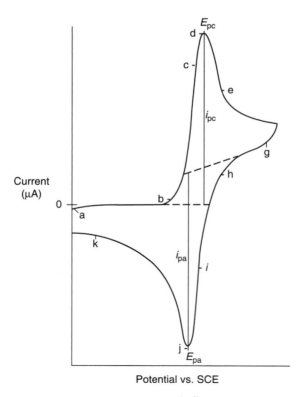

Figure A.2.1   Cyclic voltammogram of the $Fe(CN)_6^{4-/3-}$ (1 M KNO_3) couple. Descriptions of letters are given in the text in Experiment 4.4.

conditions (Nernstian conditions) required for obtaining data needs more explanation. We will look at this relationship qualitatively here and the plot will be detailed in Experiment 4.5; the corresponding quantitative explanation of concentration changes at the electrode surface can be found in Bard and Faulkner.[6]

Nernstian Conditions—How it All Works

In cyclic voltammetry, both the oxidation and reduction of the metal complex (called the *analyte* from now on) will take place in one electrochemical cell. This cell houses the analyte solution as well as three electrodes, the working electrode, the auxiliary electrode and the reference electrode. Electron transfer to and from the metal complex takes place at the *working electrode* surface (Fig. A.2.2) and does so in response to an applied potential, E_{app}, at the electrode surface. During the experiment, current develops at the surface as a result of the movement of analyte to and from the electrode as the system strives to maintain the appropriate concentration ratio (Q), through electron transfer, as specified by the Nernst equation.

Note then that E of equation (A.2.1) is a known quantity (i.e., we dial in a specified E_{appl}). Through the processes of mass transfer and diffusion of the analyte, Q will be properly maintained. The establishment of Q in response to varying E_{appl} will be reflected in the current (e^- flow to and from the working electrode) response. A cyclic voltammogram is simply the current measured as a function of the applied potential (E_{appl}). From the voltammogram we can extract what we are after, E^0 for the $[Fe^{3+}(CN)_5L]^{n-} / [Fe^{2+}(CN)_5L]^{(n+1)-}$ half reaction.

Equation (A.2.5) indicates that the establishment of the Nernstian concentrations (Q) at the electrode requires that mass transfer of analyte to and from the electrode be controlled and limited by diffusion due to a concentration difference only, called complete *concentration polarization*. Once the analyte reaches the electrode surface, the rate of electron transfer must be rapid (i.e., mass transfer not electron transfer limits the rate of

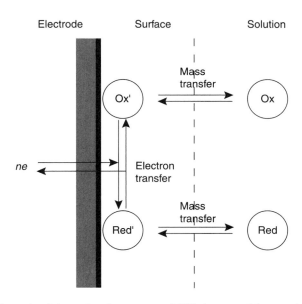

Figure A.2.2   Schematic of the molecular process of diffusion to and from and electron transfer at the electrode surface for an analyte.

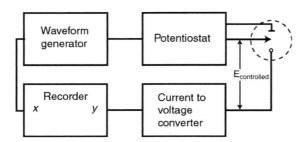

Figure A.2.3   Diagram of the cyclic voltammetry analyzer.

current—electron flow to and from the electrode). Under these "Nernstian conditions" our technique for obtaining E^0 is valid.

There are a couple of ways to maximize for the condition of complete concentration polarization experimentally:

1. The solution must not be stirred. Stirring disturbs mass transfer of the analyte by diffusion and also the concentration ratio at the electrode surface.

2. An electrolyte is added to minimize *migration* or transfer of the analyte to the electrode as a result of an electric field. When the supporting electrolyte is 50–100-fold excess over the analyte, the current in the solution due to analyte movement is minimal and the analyte's movement toward the oppositely charged electrode becomes independent of the applied voltage.

Cyclic Voltammetric Analyzer—The Instrumentation

A schematic of the cyclic voltammetry analyzer is shown in Figure A.2.3. There are different ways that a potential (excitation signal, E_{appl}) may be "dialed in" or applied to our system using a *waveform generator*. For cyclic voltammetry, this is a triangular signal as shown in Figure A.2.4. E_{appl} arbitrarily starts at a positive potential, scans more negative potential and after 20 s (point b) is reversed back to more positive potentials. At $t = 40$ s, one scan cycle is complete. The time it takes for one complete cycle is called the scan rate (v). The *potentiostat* ensures that the working electrode potential is not influenced by any surrounding reactions, and the *recorder* (<500 mV s^{-1} scan rate allowed for an x–y recorder, up to 20,000 V s^{-1} for an oscilloscope/computer and 50,000–100,000 V s^{-1} for the BAS CV50 W and CV100 W systems, respectively) reports the current measured as a function of time. The output is given as a current vs potential scanned (V) plot (see Fig. A.2.1).

The *electrochemical cell*, Figure A.2.5, houses the analyte solution and the three electrodes. The *working electrode* is the electrode through which the potential (EMF, E_{appl}, or excitation signal) is "cycled." It is where the *electrolysis*[†] (electron transfer by means of current) reaction takes place and, therefore, where the current (response to E_{appl}) is also measured. The working electrode dimensions are kept small to

[†]A more rigorous definition of electrolysis is the decomposition of water and other inorganic compounds in aqueous solution by means of an electric current, the extent being proportional to the quantity of electricity passing through the solution. The positive and negative ions formed are carried by the current to the oppositely charged electrodes (auxiliary electrode), where they are collected (if wanted) or released (if unwanted).

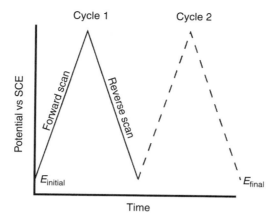

Figure A.2.4   Triangular waveform signal.

enhance concentration polarization. Common working electrodes and their properties and ranges are given in Table A.2.1.

The excitation signal (E_{appl}) is simultaneously applied across the working and *reference electrode*. The reference electrode circuit contains a large resistance ($>10^{11}$ Ω) such that, practically speaking, no current is present in the electrode,

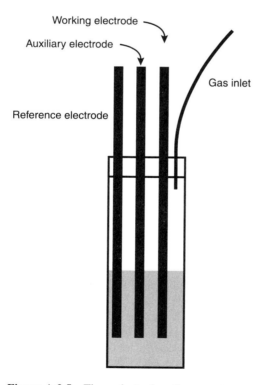

Figure A.2.5   Three electrode cell compartment.

TABLE A.2.1   Description of Common Electrochemistry Electrodes[4]

Electrode type	Properties (V vs SCE)	Approximate useful range
Carbon paste	High reproducibility	1 M $HClO_4$ (0 to + 1.8)
	Easy to prepare and maintain	0.1 M KCl (-1 to $+1.25$)
	Good anodic potential range with small background (residual) currents	
	Generally cannot be used with nonaqueous solvents	
	Periodic repacking is needed	
Glassy carbon	Are chemically resistant, highly conductive, and mechanically rigid	1 M $HClO_4$ (0 to + 1.8)
	Sustain high fluid velocity and/or totally nonaqueous solvent conditions	0.1 M KCl (-1 to $+1.25$)
	A little more noise is obtained than on a carbon paste and resurfacing by polishing with alumina is needed periodically	
Platinum	Limited applications in aqueous solution but are great for non-aqueous solutions	1 M H_2SO_4 (-0.2 to $+1.5$)
		pH 7 buffer (-0.7 to $+1.2$)
		1 M NaOH (-0.8 to $+0.8$)
	Must be surfaced and polished regularly, often in between scans	
Mercury		1 M H_2SO_4 (-0.9 to $+0.5$)
		1 M KCl (-1.75 to $+0.25$)
		1 M NaOH (-1.8 to $+0.2$)
		0.1 M Et_4NOH (-2.1 to $+0.25$)

therefore, providing a stable, reproducible voltage to which the working electrode potential may be referenced, equation (A.2.6). The current that is measured in response to E_{appl} is then proportional to the potential between the working and the reference electrodes.[a]

$$E_{appl} = E^0 - 0.0592/n \log Q - E_{ref} \qquad (A.2.6)$$

Commonly used reference electrodes, along with the reactions responsible for generating the reference potential and their potential values vs the normal hydrogen electrode (NHE) are given below. The reference electrode potential must be added to the experimentally observed value when reporting a value vs NHE.

For the *silver–silver chloride* electrode, equations (A.2.7)–(A.2.8), E_{ref} depends on [Cl⁻] in the electrolyte (3 M NaCl); the electrolyte should be replaced every once in a

---

[a]Carrying out a *linear scan* allows you to calculate $E_{1/2}$, the half-wave potential. $E_{1/2}$ is closely related to E^0, the standard reduction potential for the half-reaction, but not always equal. $E_{1/2}$ values are often reported for systems that are very irreversible.

while. The vycor plug between the internal and external solution should never be allowed to dry (and crack). The electrode should be stored in 3 M NaCl.

$$AgCl + e^- = Ag^0 + Cl^- \quad 0.199 \text{ V vs NHE} \tag{A.2.7}$$

$$E_{ref} = E^0_{Ag/AgCl} - 0.059 \log[Cl^-] \tag{A.2.8}$$

The electrolyte for a *calomel electrode*, equations (A.2.9)–(A.2.10) is a saturated KCl solution.

$$Hg_2Cl_2 + 2e^- = 2 Hg^0 + 2 Cl^- \quad \underset{\text{(temp dependent)}}{0.242 \text{ V vs NHE}} \tag{A.2.9}$$

$$E_{ref} = E^0_{Hg/Hg_2Cl_2} - 0.118 \log[Cl^-] \tag{A.2.10}$$

The *auxiliary electrode* provides the current required to sustain the electrolysis at the working electrode. It completes the circuit by conducting electricity (via movement of ions) from the signal source (waveform generator) through the solution to the working electrode. It is also used to prevent large currents from passing through the reference electrode.

REFERENCES

1. Sawyer, D. T., Heineman, W. R., Beebe, J. M. Controlled potential methods (voltammetry). In: *Chemistry Experiments for Instrumental Methods*, Chapter 4. Wiley: New York, 1984.
2. Kissinger, P. T., Heineman, W. R. *J. Chem. Educ.* **1983**, *60*, 702–706.
3. Jackson, B. P., Bott, A., W. http://www.bioanalytical.com/cursep/cs151f4.html.
4. *BAS CV-1B Operation/Instruction Manual and CV-27 Operation/Instruction Manual*, Bioanalytical Systems, West Lafayette, IN, 1984.
5. Skoog, D. A., Holler, F. J., Nieman, T. A. *Principles of Instrumental Analysis*, 5th edn, Chapters 22 and 25. Saunders: Philadelphia, PA, 1998.
6. Bard, A. J., Faulkner, L. R. *Electrochemical Methods—Fundamentals and Applications*. Wiley: New York, 1980. [2nd edn, 2000.]

States and Term Symbols for Polyelectronic Systems

Basic Definitions

1. *Term symbol* describes an arrangement of e^-s (e.g., how e^- in a p^n configuration are distributed amongst the p orbitals).

2. $s, p, d, f, \ldots$ (atomic states 1 e^- system) $\rightarrow S, P, D, F, \ldots$ (poly e^- states); $l = 0, 1, 2, 3, \ldots \rightarrow L = 0, 1, 2, 3, \ldots$ where L (total angular momentum) $= m_{l1} + m_{l2} +$, $\ldots; m_s \equiv S$ (total spin) where $S = M_s = \sum m_s$ where $m_s = \pm 1/2$. Spin multiplicity (allowed transitions in spectroscopy) $= 2S + 1$:

$$\text{if} \quad S = 0, SM = 1 \quad \text{singlet}$$
$$S = 1/2, SM = 2 \quad \text{doublet}$$
$$S = 1, SM = 3 \quad \text{triplet}$$

3. Total spin-orbit angular momentum quantum number,

$$J = |L + S|, |L + (S - 1)|, \ldots 0 \ldots, |L - S|$$

Determined *only* for $S \neq 0$, and $L \neq 0$.

Determining Term Symbols $^{SM}L_J$

1. *Ground state only* (follow steps in order):
   a. Maximize S (Hund's rule)   e.g., $3p^2 \uparrow \uparrow \underline{\quad}$ $S = 1, SM = 3$ ^3P
   $\uparrow\downarrow \underline{\quad}\underline{\quad}$ $S = 0, SM = 1$ ^1P
   b. Maximize L, i.e., fill highest m_l values;
   c. Select maximum J value for *GS* if subshell is $>1/2$ filled, and minimum if subshell is $< 1/2$ filled,

$$J = L - S = 1 - 1 = 0 \qquad ^3P_0(GS)$$

Integrated Approach to Coordination Chemistry: An Inorganic Laboratory Guide. By Rosemary A. Marusak, Kate Doan, and Scott D. Cummings

TABLE A.3.1 **Determination of the Ground and Excited States for the p^2 Electron Configuration**

	m_l	$+1$	0	-1
	$L = M_L = +1$	−	−	
	0	−		−
	-1		−	−
$SM = 2S + 1 = 3$	^{3}P			
	$L = M_L = +2$	−		
	$+1$	¬	¬	
	^{1}D $\quad 0$	−		¬
	^{1}S $\quad 0$	−	−	
	-1	¬		−
	-2		−	¬
$J = 2, 1, 0\ ^3$P$_0$ (GS)				

2. *All states* (i.e., ground and excited)
   a. Construct a table such as that shown for p^2 electron configuration in Table A.3.1.
   b. Write all combinations of spins (unpaired) for the ground state determination.
   c. Write all combinations of spins (paired) for determination of the excited states.
   d. Determine the *SM* and *J* values as above.

Setting up an Maintaining CHO Cell Culture

Equipment[†]

Water/air jacketed CO_2 incubator (Nuaire)

Laminar flow hood (Nuaire)

Centrifuge 2000 rpm/500*g*

Autoclave

37°C water bath

Dewar and cane for CHO storage

Microscope (10×) (Olympus) and hemacytometer (Hausser Scientific, Horsham, PA) for cell counting

Eppendorf automatic filler and dispensor

Pipetmen (P20, P200, P1000)

100, 500 and 1000 ml autoclavable storage bottles (500 ml bottles are most convenient for everyday use but 100 ml bottles are used to store fetal calf serum)

Disposables

Six-well plates

25 and 75 cm^2 T-flasks (for cell culture)

1–2, 5, 10 and 25 ml graduated sterile pipets

Filter sterilization apparatus [0.2 mM (large) filter—VWR catalog no. 28198–514, box of 12]

50 and 15 ml sterile conical vials

Biohazard waste bags

Reagents

Chinese hamster ovary (CHO) cells (type AA8; ATCC CRL-1859) American Type Culture Collection (Rockville, MD, USA)

α-Minimum essential medium (α-MEM) from Gibco/BRL, Burlington, Canada

[†]Some product brands and companies are given as guides for purchasing.

Integrated Approach to Coordination Chemistry: An Inorganic Laboratory Guide. By Rosemary A. Marusak, Kate Doan, and Scott D. Cummings
Copyright © 2007 John Wiley & Sons, Inc.

HEPES buffer

$NaHCO_3$—enzyme grade from Fisher

Penicillin G—streptomycin solution (Gibco)

Fetal bovine serum (Gibco, Invitrogen Corp., catalog no. 16000-044)

PBS (Dulbecco's phosphate buffered saline)

Trypsin–EDTA ($1\times$) 0.25% trypsin, 1 mM Na_4EDTA (Gibco)—upon arrival, this should be partitioned into 10 ml aliquots in 15 ml sterile conical vials.

Cell culture freezing medium–DMSO (Gibco)

DMSO

MTT (methylthiotetrazolium)—for cytotoxicity study Experiment 6.5

30% bleach

Sterile Technique[†]

Cell cultures will readily grow fungus if contaminated. You must work in a sterile environment at all times.

1. Use only sterilized equipment in the hood. Autoclave when necessary.
2. Wear gloves and rinse down with 70% EtOH periodically.
3. Wipe hood down with 70% EtOH before and after experimentation.
4. Rinse down all pipetmen with 70% EtOH prior to using in hood.
5. Always rinse with 70% EtOH underneath and around the screw cap of a bottle prior to opening in hood.
6. Keep a 70% EtOH-saturated paper towel in the hood. If needed, place any bottle caps (open end down) on the paper towel when not on reagent bottle.
7. Only work inside the hood with sterile utensils/reagents (do not breathe or sneeze inside hood!).
8. When pipetting, avoid touching the pipette to any surface, including the inside necks of reagent bottles (this requires a steady hand).

Cell Recovery and Cryopreservation

Your CHO cells will arrive frozen with instructions for bringing up the culture.

1. Warm your α-MEM growth medium to 37°C.
2. Rapidly thaw the cells by agitating in a 37°C water bath (within 40–60 s).
3. In the hood, gradually add 0.5 ml of pre-warmed α-MEM.
4. Resuspend cells with a pipette and transfer to a sterile 15 ml centrifuge tube containing warm 10 ml α-MEM.
5. Centrifuge at 800–1000 rpm for 5 min.
6. Pour off the medium and resuspend the pellet in 30 ml fresh α-MEM and transfer to a 75 cm^2 T-flask. Loosen the lid of the T-flask to allow CO_2 exchange.

[†]Minimum technique for sterility is given.

7. Incubate at 37°C for 24 h.

8. After 24 h cells will have attached to the bottom of the flask. Pour off the medium and replace with 30 ml of fresh warm medium.

9. When the cells have become confluent (single layer of cells almost completely covering the bottom of the T-flask—this can be visualized using an inverted microscope or for a rough estimate by eye, the flask appears uniformly cloudy on the bottom) pass them into a 25 cm^2 T-flask (see instructions below).

After your cells have recovered and are growing normally (doubling time is 24 h for first day of passing and 12 h afterwards), you will want to freeze several portions for future use (in case of contamination or longevity—a cell line should be terminated after ~1 year).

1. After passing your cells, remove a small aliquot from the suspension for counting (see *counting cells* below) and centrifuge the rest.

2. Pour off the supernatant and resuspend the cells to give $3 \times 10^6 - 1 \times 10^7$ cells ml^{-1} in Gibco freezing medium. Aliquot 0.5–1 ml into cryogenic vials (Nalgene) and store at $-80°C$ for at least 24 h prior to transferring to $N_{2(l)}$ dewar. Alternatively, cells may be stored at $-80°C$ for up to 1 year.

Preparation of α-MEM (50 mM HEPES, 10% Calf Serum, pH 7.4)

1. (Non-sterile.) To ~800 ml ddH$_2$O in a 2 l beaker equipped with a stir bar, add one packet of α-MEM powder, 2.2 g NaHCO$_3$, and 11.92 g HEPES.

2. Titrate in 5 M NaOH to reach pH 7.2 (note that over time the pH will slowly rise).

3. In the flow hood, add 10 ml of the penicillin–streptomycin antibiotic.

4. Bring the volume to 900 ml with ddH$_2$O.

5. In the flow hood, attach a disposable 0.2 μM filter to a sterile 500 ml bottle. (Store the bottle lid face down on an EtOH saturated paper towel.)

6. Once the filter is on the bottle, you can bring the assembly outside the hood to filter sterilize 450 ml of your α-MEM solution by vacuum filtration.

7. After filtering, bring the assembly once again into the flow hood. Remove the filter and place it on a second 500 ml bottle for filtering the last 450 ml α-MEM solution. Replace the screw cap of the first bottle.

8. Repeat steps 6–7 with the second portion of solution.

9. Label and date the solutions.

10. To only one of the 500 ml solution bottles, add 50 ml fetal calf serum in the hood and swirl to mix. This brings the total volume of α-MEM solution to 500 ml with 10% calf serum. Note on the label that you have added the calf serum.

11. Store both solutions at 4°C until needed.

Medium with calf serum is good for one month and that without is good for much longer.

Counting Cells

1. After you have passed your cells, remove 150 μl of cell suspension and add this to a 1 ml conical vial containing 150 μl of α-MEM. Flick the flask to thoroughly mix.

2. Withdraw ∼50 μl of this new cell suspension and pipette onto a hemacytometer for cell counting.

3. Count the total number of cells in all squares of the 16 square grid (1 × 1 mm total area) using the 10× lens of a microscope. Repeat this in a second 16 square grid and take the average number of cells.

4. Calculate the cell density, CD, or number of cells ml^{-1} as in equation (A.4.1).

$$CD = \text{no. cells counted mm}^{-2} \times 2(\text{dilution factor}) \times 10 \text{ mm}^{-1}$$
$$\times 1000 \text{ mm}^3 \text{ ml}^{-1} \tag{A.4.1}$$

Cell Passage

CHO cells should be passed (a portion of the culture removed and reseeded) every 4 days.

1. Prepare flow hood (see *Sterile Technique* above) and supply one 15 ml conical vial (labeled CHO cells), approximately five 10 ml graduated pipettes, 1000 and 200 μl pipetmen and a waste beaker.

2. Incubate in a 37°C water bath, the bottles of α-MEM (with 10% calf serum) and PBS and one 15 ml conical vial containing trypsin–EDTA solution.

3. When at temperature, rinse down with EtOH and put in flow hood. Loosen all lids.

4. Decant into the waste beaker the supernatant from the cells in the 25 cm^2 T-flask.

5. Add 10 ml PBS to the T-flask, gently swirl and decant.

6. Add 1 ml of trypsin–EDTA solution. Gently agitate for 2–3 min until the cells detach (note disappearance of cloudy bottom). If trypsin is left in too long, cell death can occur.

7. Quench the trypsin with 9 ml α-MEM solution for a 10 ml total volume.

8. Mix and resuspend cells by gently pipetting the solution up and down two or three times.

9. Transfer the cell suspension to the labeled 15 ml conical vial and centrifuge for 5 min between 760 and 780 rpm (do not exceed 1000 rpm).

10. Label and date a new 25 cm^2 T-flask and place in hood.

11. After centrifugation, in the hood pour off, in one fell swoop, the supernatant (this will leave the cell pellet and ∼250 μl solution at the bottom).

12. Add 10 ml α-MEM and resuspend cells as previously done.

13. Remove 150 μl for counting cells (go to *Cell Counting* procedure above) and determine the cell density.

14. Calculate the volume of cell suspension that you will need to reach confluency (10 × 10^6 cells) after 4 days. Remember that the doubling time is initially 24 h and 12 h thereafter. If you pass your cells on Monday, then by Friday the cells will have undergone 2 × 4 × 4×4 or 128 duplications.

You therefore want 10×10^6 cells after 128 duplications. Back-calculating:

10×10^6 cells/128 duplications $= 78{,}125$ cells

$78{,}125$ cells/CD (cells ml^{-1}) $\times 1000\,\mu l$ ml$^{-1} = \#\,\mu l$ cell suspension to be added to the new T-flask

15. You will want a total of 10 ml in your new T-flask. Determine the amount of α-MEM needed to make a total 10 ml volume with the amount of cell suspension calculated above. Add this to the T-flask.

16. Pipette in the $\#\,\mu l$ cell suspension determined in step 14. Loosen cap for CO_2 ventilation and incubate for 4 days (incubation may be longer if confluence is not reached in 4 days).

17. Bleach (30%) all cell waste before discarding.

Setting Up and Maintaining Yeast Culture[1]

Reagents

1. Agar
2. YPD medium (available from Difco Laboratories via Fisher)—this culture medium provides the nutrients necessary for yeast growth. The P stands for peptone, which is a protein digest that contains amino acids, other nitrogen sources, vitamins, etc. YTD (T stands for tryptone) medium is also used with *S. pombes*. The medium is prepared as directed by the company (usually 5 g 100 ml^{-1} ddH$_2$O), usually in an Erlenmeyer flask, the top covered with foil and sterilized using an autoclave prior to use.
3. *Schizosaccharomyce pombe* Lindner, teleomorph 972 h (available from American Type Culture Collection, Rockville, MD, USA). This is rated at a bio-safety level (BSL) of −1, i.e., this is not known to cause disease in humans. Cultures may be stored for short periods of time at 0–5°C or for longer periods at −80°C after freezing in liquid N$_2$. It is best to lyophilize your sample, which enables storage for extended periods of time.
4. 10 ml of 100 mM Cd^{2+} (usually as SO$_4^{2-}$ salt).
5. Tris (buffer)—pH 7.8 containing 1 mm PMSF.
6. Phenylmethylsulfonyl fluoride (PMSF).

Apparatus

1. Vortex mixer.
2. Temperature controlled incubator—shaker (for growing the yeast).
3. Petri plates.
4. 500–1000 ml Erleneyer flasks.
5. 50 ml or larger centrifuge tubes.
6. Autoclave-sterilized flat end toothpicks or metal inoculating loop.
7. Acid-washed glass beads (Sigma).

Integrated Approach to Coordination Chemistry: An Inorganic Laboratory Guide. By Rosemary A. Marusak, Kate Doan, and Scott D. Cummings
Copyright © 2007 John Wiley & Sons, Inc.

Techniques

Aseptic Techniques

1. Avoid contact of pure yeast culture and sterile apparatus with contaminating microorganisms (found on anything that is not sterilized!).
2. Clean and disinfect your work area.
3. Flame sterilize any instruments (e.g., inoculating loop) used for transferring your culture before and after transfer.
4. Flame sterilize the mouths of culture flasks before and after transfers.
5. Keep all sterilized reagents sealed before and after transfers.
6. Work quickly.

Pouring YPD Agar Plates

1. Add 7.5 g agar per 500 ml medium.
2. Autoclave at 20 psi for 20 min.
3. Cool to 50–56°C and pour into Petri dishes. Cover and allow to gel. This is enough for 20 plates.

Streaking your Plate with Yeast Culture   The goal is to grow discrete yeast colonies from isolated individual cells. There are multiple variations on this technique. Consult a microbiology laboratory book for further options.

1. With your sterile inoculating loop or the flat end of a toothpick, remove a very small drop of yeast and place it at the top of the agar medium. (If yeast is in a glycerol medium, allow the drop to melt on the medium.)
2. Re-flame sterilize your loop or pick up a new toothpick.
3. With the flat surface against the agar, streak across the surface to the end of the plate creating the pattern shown in Figure A.5.1 (do not lift your tool off the agar during the process).

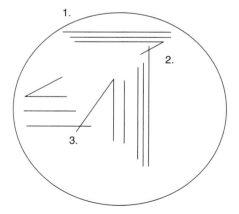

Figure A.5.1   Schematic of a streaked agar plate.

4. Reflame the loop or use the opposite side of the toothpick and touch the end of the yeast pattern and streak back and forth across the plate at a right angle to your original streak.

5. Repeat step 4 in a new direction as shown in Figure A.5.1 if more cell dilution is needed.

REFERENCE

1. Atlas, R. M., Brown, A. E. *Experimental Microbiology: Fundamentals and Applications.* Prentice Hall: Upper Saddle River, NJ, **1997**.

A Brief Guide to Writing in Chemistry

To avoid criticism, do nothing, say nothing, be nothing (Elbert Hubbard)

This document is a guide to assist students in chemistry courses with writing and formatting laboratory reports and research reports. An important goal of an undergraduate curriculum is for our students to organize and communicate research results effectively and to write with acceptable scientific style. By providing many of the common stylistic, grammatical and organizational points, this document provides students with a reference to use throughout their study of chemistry. However, an effective way to improve scientific writing is to read primary research papers. Students are encouraged to consult papers in leading chemistry journals such as *The Journal of the American Chemical Society* in order to appreciate the formatting and writing style of research reports in the field of chemistry.

Formatting a Report

Layout   Use 12-point Times New Roman font and double spacing to allow space for comments and corrections. Number all pages, including those in appendices.

Organization   A standard lab report or research paper should be formatted with sections.

1. *Title*—give the title of your experiment or research report a meaningful name.

   poor: "The Synthesis of a Coordination Compound"
   better: "The Synthesis of a *cis*-Platin"

The title is followed by your name and the date submitted. If you worked with partners, list their names next to yours, but put an asterisk after your name to indicate that you wrote the report.

2. *Abstract*—the abstract should be able to "stand alone." This means that someone should be able to read *only* the abstract and understand the basic nature of your report. For this reason, a good abstract clearly identifies the purpose of the

Integrated Approach to Coordination Chemistry: An Inorganic Laboratory Guide. By Rosemary A. Marusak, Kate Doan, and Scott D. Cummings

experiment *and the important results; a good abstract contains a summary of your results.* Avoid pedagogical comments such as "this experiment helped us learn about the nature of chemical kinetics" or "the goal of this experiment was to learn about metal ions." Although those are important aspects and goals of a lab experience, the lab/research report should focus on the data and results. Avoid starting your abstract with "The purpose of this experiment was ..." Background information on the theory or applications of your experiment belongs in the Introduction section. Avoid referencing other parts of the report, because the abstract should be able to "stand alone." Be specific about what was done: name the reagents or types (not models) of instruments that were used, identify the products of a reaction, and indicate numerical values that were measured or calculated, etc. Avoid vague statements such as "a metal complex was prepared and the percentage yield was calculated." A better abstract would read "hexaammine cobalt(III) chloride was prepared from cobalt (II) chloride, ammonia, ammonium chloride, and hydrogen peroxide. The yield was 8.45 g (64% based on cobalt)." *Tip*: when writing a full report, write the abstract last.

3. *Introduction*—the Introduction section explains to the reader what basic scientific question is being addressed. This can be offered through general background material or a brief historical perspective on the topic being investigated followed by brief summaries, with references, of previous related work. An effective introduction guides the reader from the larger research area, through specific examples of progress in the field to a clear statement of the research problem or approach being addressed in the current report.

4. *Experimental*—this section includes a description of your experimental procedure, and names of instruments used. For lab courses, the procedure can simply *reference the lab manual*, listing any changes to the published procedure. *Do not rewrite the lab manual.* For advanced labs or independent research projects, the experimental section should provide all the necessary detail for someone to be able to reproduce your work. Often, an Experimental section is subdivided into Materials (sources and purity of reagents used), Preparation of Compounds (with procedure, and summary of characterization by NMR, IR, UV–vis spectroscopies, melting point, chromatography, or elemental analysis) and Instrumentation (manufacturer, description of any adaptation or sample preparation) sections. Consult chemistry journals to see examples of Experimental sections for various types of reports. *Tip*: a good experimental section should allow another person, using what you have written and a lab manual, to completely reproduce what you did in the lab.

5. *Results and Discussion* (may be single or separate sections)—the Results section is a summary of your experimental results, important observations and numerical data (preferably in tabular form, when possible). Avoid including especially long tables of raw data, instead presenting the results of calculations or averaged values. A sample calculation may be included in this section or in an Appendix, and a description of equations used in your calculations should be presented. Hand-written calculations may be acceptable for lab reports.

A Discussion section should take the form of an analysis of your results. Comment on the purpose of the experiment. What do the results indicate? What are sources of error (experimental uncertainty/precision)? What additional experiments could help address any dangling ends? Do the results agree with what others have found? Do the results support a model or hypothesis? For some lab courses, you can use this section to answer any questions presented in the manual or in class. Although you should answer the questions in the lab manual, this section should have the style of flowing prose, not simply answers to numbered questions.

6. *Conclusion*—summarize your results and discussion with a short conclusion that is more than simply a reiteration of your results. Phrase it in terms of the broader questions addressed in the Introduction.

7. *References*—citations of the literature used in the previous sections (see below).

8. *Appendix*—graphics may appear here, along with lengthy calculations or additional material not needed by the reader when reading the report.

9. *Graphics*—graphics include tables, figures, schemes and chemical structures. Tables are columns of measured and/or calculated values or observations. All quantities should have units and be expressed using proper significant figures and scientific notation. Important experimental conditions should be listed as footnotes, especially when the table includes data obtained under different experimental conditions. Figures include: spectra, graphs, cartoons of experimental set-up or other drawings intended to show an *object*. Schemes include: reaction mechanisms, experimental flow charts or other drawings that are intended to show a *process*. All tables, figures and schemes should be numbered sequentially and must be mentioned in the text. Graphics are best when large, and should be at least half-page in size; they can be inserted into the text or included at the end of the manuscript in the Appendix. Chemical structures can appear in the text and should be labeled with the same name, formula or compound number that appears in the text.

Sentence Structure and Writing Style

1. *Beginning a sentence.* Avoid beginning a sentence with a symbol, numeric value or equation.

   incorrect: 315.6 mg of ammonium chloride was added to the solution, which was then heated to 50°C.

   correct: After the addition of 315.6 mg of ammonium chloride, the solution was heated to 50°C.

   incorrect: v is both the vibrational frequency and the IR radiation frequency.

   correct: The frequency v refers to both the vibrational frequency and the frequency of IR radiation.

2. *Dangling modifiers and illogical construction*—check that a modifier phrase or the pronoun "it" actually refers to the intended subject (see also: Subject–verb agreement).

   incorrect: Being slightly wet, I first dried the solid in a dessicator.
      Was I wet or was the solid?

   correct: Because the solid was slightly wet, it was dried in a dessicator.

   incorrect: After transferring to a larger flask, the solution was heated to a boil.
      Did the solution transfer itself?

   correct: The solution was transferred to a larger flask and heated to a boil.

incorrect: A diagram of the SARS virus is now available. We obtained it from an online source. *The SARS virus is available online?*

correct: A diagram of the SARS virus is now available from an online source.

3. *Equations*—equations typically appear as a separate line from the text and are numbered sequentially throughout the manuscript. Equations can then be referred to by number.

Example: The quenching rate constant can be calculated using the Stern–Volmer equation:

$$\Phi_0/\Phi_q = 1 + k_q \tau_0 [Q] \qquad\qquad (A.6.2)$$

4. *Spaces*—there should be a space between a quantity and its units and between a quantity or word and subsequent parenthetical phrase.

Examples:

6.626 J s

25.15 K = 298.15 °C

45 ml

456 nm (34,000 M^{-1} cm^{-1})

5. *Personal pronouns*—by tradition, scientists tend to avoid using the personal pronouns "I" and "we" and "you" in most technical communications. The use of third person instead of first person is standard for research results (see also: Active voice).

first person: I heated the solution at 100 °C for 1 h and I noticed that it turned blue.

third person: When heated at 100 °C for 1 h, the solution turned blue.

6. *Pedagogical comments*—avoid including pedagogical comments in a report or scientific communication. Phrases such as "this experiment helped us learn about ligand field theory" or "the goal of this experiment was to learn about ligand substitution kinetics" are addressing the process of learning not the science of the experiment. Although those *are* important aspects and goals of the lab experience, the lab report should focus on the data and results.

7. *Personification*—molecules and equipment are not people, so do not personify them in your writing.

incorrect: Potassium nitrate really wants to dissolve in water.

correct: Potassium nitrate is very soluble in water.

incorrect: Sodium wants to lose one electron to form Na^+.

correct: Oxidation of Na to Na^+ is thermodynamically favorable.

incorrect: The spectrum shows two bands of equal intensity

correct: Two bands of equal intensity appear in the spectrum.

8. *Plural nouns*—"Data" is plural for "datum," "spectra" is plural for "spectrum," "phenomena" is plural for "phenomenon," and "formulae" is plural for "formula." The amount of chemical reagent is singular, so use the correct verb tense.

incorrect: Data was acquired and a spectra is in the appendix.

correct: Data were acquired and a spectrum is in the appendix.

    incorrect: While the solution boiled, 5.0 g of KBr were added.

    correct: While the solution boiled, 5.0 g of KBr was added.

9. *Prepositions*—Do not forget "of" between quantities and substance name.

    incorrect: ... and 10 ml MeOH was added.

    correct: ... and 10 ml of MeOH was added.

10. *Redundant or unnecessary phrases*

    incorrect: A photon of light having a wavelength of 530 nm ... *if not "of light," what was the photon made of?*

    correct: Light having a wavelength of 530 nm ...

    incorrect: In this experiment, *cis*-platin was prepared from tetrachloroplatinate. *If not this experiment, then in which experiment?*

    correct: *cis*-Platin was prepared from tetrachloroplatinate.

11. *Subject–verb agreement*—are you stating that an inanimate object is drawing a conclusion, or suggesting a strange cause and effect? (See also: Dangling modifiers.)

    incorrect: The IR spectrum implies that water is in the sample of $Cr(phen)_3^{3+}$. (*spectra do not imply, people do*)

    correct: The presence of water in the sample of $Cr(phen)_3^{3+}$ is inferred from the IR spectrum. (*the presence is inferred*)

    incorrect: Water was present in the product because of the peak at $3200 \ cm^{-1}$ in the IR spectrum. (*the peak in the spectrum did not cause water to be present*)

    correct: The peak at $3200 \ cm^{-1}$ in the IR spectrum indicates that water was present in the product. (*water caused the peak in the spectrum*)

Verbs

1. *Active voice*—by avoiding personal pronouns, scientists often depend excessively on the passive voice, which can weaken the writing style. *When possible*, replace passive voice with active voice.

    passive voice: A color change was observed when the solution was heated.

    active voice: A blue color formed as the hot solution.

    passive voice: There was some solid $CrCl_3$ that did not dissolve.

    active voice: Some $CrCl_3$ solid did not dissolve.

2. *Subject–verb agreement*—based on whether the subject is singular or plural, use the correct verb tense. A quantity used is a singular subject, even when that quantity is in a plural form of units.

    incorrect: 45 mg **were** added

    correct: 45 mg **was** added

3. *Verb tense*—past tense is used to describe a procedure that you followed in an experiment. Present tense is used to describe a scientific fact, such as the properties of a molecule.

    Examples: Potassium tetrachloroplatinate was used as the starting material for the synthesis of *cis*-Platin. When dissolved in water, K_2PtCl_4 forms a cherry-red solution.

4. *"Verbing" a noun*—do not turn nouns into verbs.

incorrect: ammonia complexes to cobalt ions

correct: ammonia forms complexes with cobalt ions.

incorrect: the mixture was centrifuged to separate the solid.

correct: The solid was separated from the mixture using a centrifuge.

incorrect: The solution was rotovapped to dryness

correct: The solvent was removed by rotary evaporation

Abbreviations, Formulae, and Numerals

1. *Standard abbreviations*—use standard *JACS* abbreviations (*note*, not all journals use exactly the same abbreviations):

    Examples: ml = milliliter; μg = microgram; nM = nanomolar

    h = hour; min = minute; s = second

    K = degrees Kelvin, °C = degrees Celsius

2. *Chemical formulae*—use subscripts, superscripts, parentheses, and symbols appropriately in chemical formulae.

    Examples: $Cr^{3+}(aq)$

    $K_2[PtCl_4]$

    $[Ru(bpy)_3^{2+}](PF_6)_2$

3. *Compound numbers*—compounds can be numbered if repeated long compounds names become cumbersome. The number should be defined (usually in bold or underlined) somewhere early in the manuscript, often when it is first presented.

    Example: "Investigations into the fluorescence of 8-hydroxyquinoline (**1**), 4-iodo-8-hydroxyquinoline (**2**) and 2-methyl-4-iodo-8-hydroxyquinoline (**3**) are described in this paper. Recrystallization of **1** and **2** afforded analytically pure samples, but vacuum sublimation of the methyl derivative (**3**) was necessary to remove fluorescent impurities."

4. *Decimal places*—for values less than unity, use a leading zero. Avoid writing values having too many zeroes; use scientific notation.

    Examples: "0.15 μL" not ".15 μl"

    "2.3×10^{-5} M" not "0.000024 M"

5. *Defining abbreviations*—abbreviations for chemical compounds, ligand, instruments or methods should be defined in the text before using throughout the manuscript.

    Examples:

    "The complex cation $Ru(bpy)_3^{2+}$, where bpy = 2,2'-bipyridine, is luminescent ..."

    "Surfactants such as sodium dodecyl sulfate (SDS) form micelles ..."

    "Chelate structures were minimized using the empirical force field (EFF) method."

    "Results were analyzed using both molecular orbital (MO) and ligand field (LF) theories"

6. *Organic abbreviations*—standard organic abbreviations can be used in text and formulae.

> *Examples*:
>> Me = methyl
>>
>> Et = ethyl
>>
>> iPr = *iso*-propyl
>>
>> tBu = *tert*-butyl
>>
>> Ch = cyclohexyl

7. *Reagents and solvents*—use chemical formulae for standard reagents and solvents, but not when the name is shorter or more precise.

> *Examples*: NaOH (*aq*) in place of "sodium hydroxide"
>
> H_2SO_4 (*aq*) in place of "sulfuric acid"
>
> CH_2Cl_2 in place of "dichloromethane"
>
> "caffeine" in place of $C_8H_{10}N_4O_2$

Chemical Terms and Expressions

1. *Chemical names*—the names of chemicals are not capitalized, unless they are trade names such as "Tylenol" or "Viagra."

> *incorrect*: The reaction of aqueous Cobalt(II) with Aspirin was investigated.
>
> *correct*: The reaction of aqueous cobalt(II) with aspirin was investigated.

2. *Create*—chemistry involves "synthesizing" new compounds, "preparing" solutions,7 "characterizing" products. Avoid using phrases such as "products were *created*."

3. *Measurements*—spectra are measured "with" or "using" a spectrometer, not "on" a spectrometer.

4. *Machines*—spectrometers (UV–vis, IR, NMR, etc.) are "instruments," not "machines."

5. *React*—as an intransitive verb, "react" should not have an object and should not have a passive voice. Chemical regents react with each other, they are not reacted.

> *incorrect*: Hydroxyquinoline and sodium hydroxide were reacted to form sodium hydroxyquinolinate.
>
> *correct*: The reaction of hydroxyquinoline and sodium hydroxide produced sodium hydroxyquinolinate.

6. *Tested*—a hypothesis can be "tested" and a student can be "tested." For most laboratory work, the terms "measured," "investigated," "determined," "calculated", or "obtained" often work better.

> *incorrect*: The absorbance of the solution was tested using the UV–vis machine.
>
> *correct*: The absorbance of the solution was measured using a UV–vis spectrophotometer.

REFERENCES

There are numerous styles for formatting references. Unless otherwise instructed, citations should be formatted in the *JACS* style and appear as endnotes. Alternatively, article titles can also be included. Most important is to prepare citations with a uniform style.

Last name, initials, Last name, initials. *Journal Title* **year**, *volume (issue)*, starting page.
or

Last name, initials, Last name, initials. "Article Title" *Journal Title* **year**, *volume (issue)*, starting page.

Examples:

Schlabach, M., Limbach, H.-H., Shu, A., Bunnenberg, E., Tolf, B., Djerassi, C. *J. Am. Chem. Soc.* **1993**, *115*, 4554.

Additional Materials for Writing Lab/Research Reports

1. Davis, M. *Scientific Papers and Presentations* Academic Press: San Diego, CA, 1997.
2. Dodd, J. S. (ed.). *The ACS Style Guide: a Manual for Authors and Editors*. ACS: Washington, DC, 1997.
3. Eisenberg, A. "Strategies five productive chemists use to handle the writing process." *J. Chem. Educ.* **1982**, *59*, 566.
4. Potera, C. "The Basic Elements of Writing a Scientific Paper: The Art of Scientific Style." *J. Chem. Educ.* **1984**, *61*, 247.
5. Spector, T. "Writing a Scientific Manuscript: Highlights for Success." *J. Chem. Educ.* **1994**, *71*, 47.